Between 1920 and 1960 astronomers began working with scientists in other fields to improve their understanding of the nature of the solar system. Well before the launch of *Sputnik,* researchers made wide-ranging attempts to solve such problems as the nature of lunar and terrestrial craters, the origin of comets and meteors, and the birth of the solar family. Their achievements included the Oort cloud and Kuiper belt. Cooperation among specialists often dissolved in controversy – including the famous Kuiper–Urey conflict over the Moon's history – yet their work provided the foundation for planetary science in the space age.

Exploiting previously unused archival material, Ronald E. Doel investigates the emergence of this interdisciplinary scientific community and its influence on research in astronomy, meteorology, geology, and geophysics. He examines how studies in planetary science were influenced by shifts in institutional mandates, new research techniques, and government–military funding during the cold war. One example analyzed is the challenge to the geological doctrine of uniformitarianism that emerged in light of cold-war weapons research. Above all, this book explores an important branch of earth science, central to what we now call the environmental sciences.

Solar System Astronomy in America

Solar System Astronomy in America

Communities, Patronage, and Interdisciplinary Science, 1920–1960

RONALD E. DOEL

CAMBRIDGE UNIVERSITY PRESS

CAMBRIDGE UNIVERSITY PRESS
Cambridge, New York, Melbourne, Madrid, Cape Town, Singapore, São Paulo, Delhi

Cambridge University Press
The Edinburgh Building, Cambridge CB2 8RU, UK

Published in the United States of America by Cambridge University Press, New York

www.cambridge.org
Information on this title: www.cambridge.org/9780521115681

First published 1996
This digitally printed version 2009

A catalogue record for this publication is available from the British Library

Library of Congress Cataloguing in Publication data
Doel, Ronald Edmund.
Solar system astronomy in America : communities, patronage, and
interdisciplinary science, 1920–1960 / Ronald E. Doel.
p. cm.
Includes bibliographical references and index.
ISBN 0-521-41573-X
1. Solar system. 2. Astrophysics – United States – History.
3. Science – United States – History. I. Title.
QB501.D64 1996
523.2′0973′0904 – dc20 95–23112
CIP

ISBN 978-0-521-41573-6 hardback
ISBN 978-0-521-11568-1 paperback

To my parents,
Edmund C. Doel
and
Ella Doel

and to
Katharine

"The big opportunities are obviously not in pure astronomy or physics or chemistry or geology, but in mixtures."

– Harlow Shapley to Hugo Benioff, February 9, 1923

"Coordination of knowledge in a broad field of science is just as important as building a new observatory."

– Gerard P. Kuiper to Peter van de Kamp, April 26, 1954

"Whenever I lecture . . . I always admit that the origin of the solar system involves miracles. I am afraid that astronomers ask for chemical miracles and chemists ask for astronomical miracles."

– Harold C. Urey to Otto Struve, January 15, 1953

Contents

Illustrations

Preface

In 1908 the founder of Lowell Observatory, Percival Lowell, published a new book. In *Mars as the Abode of Life,* which drew on fifteen years of telescopic observations and theoretical work, Lowell boldly proclaimed that fine weblike markings he and others observed on Mars were irrigation canals, built by an intelligent race cultivating a parched, dying planet. Well received by popular audiences, Lowell's interpretations caused an intense furor among American astronomers. Many agreed with Lick Observatory astronomer Edward S. Holden that Lowell's conclusions were deeply biased, reflecting nothing more than what he had set out to prove.[1]

This embarrassing, protracted controversy – and the building of large, new telescopes in the American Southwest – led astronomers to abandon Mars and the solar system early in the twentieth century in favor of studying stars and distant nebulae; or so it has been claimed. In this book I argue that solar system research, although not a core program of American astronomy through the first half of the twentieth century, nevertheless was far from neglected. Long before *Sputnik*'s 1957 launch ignited the space race, numerous astronomers had begun important investigations of the origin and evolution of the planetary system. Dedicated institutes of planetary science were lacking, and many researchers contributed to these investigations on a part-time basis. Nevertheless, American astronomers played leading roles in resolving some of the great challenges of solar system astronomy by the mid-1950s, including the origin of meteors, the origin and physical structures of comets, and the nature of craters on the Moon and Earth.

Solar system studies research was also *interdisciplinary.* The key to understanding how this broad field of science grew during the first half of the twentieth century is to look not at astronomical research alone, but rather at the wide range of disciplines that comprised the earth sciences. To study planetary atmospheres and the geochemical evolution of planets, astronomers forged new and untested liaisons with chemists, meteorologists, geochemists, geophysicists, and geologists. These relationships proved difficult to maintain; frequently they fell victim to intense disputes over

[1] Lowell (1908); on the reception of Lowell's claims, see DeVorkin (1977) and Hoyt (1976, 87–102, 200–36).

methodology and professional standing. Yet endure they did. Although this emerging interdisciplinary community was not yet fully formed by 1960 – and in some respects remains incomplete today – it provided the foundation on which the scientific component of the early U.S. space program was built. The history of solar system astronomy is thus an important dimension of the emerging earth sciences and the broader tapesty that we now call the environmental sciences.

This book traces the emergence of solar system research in America from 1920 to about 1960 – that is, from the time astronomy in the United States assumed its modern cast until the era of direct planetary exploration, when American astronomy in general and solar system research in particular became exemplars of Big Science undertakings. It is not a complete history of solar system research during this period. By following the main community of American astronomers, for example, I have focused principally on astronomers who employed optical rather than radio and radar techniques. However, this is the first book to examine the broad field of solar system studies prior to the era of space exploration; by so doing, it illuminates little-explored aspects of the history of American astronomy, meteorology, geology, and geophysics. It is based on primary and secondary sources, as well as on an extensive examination of archival collections of individuals and institutions, spanning several disciplines and four countries.

A central theme of this work is that the history of solar system research – indeed, of any science – cannot be separated from its social, cultural, institutional, or political context. This point is particularly telling for interdisciplinary research. The boundaries of scientific disciplines are not set by nature but by scientists, and in turn by national traditions and patterns. Opportunities for cooperative research, which transcend the institutional and patronage structures of traditional disciplines, often require extramural support. In the United States, such funding came from a mixture of private, federal, and military patrons, including the General Education Board and the Carnegie Institution of Washington before World War II, and the National Science Foundation and military agencies in the cold-war era that followed. The willingness – or hesitation – of patrons to fund interdisciplinary collaborations profoundly shaped the development of this field. For this reason, readers will find patronage a recurring concern.

Readers accustomed to "planetary astronomy" may find my use of "solar system astronomy" startling, and a word of explanation is in order. I define solar system astronomy as studies of the Moon, planets, planetary satellites, and asteroids, as well as smaller bodies now considered to be members of the Sun's family, including comets, meteors, and meteorites. Since the early 1960s, *planetary astronomy* has typically been employed to refer to investigations of the planets alone; but through the first half of the twentieth century, astronomers used *planetary* and *solar system* interchangeably to refer to objects gravitationally bound to the Sun, for them a common field of study. As late as 1956, one of the leaders of planetary astronomy, Gerard P. Kuiper, mailed a grant application to the National Science Foundation that concerned planetary atmospheres, asteroids, and the lunar surface, entitled "Solar System Studies." This broad perception was similarly shared by American astronomers in other fields.[2] It is a reminder that we must not impose later definitions on the science of an earlier time.

Fittingly, this study itself is a product of diverse influences. My research and writing, in chronological order, were supported by Princeton University, the Smithsonian Institution, and the Center for History of Physics of the American Institute of Physics. At Princeton, Charles Coulston Gillispie supervised this study as dissertation advisor; he provided consistent encouragement and constructive criticism. My analysis of professional and institutional issues facing American astronomers was sharpened by discussions with David H. DeVorkin and Robert W. Smith, while my treatment of patronage reflects fruitful exchanges with Paul Forman, Robert E. Kohler, and particularly Stanley N. Katz. Steven J. Dick, LeRoy Doggett, James R. Fleming, Julian Loewe, Paul Lucier, Michael S. Mahoney, Frederik Nebeker, Allan A. Needell, Michael J. Neufeld, Naomi Oreskes, Joseph N. Tatarewicz, and Spencer R. Weart joined DeVorkin and Smith in reading all or part of the manuscript. Their suggestions, comments, and criticisms have greatly strengthened this work. Additional support and advice came from Faye Angelozzi, Gary Braasch, Pete Daniel, A. Hunter Dupree, Lawrence J. Friedman, Richard Jarrell, Peggy Kidwell, John Lankford, Cathy Lewis, John Mauer, and Mandy Young. My thanks also go to the staff of the Jet Propulsion Laboratory's Public Information Office, particularly Jurrie van der Woude. Finally, I am grateful to librarians and archivists at several universities and institutions, particularly at Arizona, Chicago, Harvard, Princeton, UCSD, Lowell Observatory, the Rockefeller Archive Center, the U.S. Naval Observatory, the Smithsonian Institution, the U.S. Geological Survey, and the American Institute of Physics.

One advantage of writing recent history is that one can sometimes talk directly with participants in the drama. Those (some now deceased) who gave freely of their time and recollections in oral history interviews include Ralph B. Baldwin, Harmon Craig, Gérard de Vaucouleurs, Audouin Dollfus, M. King Hubbert, Gordon MacDonald, Harold Masursky, S. Keith Runcorn, Carl Sagan, Robert P. Sharp, Eugene M. Shoemaker, Athelstan Spilhaus, John Verhoogen, Fletcher G. Watson, Jr., Don E. Wilhelms, and J. Tuzo Wilson. Transcripts of these interviews have been deposited at the Niels Bohr Library of the American Institute of Physics (now in College Park, Maryland). They were made possible by grants-in-aid from the Friends of the Center for History of Physics and direct support during my term as postdoctoral historian at the center, and by a grant from the National Science Foundation; assistance from these sources is gratefully acknowledged.

[2] G. P. Kuiper, "Solar System Studies," draft proposal, Nov. 12, 1956, Box 33, GPK; Aller to P. Oosterhoff, Dec. 20, 1957, Box 10, OS (see the List of Abbreviations for archival sources); and Whipple (1956c, iii–ix). A quarter century earlier, Harlow Shapley (1930, 81) had similarly defined planetary systems as a "permanent entourage of gravitationally-controlled particles." By contrast, post-*Sputnik* definitions of planetary astronomy emphasized the major planets; see Tatarewicz (1990a, xi).

Other individuals and institutions assisted me in various ways. Dale P. Cruikshank, Ian Halliday, and the late Jan Oort discussed with me the development of solar system astronomy, and offered valuable insights. A Goethe-Institut language fellowship from the Deutsche Akademischer Austauschdienst indirectly permitted me to explore significant archival collections in Europe; I also benefited from a travel grant from the National Endowment for the Humanities. In later stages my research was ably assisted by Kevin J. Downing; Katharine Doel prepared the index.

At Cambridge University Press, Helen Wheeler first championed this manuscript; John X. Kim and Alex Holzman then cared for it in turn. Careful editing by Michael Gnat saved me from numerous mistakes; any that remain are my responsibility alone.

My parents, Edmund and Ella Doel, were supportive through the years this project expanded and matured. To Katharine, who has lived with this book longer than she would have liked – but nonetheless offered steadfast encouragement – I offer my love and heartfelt appreciation.

Armin O. Leuschner and family, Berkeley, California, 1920s. An expert in the theory of orbital computations and a leader of asteroid studies, Leuschner led the largest American graduate program in astronomy in the early twentieth century, and introduced many graduate students to his research. (Courtesy the Bancroft Library)

Samuel Boothroyd *(left)* and Vesto M. Slipher *(right)* at a station of the Harvard Arizona Meteor Expedition, 1932. The largest meteor expedition then undertaken in North America, the project reflected the hopes of Harlow Shapley to employ meteors to study the interstellar medium and stellar spectra. (Lowell Observatory photograph.)

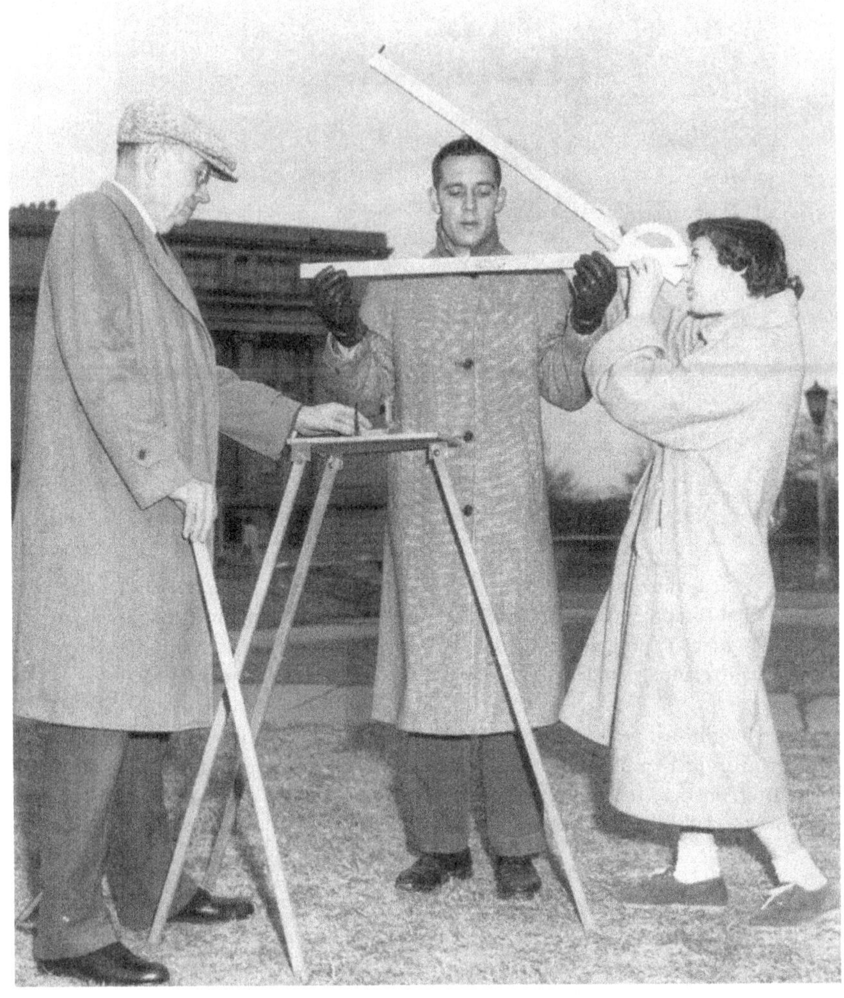

Charles C. Wylie *(left)* at the University of Iowa. Wylie's studies of the paths and durations of meteors reported by visual observers helped convince American astronomers that these velocities were too high, and that meteors are not interstellar in origin. (F. W. Kent Collection, University of Iowa Archives.)

(Facing) Fletcher G. Watson, Jr., and Alice Watson, Massachusetts, ca. 1935. Watson's 1938 doctoral dissertation on small bodies in the solar system reflected the interests of Harvard astronomers in meteor and meteorite astronomy, and included the first modern estimate of the collision rate between asteroids and Earth. (Courtesy Fletcher G. Watson, Jr., and the Niels Bohr Library, American Institute of Physics.)

Harlow Shapley and Henry Norris Russell en route to the 1938 International Astronomical Union meeting in Copenhagen. Shapley's support of meteor astronomy and Russell's strong interest in cosmogony encouraged American astronomers to take up problems in solar system research early in the twentieth century. (Photograph by Dorothy Davis Locanthi, W. F. Meggers Collection, courtesy Niels Bohr Library, American Institute of Physics.)

Gerard P. Kuiper at the Yerkes Observatory, Williams Bay, Wisconsin, ca. 1947. After switching from stellar to solar system astronomy in the mid-1940s, Kuiper became the leader of this field in America, initiating studies of the Moon, planets, and cosmogony. His edited volumes on solar system phenomena provided an important framework for interdisciplinary studies. The Planetary Atmospheres conference of 1947, which Kuiper organized, was one of the first major interdisciplinary gatherings in solar system astronomy after World War II. (University of Chicago Archives.)

The 82-inch telescope of the McDonald Observatory. Kuiper employed this instrument – the second-largest reflecting telescope in the United States until the completion of the 200-inch telescope in 1948 – through the late 1950s. With it he discovered the atmosphere of Titan and carbon dioxide in the atmosphere of Mars, studied asteroids, and made an extensive visual survey of the surface of the Moon. (Yerkes Observatory photograph.)

The Conference on Planetary Atmospheres, Lowell Observatory, March 1950. The Lowell project marked the first sustained interdisciplinary collaboration between meteorologists and astronomers on such problems as global climatology and planetary meteorology. *From left to right: (front row)* Roger Putnam, Vesto M. Slipher, Edison Pettit, Gerard Kuiper, Seymour L. Hess; *(middle row)* Earl C. Slipher, Rudolf Penndorf, Arthur Adel, James B. Edson, George Herbig, Christian T. Elvey; *(back row)* William Putnam, Roger L. Putnam, Jr., Franklin E. Roach, Henry Giclas, and Carl Otto Lampland. (Lowell Observatory photograph.)

Fred L. Whipple beside a Baker Super-Schmidt camera, 1952. The Super-Schmidt camera attested to the willingness of American military officials to fund studies of the upper atmosphere after 1945, and Whipple's desire to employ meteor observations to investigate the upper atmosphere as well as small particles in the solar system. Whipple's Navy contracts for meteor research were among the largest granted in American astronomy before the mid-1950s. (Courtesy Fred L. Whipple and the Niels Bohr Library, American Institute of Physics.)

Dirk Brouwer of the Yale University Observatory. By 1950 Brouwer was a dominant figure in celestial mechanics, and gained wide recognition for his investigations of planetary orbits and asteroid families. (Courtesy Yale University Archives, Manuscripts and Archives, Yale University Library.)

Eugene Rabe and Paul Herget at the University of Cincinnati Observatory, ca. 1955. After World War II Herget directed the Minor Planet Center of the International Astronomical Union, the international clearinghouse for asteroid observations. American interest in asteroid research rose after 1945 when it became clear that leading European centers for celestial mechanics, particularly the Berlin Rechen-Institut, had been incapacitated during the war. (Reprinted from *Scientific Monthly* 70[6] [January 1950]; courtesy Archives and Rare Book Department, University of Cincinnati Libraries.)

Harrison Brown in his laboratory at the California Institute of Technology, ca. 1955. Brown applied the technique of mass spectroscopy to meteorite analysis, initiating new work in solar system geochemistry. His student Claire Patterson became the first to determine the modern figure for the age of Earth of roughly 4.5 billion years. (University of Chicago Archives.)

The Rancho Sante Fe conference, California, Jan. 23–25, 1950. One of the most influential interdisciplinary conferences in the mid-twentieth century, its attendees debated the formation of Earth, the structure of Earth's core, and the process of mountain building. Participants included Louis Slichter, Beno Gutenberg, Urey, Linus Pauling, Edward Teller, Fred L. Whipple, John Verhoogen, Harrison Brown, and David T. Griggs. (Courtesy Leon Knopoff and the Niels Bohr Library, American Institute of Physics.)

Introduction

August 1955 – Gerard P. Kuiper, anxious and exhausted, arrived in Ireland. One of over five hundred delegates bound for the ninth General Assembly of the International Astronomical Union in Dublin, Kuiper, a stocky, intense-looking man, was distracted by his escalating conflict with Harold C. Urey. Three years earlier, as president of the IAU's section on planetary research, Kuiper had invited Urey, the physical chemist and Nobel laureate, to join the IAU. At the time Kuiper and Urey had collaborated for more than two years on the chemical evolution of planets and their atmospheres. Working with Urey, Kuiper had written a colleague, would allow him to resolve important aspects of planetary formation, since Urey knew the "necessary chemistry" needed to advance knowledge on this problem.[1]

Now all had changed. Furious with Kuiper for what he regarded – not without justification – as a pattern of misleading citations to his work, Urey had mailed an inflammatory denunciation of Kuiper to nearly a hundred members of the IAU, indirectly calling for Kuiper's ouster as president of the planetary commission. Although Kuiper's briefcase bulged with new results and ambitious plans for expanding lunar and planetary research, he spent considerable time defending himself against public charges from a noted member of his commission, an embarrassing episode for the IAU and one without precedent in the organization's history. Moreover, Urey had condemned Kuiper's recent lunar work as chemically naïve in the *Proceedings of the National Academy of Sciences,* ensuring that their controversy would come before the entire U.S. scientific community.[2] Kuiper typed a heated rejoinder to Urey's article after he departed Dublin. Edwin B. Wilson, the Harvard zoologist and *Proceedings* editor, urged Kuiper to retract his bitterest charges. "This sort of situation," Wilson lamented, "always occurs in these complicated fields."[3]

Kuiper did mute his published reply and, with the strong backing of IAU leaders, retained his presidency of the planetary commission. After a time he resumed his fruitful studies of the the solar system, later with new collaborators; but the incident

[1] Kuiper to Bengt Strömgren, Nov. 13, 1950, Box 14, GPK.

[2] Urey (1955b).

[3] Kuiper to O. Struve, Aug. 15, 1955, Box 14, and Wilson to Kuiper, Sept. 6, 1955, Box 18, GPK.

revealed stresses characteristic of interdisciplinary research. Solar system studies were not contained within the discipline of astronomy alone, but instead were pursued at the intersection of existing disciplines. Researchers interested in such fundamental questions as the age, origin, and evolution of Earth and other planets found it necessary to work with members of a wide range of disciplines to achieve trustworthy results. During the 1920s and 1930s solar system astronomy was praised as a promising example of "cooperative research," a catchall phrase for the interdisciplinary programs that many leading researchers regarded as the strength of U.S. science. After 1945, solar system studies benefited from federal and military interest in global climatology and weather modification, isotope geochemistry, and high-pressure impact phenomena. In the mid-1950s, rapidly escalating federal patronage for the International Geophysical Year, the National Science Foundation (NSF), and the nascent space program caused important social, instrumental, professional, and disciplinary transformations in this field, making solar system research a distinct discipline. How solar system studies evolved in the twentieth century, and how they helped to shape the earth sciences in America, are the subjects of this book.

Astronomy in America grew ever more complex in the twentieth century. A tiny but thriving community in 1900, astronomers expanded their intellectual and institutional horizons by building new telescopes, creating centers of graduate training, and by addressing an ever wider set of theoretical and practical problems. Already by 1910, aided by large telescopes located under the dark skies of Arizona and California, American astronomers became leaders of observational astrophysics, discovering that spiral nebulae are themselves galaxies and that the universe is expanding, two of the most significant concepts in early-twentieth-century astronomy. Until the launch of *Sputnik* in 1957, U.S. astronomy remained one of the smallest of the physical sciences, and stellar and galactic astronomers among the last affected by the flood of government funding for research that followed World War II. These developments have received considerable attention from scientists and historians.[4]

Far less attention has been devoted to studies of the solar system during this time. Compared to stellar astrophysics, solar system research was a relatively modest enterprise. Only a handful of individuals worked on these problems full-time, joining other astronomers who occasionally contributed to the field while pursuing long-term investigations of solar flares, variable stars, or stellar atmospheres. Yet while these studies were distinct from the core problems of American astronomy, they led to startling reconceptualizations about the composition and dynamics of planetary atmospheres, the origin of comets and meteors, the influence of celestial impacts on planetary evolution, and the birth of the solar system. These studies, often accomplished jointly with geochemists, geophysicists, geologists, and meteorologists, challenged prevailing estimates of the age of the universe, led to modern age estimates of Earth, and entered British–U.S. debates over the superiority of the steady-state versus

[4] North (1990), Pannekoek (1961), and Smith (1982). For a historiographic overview, see Rothenberg (1985).

Big Bang cosmologies. Their evidence for high rates of impact collisions over geological time reopened fundamental debates over uniformitarian and catastrophic processes in geology. Distinct research schools for these studies also took root, albeit fitfully, at Harvard, Berkeley, Chicago, the Dominion Observatory of Canada, and within the U.S. Geological Survey.

What makes solar system astronomy particularly interesting and worthy of study is its interdisciplinary character. In recent years historians have begun to focus on the increased specialization of twentieth-century science, and the challenges that scientists face when addressing broad problems involving Earth and other planets as a whole – difficulties that persist despite considerable scientific and public interest in such issues as climate change and mass extinctions of species. A key characteristic of modern science has been the creation of integrated scientific programs, designed to counterbalance the continued fragmentation of the natural sciences and the entrenched authority of scientific disciplines. Scientists often find disciplinary boundaries difficult to transcend; as John Servos has noted, "the successful discipline affords its practitioners social and intellectual security, institutional support, and a sense of direction and opportunity. . . . Essential to the sustained accumulation of facts, the elaboration of ideas, and the transmission of technique, disciplines are at least partly responsible for giving modern science its cumulative and progressive character."[5] Efforts to establish interdisciplinary structures have thus required specific intellectual stimuli. In the earth sciences, for example, the argument that impact collisions precipitated past episodes of mass extinction has caused heightened interactions between geologists, physicists, astronomers, and paleontologists, just as recognition of emergent properties of complexes has led to increased collaboration between chemistry and physics. Social factors outside science proper also influence the development of interdisciplinary fields, however. The unity of the "environmental sciences," as Peter J. Bowler points out, was "not created by the sciences themselves," but rather by public perceptions of potential losses resulting from human activities.[6]

Solar system research in America was similarly shaped by both intellectual and social pressures. Although several challenges that researchers faced – such as the origin of meteors and comets – were largely resolved by astronomical instruments and methods, many others required direct and sustained collaboration with members of other disciplines. Investigations of possible global circulation in the atmospheres of Mars and Jupiter were not undertaken until astronomers and meteorologists came together to analyze planetary atmospheric and radiometric data. The efforts of astronomer Carlyle S. Beals to test Ralph B. Baldwin's predictions about the structure of terrestrial impact scars during the 1950s succeeded only because Dominion Observatory geophysicists willingly participated in these studies. However, this should not be interpreted to mean that such collaborations were motivated and enabled

[5] Servos (1990, xiv); see also DeVorkin (1993), Kevles (1977), Olesko (1991), Pauly (1984), and Whitley (1976).

[6] Bowler (1992, 2); see also Nash (1982, 182–99) and Schweber (1993).

solely by scientific considerations. It was the fascination of John C. Merriam, president of the Carnegie Institution of Washington from 1920 to 1938, in interdisciplinary science – and the CIW's enormous resources – that created the Moon Committee, a significant but failed attempt to link astrophysics with geology and geophysics. National security concerns during the early years of the cold war had no less profound influence on interdisciplinary research. Large contracts for studies of planetary meteorology and the upper atmosphere arose from military interest in the atmosphere as the medium through which ballistic missiles would travel, and the nascent hope that weather and climate could be modified for military advantage; it was the latter issue that drew military scientists to John von Neumann's famous meteorological computation project at Princeton.[7] Atomic Energy Commission anxieties over locating or creating additional supplies of fissionable elements made them patrons of Eugene M. Shoemaker's studies of atomic explosion craters at the Nevada Test Site, a springboard for Shoemaker's critical comparative studies of terrestrial and lunar craters. The International Geophysical Year and, still more broadly, the international space race that abruptly followed the launch of *Sputnik,* were similarly shaped by overt political pressures that created as well as constrained opportunities for scientific collaborations.[8]

Writing the history of an interdisciplinary science field thus requires that we examine intellectual achievements not in isolation but within appropriate social, professional, institutional, and disciplinary contexts. We do well to remember that the borders of modern disciplines are themselves historical artifacts, not a reflection of the natural world. Moreover, these demarcations vary, sometimes in important ways, among scientifically advanced nations. For example, during the first half of the twentieth century, the disciplinary boundaries of astronomy in the United States differed from those in England. The Royal Astronomical Society's Geophysical Committee, established in 1919, had no clear counterpart in the American Geophysical Union (AGU) or the American Astronomical Society (AAS), meaning that the sort of casual after-dinner conversations between geophysicists and astronomers at the Royal Astronomical Society were typically more difficult to arrange in the United States.[9] Careful attention must also be paid to the role of key individual collaborators, methodological and professional disputes, access to new instruments, and the extent and type of patronage. Finally, one must also bear in mind that these conditions change markedly from one historical period to another. The kinds of interdisciplinary cooperation made possible by Rockefeller-funded grants in the 1920s, for instance, differed from those enabled by military contracts in the early 1950s. To assess the history of solar system astronomy in America, we must comprehend what

[7] "Note on the Chicago Conference on Problems in Meteorological Research," Dec. 9–13, 1946, Box 37, HW; further details appear in Chapter 2.

[8] Bulkeley (1991), Doel (1992), and Needell (in press-a,b).

[9] Jarrell (1988, 5) and Tayler (1987, 24–5).

Charles Rosenberg has termed the "ecology of knowledge" of the modern physical sciences.[10]

The novelty of the interdisciplinary collaborations required for solar system astronomy make social and professional factors particularly crucial for understanding its historical development. How important these factors were is readily illustrated by comparing solar system research with an older and more successful interdisciplinary field, stellar astrophysics. Born of the merger of classical telescopic observation, the recording medium of photography, and the application of chemical spectroscopy carried out by Hermann C. Vogel, Norman Lockyer, Johann Zöllner, and others in the mid and late nineteenth century, astrophysics made it possible for astronomers for the first time to examine the temperatures, chemical compositions, and velocities of the Sun and other stars. By the first decade of the twentieth century, astrophysics had become the dominant branch of astronomy in the United States. It had dedicated institutions (including Yerkes and Mount Wilson, whose spectroscopic facilities were closely linked with the nascent California Institute of Technology), a sympathetic professional society (the American Astronomical Society), and a respected publication, the *Astrophysical Journal.* One of its strongest promoters, George Ellery Hale, was simultaneously one of America's leading researchers and institution builders. In 1927 Henry Norris Russell, dean of American astronomers, jubilantly declared that the boundaries between physics and astronomy had been erased; but he had certainly overstated the case. For example, astrophysicists continued to quarrel with physicists over such fundamental issues as how stars produced energy, the physical implications of stellar classifications, and the end result of stellar evolution. Even by 1940, major differences between these communities remained unresolved.[11]

Frictions within astrophysics, however, paled in contrast to those that faced scientists attempting to forge new fields of "astro-meteorology," "astro-chemistry," "astro-geology," and "astro-geophysics." Astronomy and physics had long been close disciplinary neighbors. After 1920, many individuals who earned their doctorates in astronomy received ample training in classical mechanics, electromagnetism, nuclear physics, and spectroscopy, core fields of U.S. departments of physics; later increasing numbers received training in quantum mechanics as well. Beginning in 1942, astronomers were frequently recruited to wartime research positions advertised for physicists, and in 1946 a physicist and spectroscopist, Ira S. Bowen, was appointed to direct the Mount Wilson–Palomar Observatories, then home to the world's largest telescopes. Such links were largely absent between astronomers and members of more distant disciplines. Ph.D. programs in astronomy did not require graduate students to study meteorology or planetary chemistry – and few students found their way to these subjects on their own. Moreover, astronomers occasionally

[10] Rosenberg (1979); see also Fleming (1990, ix–xi), Kohler (1982, 1–23), and Servos (1990, xiii–xvii).

[11] DeVorkin (1984, 48), DeVorkin and Kenat (in press), Hufbauer (1991), Pliotkin (1978), and Smith (1982, 170–94).

scorned the chemical and earth sciences as excessively descriptive and less esteemed than the disciplines of physics and mathematics; as one astrophysicist recalled, astronomers felt "a natural scientific prejudice that links between astronomy and physics were more important than links between astronomy and chemistry or geology."[12] In contrast to the spectrometer, the defining instrument for astrophysics, there were few common instruments in solar system research, and clashes over methodologies, worldviews, and disciplinary cultures were commonplace. It is hardly surprising that many researchers who tried to work at the periphery of their disciplines felt socially and intellectually isolated.[13]

Despite these impediments, many astronomers, meteorologists, and earth scientists (and their patrons) still found interdisciplinary research alluring. One powerful magnet was the professional and individual rewards for opening up new spheres of knowledge. The possibility of forging a new field of "astro-chemistry," intended to yield fundamental new insights about planetary compositions and the solar system's origin, was what drew Kuiper and Urey toward their ill-starred collaboration in the first place. The intensity of their later clash became a reverse image of their initial exhilaration. Similar motivations drove Harlow Shapley, director of the Harvard College Observatory, to plan an ambitious institute of cosmogony at Harvard in the 1920s, intended to embrace physics, astronomy, chemistry, and geophysics; he pursued this hope as vigorously as his more widely known efforts to chart the structure and size of the Milky Way. Interdisciplinary programs in astronomy were also attractive to some astronomers because the discipline was small and intellectually unified, at least until the mid-1950s. With only slightly over two hundred active members, the American Astronomical Society remained an integrated body; not until the 1960s, when the *Sputnik* crisis precipitated massive increases in astronomy budgets, did it finally add separate divisions to accommodate its swelling ranks of solar physicists and planetary scientists, whose professional identities had grown increasingly distinct.[14] This circumstance allowed leaders of solar system astronomy, including Kuiper, Harvard astrophysicist Fred L. Whipple, and mathematical astronomer Dirk Brouwer of Yale, to maintain high status within the *discipline* of American astronomy as well as within this *field*. For example, Kuiper was an AAS Councilor between 1949 and 1952, while Whipple (1948–50) and Brouwer (1949–51) served as AAS Vice-Presidents. No less significantly, Kuiper led the Yerkes–McDonald observatories between 1958 and 1960, Whipple Harvard's Department of Astronomy between 1949 and 1956 (and thereafter the Harvard-based Smithsonian Astrophysical Observatory), and Brouwer Yale's Department of Astronomy between 1941 and 1966.

A third and final factor that aided interdisciplinary research in early-twentieth-century America was the ideal of cooperative research. Borrowed from and in-

[12] Greenstein (1986, 18).
[13] Abir-Am (1987; 1988), Coben (1971), and Kohler (1991, 303–6).
[14] Tatarewicz (1990a, 122).

formed by progressive social philosophy ("cooperation" appeared as frequently in diplomatic cables of the 1920s and business rhetoric as in scientific proposals), cooperative research became a strategy for maximizing return from the nation's young but rapidly maturing science institutions. Cooperative research sometimes implied interinstitutional collaborations, but most scientists understood it to mean interdisciplinary research. Blessed by Hale, Merriam, and other leaders of early-twentieth-century American science, interdisciplinary research became a cornerstone of the natural sciences in America during the 1920s and 1930s. The Rockefeller Foundation unapologetically supported the interdisciplinary agendas of its leading managers of science, Wickliffe Rose and Warren Weaver, throughout this period.[15] Although the Rockefeller Foundation turned to the social sciences after World War II, and the CIW became a much smaller player in the political economy of postwar U.S. scientific institutions, favorable attitudes toward interdisciplinary cooperation persisted into the post-1945 era, where they continued to be articulated by scientists and promoted by military and federal patrons. Navy funds supported the meteor research program that Whipple developed at Harvard, straddling the border between meteorology and astronomy and one of the largest postwar U.S. astronomy projects, while the Office of Naval Research and the NSF supported Kuiper's efforts to find common ground between astronomy and the earth sciences.

No institutions solely devoted to solar system research were created before the dawn of the space age in the late 1950s, which some have interpreted as a sign of limited activity in this field. What this argument ignores, however, is the founding of numerous short-lived interdisciplinary communities, or "transient" institutions.[16] Typically funded by patrons outside traditional disciplinary structures, transient institutions permitted intense collaborations among individuals from different disciplines on problems of mutual interest. The 1940s and 1950s saw the creation of such transient institutions as the Planetary Atmospheres Project at Lowell Observatory and the Rancho Santa Fe conference in 1950, both of which had considerable impact on solar system astronomy in America. These agencies gave participants a context to reach consensus on research problems, to plan new instruments or observations, and to hold extended, informal discussions; no less important, they allowed participants to evaluate the data, methodological criteria, and tacit assumptions of scientists in neighboring fields. In few instances were they progenitors of permanent interdisciplinary centers. More frequently, disputes over instrumental techniques and interpretations caused them to dissolve after active periods lasting days, months, or years. Nevertheless, transient institutions were prolific in solar system astronomy at a time when permanent institutions were not; their structure and

[15] Bugos (1989), Cain (1993), and Doel (in press-a), Genuth (1987), Kargon (1977), Kohler (1991), and Servos (1983). In the decentralized, "associative" U.S. political economy after World War I, the ideal of "cooperation" had broad resonance; see Hawley (1992) and Rosenberg (1982).

[16] I use the term "transient institution" much as Spencer R. Weart (1992b) does in his examination of solid-state physics, although this field was largely confined *within* the discipline of physics.

aims reflected the ambitions of interdisciplinary-minded scientists and the goals of their patrons. They are equally revealing of both.

Evidence of strong interest in planetary questions in early- and mid-twentieth-century America, I hope, will stimulate reassessments of the emergence of the planetary sciences, which have long been regarded as neglected or abandoned fields of U.S. science prior to the start of the space age.[17] It is true that the difficulty of making trustworthy spectroscopic observations of the planets constrained studies of these objects; their cool temperatures made it far more difficult to record and interpret spectroscopic information about them than the far hotter (and vastly more numerous) stars. Nor can it be denied that, in an obvious way, our knowledge of the planets has expanded by several orders of magnitude since direct lunar and planetary explorations began in the late 1950s, which Stephen J. Pyne has termed the start of a "Third Great Age" of exploration.[18] However, instrumental difficulties and the smaller scale of an earlier scientific era does not add up to neglect. More important, careful examination of manuscript sources reveals that astronomers first voiced anxiety about possible "abandonment" of their field only in 1956, precisely when profound shifts in federal patronage for astronomy and geophysics fractured American astronomy into distinct subcommunities and made it likely that new, *distinct* institutions for solar system research would be needed. This was a disturbing change for old-line astronomers like Kuiper, at home in traditional astronomical facilities and their stable patronage structures. It is also helpful to bear in mind Nathan Reingold's reminder that "strongly committed scientists [are] very conscious of the gap between needs and actual support. . . . We can even speculate that complaints are a sign of a healthy, lusty scientific community. Their absence may denote dangerously smug self-satisfaction, stagnation, or actual decline."[19] By using original manuscripts and letters whenever possible, and by ending this account at the birth of the space age, I have endeavored to illuminate a field of science formed under conditions remarkably different from those of the Big Science era that followed. At a time when the planetary sciences have become more important for society, it is all the more imperative to understand the history of solar system astronomy, and its principal architects, on their own terms.

[17] Doel (1990); Donald E. Osterbrock (1991) independently voiced similar conclusions. Past accounts by planetary scientists stressing neglect include Cruikshank (1991) and Sagan (1974). These views are uncritically reported in Tatarewicz (1990a, 1–2); see also Brush (1978b).

[18] Pyne (1986, 82–5, 108–15).

[19] Reingold (1972, 50; 1994).

I

American Astronomy in the Foundation Era, 1920–1940

For many Americans who came of age during the Great Depression, solar system astronomy is indelibly associated with a young Kansas astronomer at a famous western observatory. In 1929, Clyde Tombaugh, a tall, gangling, 23-year-old amateur astronomer, left his home farming community of Burdett for the Lowell Observatory, set on a hill overlooking Flagstaff, Arizona. Tombaugh had been hired to help locate "Planet X," whose existence had been hypothesized by Percival Lowell in 1905. A former diplomat and writer, brimming with restless energy and large ambitions, Lowell had founded this hilltop observatory in 1894 in then-remote Arizona to study Mars and what he insisted was its network of artificial canals; but he also investigated the gravitational attraction of planets, finding discrepancies in the orbit of Neptune that led him to predict an undiscovered, more distant world. Within a year, Lowell astronomers announced that they had found a ninth planet. Subsequently named Pluto, its identification made Tombaugh famous worldwide. It also gave the Lowell Observatory – already an icon in American popular culture – renewed fame and attention.[1]

The discovery of Pluto, the only planet found in the twentieth century, figures prominently in past histories of American astronomy. It is often viewed as an isolated achievement, made when professional astronomers devoted almost all their energies toward stellar and galactic research.[2] Yet like many often-told accounts, this story is only partially correct, and hides a larger, more interesting history. Lowell Observatory certainly occupies an important place in the history of planetary astronomy and American astrophysics. After Lowell's death in 1916, however, staff astronomers restricted their planetary research to photovisual studies of planetary atmospheres and the search for Planet X. Moreover, Lowell Observatory had no direct university links, heightening its geographic and intellectual isolation. It lacked the

[1] On Pluto's discovery, see especially Hoyt (1980a, 161–219). The notion of "discovery" in science is problematic; see Smith (1989).

[2] See, e.g., Cruikshank (1991), de Vaucouleurs (1960), and Sagan (1974); see also Tatarewicz (1990a). On Lowell and the canal controversy, see Crowe (1986, 502–46) and Hoyt (1976). On Tombaugh, see Levy (1991) and Tombaugh and Moore (1980).

capacity to train new researchers, and its senior staff cultivated few links with scientists in neighboring disciplines.[3]

Solar system astronomy was pursued more vigorously at other centers of American astronomy, often as extensions of stellar astrophysics. This work was limited by factors common to all fields of science, including available data and instruments, institutional mandates, disciplinary boundaries, and patronage. However, this field was neglected by neither astronomers nor their patrons. Indeed, solar system research illuminates circumstances that particularly aided interdisciplinary science fields in the United States after 1920. How these research programs emerged – and how they influenced early-twentieth-century American astronomy – is the subject of the present chapter.

THE RESOURCES OF THE FIELD

As the twentieth century dawned, astronomers in the United States and Europe remained fascinated with the origin and development of the solar system, much as they had throughout the previous century. Such problems as the evolution of planetary atmospheres, the origin and structure of comets, the creation of asteroids, and the formation of satellite systems held great appeal, largely because few satisfactory explanations for them had been achieved. Moreover, because the discipline of astronomy was not yet differentiated into component fields, astronomers worked simultaneously on problems in solar system, stellar, and nebulae research. Textbooks of the time devoted roughly equal space to stellar and solar system topics, the same ratio found in texts written half a century earlier.[4] The 1902 *Manual of Astronomy*, by Princeton's Charles Young, devoted nine chapters to the Sun's family, three more than those addressing the Sun, stars, and nebulae. An 1896 text by the British astronomer and writer Agnes Clerke similarly dedicated six of twelve chapters to new results about the Moon, planets, and comets obtained by visual, photographic, and spectroscopic techniques.[5]

Solar system astronomy at the time, and for many years afterward, comprised two distinct branches. Mathematical astronomers studied the solar system to advance the theory of gravitation. Celestial mechanics remained the main branch of astronomical research through much of the nineteenth century. British astronomer Robert Grant had proclaimed in 1852 that, through planetary studies, "the development of the Theory of Gravitation had attained a high state of perfection, and the various phenomena which had seemed irreconcilable with its principles, were all satisfactorily accounted for"; but by the 1890s mathematical astronomers realized that difficulties plagued several important theories, particularly those of the Moon and Mer-

[3] Marsden (1973) and Hoyt (1980a, 142–219).
[4] Dick (1990, 33) and Herrmann (1984, 59–60).
[5] Clerke (1896) and Young (1902).

cury. Celestial mechanics thus continued to attract leading scientists, including the distinguished French mathematician and physicist Henri Poincaré. Other astronomers placed greater emphasis on recording occasional planetary details revealed by the large refracting telescopes then standard at many observatories, an undertaking that resembled contemporary research efforts in natural history. One reason why astronomers found visual study of the Moon and planets appealing was that physical studies of stars and faint nebulae were correspondingly harder to make. As Simon Newcomb, director of the U.S. Naval Observatory, declared in 1887, "[p]roblems lying beyond the range of the solar system" until recent times "were almost unheeded, because they seemed inscrutable."[6]

Although interest in the solar system remained high after 1900, fewer American astronomers took up planetary research in the twentieth century as compared to the nineteenth. At the Harvard College Observatory, for example, where visual studies of solar system objects had been commonplace from its founding in 1843 through the 1880s, the focus of its staff increasingly became stellar research. In 1905, Harvard College Observatory director Edward C. Pickering, responding to a request from Newcomb to describe astronomical research underway at Harvard, listed only one program ("Photometric Observations of the satellites of Jupiter") that dealt with planetary phenomena. Pickering made no effort to justify the Harvard program as in any way unusual. Astronomers at the large observatories in the American Midwest and West, newly established around the turn of the century, also concentrated increasingly on stars and nebulae. By 1910, stellar research became the principal interest of astronomers at major observatories such as Yerkes (Williams Bay, Wisconsin, completed 1897), the Lick (Mount Hamilton, California, 1888), and Mount Wilson (near Pasadena, California, 1908). Since these three observatories dominated observational astronomy and astrophysics in the United States and Europe after 1920, interpreting stellar phenomena became a central concern for many astronomers.[7]

Several developments caused this shift of research emphasis from planetary to stellar research. Of critical importance was photography's rise as an astronomical technique. By the beginning of the twentieth century, time-exposure photographs could record stars several magnitudes fainter than those glimpsed by visual observers employing large telescopes, and revealed both the extent and structure of nebular objects. Photographs also provided a permanent record of stellar positions with an accuracy surpassing that of visual methods, permitting astronomers to make astrometric measurements in far less time. Astrophotography, however, provided fewer benefits to lunar and planetary study. Planets proved extremely difficult to photo-

[6] Quotes from Grant (1852, iii) and Newcomb (1887, 7); see also Herrmann (1984, 54).

[7] Newcomb to Pickering, Nov. 15, 1905, and Pickering, "Works of the Harvard Observatory, 1785–1905," n.d. [ca. 1905], both Box 3, ECP. This same list announced the discovery of a new satellite of Saturn from Harvard's Southern Hemisphere Arequipa station; on Harvard and early-twentieth-century American astronomy, see Jones and Boyd (1971) and Plotkin (1990; 1993a,b).

graph, since fine detail glimpsed during brief moments of calm "seeing" (when atmospheric turbulence is at a minimum) greatly exceeded that recorded by photographic plates, in part because required exposure times exceeded intervals of good seeing. "The invention of photography," lamented the Mount Wilson astronomer Theodore Dunham, Jr., "has not quite had the same revolutionizing effect on the study of planetary surfaces than it has had on stellar and nebular investigations and on spectroscopy."[8]

A second factor was the rise of the reflecting telescope as a serious competitor to the long-focus refracting telescope, long astronomy's standard research tool. As photography became central to the practice of astronomy, astronomers grew increasingly troubled by chromatic aberration and other problems of refractor optics, a deficiency that "photographic" refractors developed in the late nineteenth century failed to eliminate. Nebular photographs produced by Edward Keeler in 1906 with the Lick Observatory's 36-inch reflector – coupled with advances in the silvering of telescope mirrors – convinced astronomers that reflectors were better suited than refractors for astronomical research. Thereafter astronomers employed reflectors in all new American observatories, including the 100-inch instrument completed at Mount Wilson in 1918. Although reflecting telescopes also proved excellent instruments for solar system research, the traditional long-focal-length refractors they replaced were optimal for visual observations of the planets.[9]

The most significant development to stimulate stellar and nebular research, however, was the application of spectroscopy to astronomy. The linkage of physics with astronomy, well underway by the turn of the century, transformed both the practice and institutional structure of American astronomy, bringing laboratory spectroscopy into close contact with observational astrophysics; after 1900, all major new observatories were equipped to pursue astrophysical investigations. Astronomers valued astrophysics as a means to determine not simply the movements of planets and stars but also their compositions, temperatures, periods of rotation, and other physical characteristics; yet astronomers found it extraordinarily difficult to extend astrophysics to the planets. Compared to the constantly writhing Sun, for example, the Moon produced few phenomena to investigate, and spectral lines observed in planetary atmospheres proved harder to interpret than those generated by stars and nebulae. David Todd, the American astronomer, complained in 1897 that whereas the atmospheres of Jupiter, Saturn, Uranus, and Neptune all revealed "the presence of gaseous elements as yet unrecognized on the earth and in the Sun," existing techniques provided little information about them or what their presence implied. Samuel P. Langley, the American pioneer of astrophysics, similarly declared that even when astrophysical instruments were aimed at Mercury and Venus, "the great

[8] Dunham (1938, 258) and Turner (1904); see also Lankford (1984).
[9] Lankford (1983) and Van Helden (1984, 57).

telescopes of modern times cannot do much more than those of Galileo."[10] In practical terms, as debates over whether spiral nebulae are gaseous objects within the Milky Way or distant, individual galaxies grew intense in the 1920s, American astronomers concentrated on problems in stellar and galactic astronomy that seemed capable of solution, and trained students to apply astrophysical techniques to these problems.[11]

A final factor that diminished the appeal of planetary astronomy in America was the bitter turn-of-the-century controversy over the canals of Mars. This famous battle, which Lowell waged against a majority of his astronomical colleagues, was one of the most public of its day.[12] Fascinated by the reports of such European astronomers as Giovanni Schiaparelli that canallike markings were visible on Mars in moments of good seeing, Lowell used his personal resources to build the Lowell Observatory in the dark skies of the American Southwest, and undertook to educate himself in mathematical astronomy. By 1910, with its new 44-inch reflector, Lowell Observatory ranked among the largest astronomical facilities in the United States. Its research staff, including Vesto M. Slipher, were later recognized for their contributions to stellar and nebular astronomy as well as to planetary studies (e.g., Slipher discovered the large redshifts in spiral nebulae). However, during Lowell's lifetime – and indeed for decades afterward – a majority of Lowell staff astronomers worked tirelessly to find evidence supporting Lowell's claim that Mars is essentially Earthlike, its canals artifacts of an advanced civilization irrigating the parched agricultural lands of an evolved, dying world. Between 1905 and 1912, Lowell engaged in bitter polemics with leaders of other American observatories, including Lick's Wallace W. Campbell, who disagreed with his visual interpretations of canaliform markings and the implications Lowell assigned to them. Rich in cultural symbolism – the Panama Canal was then under construction – the controversy became one of the most embarrassing and disturbing episodes experienced by American astronomers, just then beginning to organize themselves as a distinct professional community.[13]

The Martian canal controversy caused several distinct problems for American astronomers. One was that it exposed the weakness of visual methods for studying planetary features just as photography became the discipline's chief standard of evi-

[10] Langley (1888, 120–1), Meadows (1984, 59), Smith (1982, 5), and Todd (1897, 350).

[11] Meadows (1984, 61) and Smith (1982, 7).

[12] Hoyt (1976, 173–186, 190–200) has argued that the emphasis staff members placed on securing precise data to rebut critics enhanced the observatory's reputation for accurate observations, helped lead to Pluto's discovery, and generally heightened interest in the planets. However, it seems more correct to argue, as Tatarewicz and Crowe have done independently, that the astronomical community paid a high price "in loss of credibility, internal discord, methodological misconceptions, and substantive errors, as well as the efforts wasted on the observation of ambiguous detail" (Crowe 1986, 545); see also Tatarewicz (1990a, 3–5).

[13] Lowell (1911); on Lowell Observatory, see Crowe (1986, 502–46), DeVorkin (1977), Hoyt (1976; 1980a, 83–219), Strauss (1994), and Webb (1983, 13–69). Tombaugh joined other staff members in sketching Martian canals; see Levy (1991, 57).

dence. Although an observer could see far greater planetary detail visually in a 10-inch telescope than on a photograph with a 100-inch instrument, visual observations quickly gained a reputation among American astronomers as unreliable.[14] Moreover, the controversy could not be resolved by experiment or further observation. Although such senior Lowell astronomers as V. M. Slipher, his brother Earl C. Slipher, and Carl Otto Lampland remained faithful adherents of Lowell's vision of an Earthlike Mars, this view was shared by only a minority of American astronomers, including Lick's Robert J. Trumpler. The canal controversy, though muted by Lowell's death in 1916, thus resembled a time bomb that threatened to split the community over one of astronomy's most highly charged philosophical issues: the existence of life on other worlds. Although the declining resources of Lowell Observatory during the Great Depression lessened anxiety among American astronomers about the canal controversy, they hardly ended them. As late as 1950, Gerard P. Kuiper privately voiced relief when at a Lowell-sponsored conference "the question of the canals on Mars did not come up." Not surprisingly, visual planetary observations in the United States, more than in France or in the Soviet Union (where the canal controversy attracted little notice), largely became the province of amateur astronomers.[15]

American astronomers also worried that the canal controversy could curtail their support from major American foundations, then the discipline's most critical patrons. This anxiety was particularly evident in the actions of George Ellery Hale. A skilled astrophysicist and an even more successful organizer, Hale's talents for fund-raising and dedication to institution building made him by World War I a leader of American astronomy and of U.S. science. By the early 1920s, Hale had developed close ties with such individuals as Wickliffe Rose, director of the Rockefeller-funded International Education Board. Already he had secured funds to build the 60- and 100-inch reflectors at Mount Wilson from the Carnegie Institution of Washington, and had employed Rockefeller support in his ambitious plan to refashion the former Throop Academy in Pasadena into the California Institute of Technology. He had relied on similar networks of patrons, including the General Education Board, to create the National Research Council (NRC), his preferred means to stimulate interdisciplinary research.[16]

For Hale, Lowell's insistence on an inhabited, canal-girded Mars – and his willingness to attack publicly such senior astronomers as Campbell and Edward E. Barnard – threatened disaster for American astronomy by marring its image as a serious, mature scientific community. Well aware that foundation patronage was essential for his postwar plans to build a 200-inch telescope on Palomar Mountain,

[14] Bobrovnikoff (1951, 303); see also Whipple (1941, 208).

[15] Quoted from Kuiper to G. de Vaucouleurs, Sept. 6, 1950, Box 28, GPK; see also V. M. Slipher to R. L. Putnam, Feb. 6, 1928, LOW; G. de Vaucouleurs OHI (Ronald E. Doel, Nov. 20 and 23, 1991, at AIP); and Crowe (1986, 524), DeVorkin (1977, 52), Hoyt (1976, 173–236), Tatarewicz (1990a, 3–6), Tombaugh and Moore (1980, 83), and Whipple (1941, 224–5).

[16] Hufbauer (1991, 75) and Kohler (1991, 90–108).

Hale wanted no reminders of the canal controversy to thwart his disciplinary and institution-building ambitions. In a famous article about the proposed telescope that he placed in *Harper's Magazine* in 1928, Hale solely extolled the excitement of galactic astronomy, never mentioning the solar system even while recounting the history of astronomy since Galileo. His omission was deliberate: As he later told Rose, he had sidestepped this field "in the hope of heading off the inhabitants of Mars!" For Hale, like many American astronomers, Lowell's shadow still loomed large. Staff members at Hale's Mount Wilson Observatory did turn to solar system astronomy on occasion during the 1920s and 1930s, typically with Hale's blessing, but took care to avoid sensational popular accounts of their work. Since Hale also held the reigns of the *Astrophysical Journal,* and had considerable influence within the American Astronomical Society (AAS), controversial aspects of solar system research received limited attention.[17]

These various factors – the advance of photographic techniques, the rapid rise of astrophysics involving stellar and galactic objects, and the embarrassment of the Martian canal controversy – made solar system research less central to the discipline of astronomy in America after 1900 than during the nineteenth century. However, it would be a mistake to see solar system astronomy as a field in sharp decline, or to identify it solely as the province of Lowell Observatory. American astronomy was not yet the differentiated discipline it would become after the late 1950s. Nor was solar system astronomy yet a distinct field; rather, the constitutive elements of what would come to be solar system astronomy by the 1950s (including comet, meteor, and planetary atmosphere research) then existed only as *problems* within the discipline. Throughout the early twentieth century, American astronomers remained fascinated with large, unresolved problems such as the origin of comets and meteors, the nature of planetary atmospheres, and the origin of the planetary system itself. These problem areas became magnets for interdisciplinary collaborations, making novel cross-field research programs possible.

Programs devoted to aspects of solar system astronomy took root at several American observatories in the early twentieth century. The remainder of this chapter examines three important branches of this field active between 1920 and 1940. The first branch was mathematical astronomy, particularly studies of asteroid orbits. The others were research programs begun as adjunct investigations of stellar phenomena: the study of meteors and the study of planetary atmospheres. These latter lines of investigation were undertaken not to understand the solar system per se, but instead to comprehend the composition and structure of the universe at large. To understand the subsequent growth of this field, and the institutional structures that sustained it, we must first examine these programs in detail.

[17] Hale to Rose, June 14, 1928, Box 21, Folder 313, RF (quoted) and Hale (1978); see also Crowe (1986, 512, 544), DeVorkin (1977, 43, 51), Kohler (1991), and Miller (1970, 179–81).

CELESTIAL MECHANICS AND ASTEROIDS

Most historical studies of early-twentieth-century astronomy make little mention of mathematical (or dynamical) astronomy, focusing instead on developments in stellar spectroscopy and galactic studies. However, as Karl Hufbauer has pointed out, the rapid rise of astrophysics did not supplant mathematical astronomy in the United States. Instead, "the two fields, though hardly on an equal footing, were generally regarded as astronomy's two principal subdivisions."[18] From the 1920s until the early 1950s, mathematical astronomers interested in studying the overall properties of orbits continued to concentrate on solar system problems. "It is only within the solar system," declared Naval Observatory astronomer and future AAS president Gerald M. Clemence, "that we are able to study gravitational phenomena in a way more or less pure and direct, and it is here that the theory has been developed to the highest degree of perfection."[19] American astronomers investigated not only the orbits of the major planets but those of comets, meteors, and particularly asteroids, seeking to discover clues to their origin and to the origin of the solar system itself.

Although mathematical astronomy declined relative to other fields after the late nineteenth century, even including those workers who labored solely to calculate practical theories and new ephemerides, celestial mechanics remained a viable field of research in the 1920s and 1930s. At smaller observatories lacking the capability for astrophysical work, such as that at the University of Cincinnati, mathematical astronomy provided a way for staff members to engage in research. Research in celestial mechanics was not confined to smaller facilities, however. Forest Ray Moulton, using advanced mathematical approaches, worked on mathematical astronomy and cosmogony at Chicago, while Seth B. Nicholson – trained in celestial mechanics as well as in astrophysics – pursued both as a staff member at Mount Wilson. The U.S. Naval Observatory, whose Nautical Almanac Office produced the annual *American Ephemeris and Nautical Almanac*, also included fundamental work on solar, lunar, and planetary theories within its mandate. When Harvard College Observatory director Harlow Shapley surveyed the institutional landscape of astronomy worldwide for Rose in 1926, he reported that four of the ten leading centers for celestial mechanics were in America. He listed Harvard, Chicago, Yale, and Berkeley.[20]

More than any others, Yale and Berkeley came to define the style and scope of asteroid research in the United States in the early twentieth century. Leaders of their research programs shared several characteristics. Yale's Ernest W. Brown and University of California's Armin O. Leuschner were Americans born during the Civil War years who had taken their Ph.D.'s overseas: Brown under George Dar-

[18] Hufbauer, "Astrophysics' Rise and Evolution into an Interpretative Subdivision, 1860–1940," paper at session on the Emergence of Hybrid Disciplines, History of Science Society Annual Meeting, 1982, MS p. 3 (in my possession).

[19] Clemence (1951, 220); see also Whipple (1941) and Hoyt (1980a, 9–59).

[20] Shapley to Rose, Appendix to Research Survey, Part II, Nov. 20, 1926, Series 1.1, Box 18, Folder 265, IEB; see also Clemence (1951), Herrmann (1984, 149–50), and National Research Council (1922).

win at Cambridge, Leuschner at Berlin. Dirk Brouwer, who joined Yale's staff in 1928 and later replaced Brown, was born one generation later in Holland, and had studied under Leiden's eminent Willem de Sitter. All three individuals trained new researchers, making American astronomy increasingly self-reliant in this field. Moreover, Brouwer, and particularly Leuschner, became skilled scientific entrepreneurs. Their success in securing extramural patronage sustained research programs far beyond what the local capacities of these observatories could afford.[21]

Most important, all three men shared an interest in asteroids. Mathematical astronomers found these bodies fascinating for two reasons. One was the dramatic increase in the number of asteroids found after the turn of the century. By 1872, just 131 asteroids had been discovered, all by visual means. By 1885, however, after the German astronomer Max Wolf began a photographic survey of the ecliptic, this number had leapt to 250 (a rare instance where photography aided solar system astronomy). In the 1930s, the Mount Wilson astronomer Walter Baade estimated that the number of asteroids within photographic reach of its 100-inch reflector was 44,000. Efforts to compute ephemerides for these tiny worlds were quickly undertaken, both to reduce priority disputes over discoveries and to lessen instances of mistaken identity. Since the orbits of asteroids are easily disturbed by the gravity of their larger planetary neighbors, a major, international program was required to track them. A second reason that astronomers turned to asteroids was precisely because their diminutive masses also magnified perturbations in their motions introduced by the larger planets. This made asteroids ideal instruments for the study of gravitational phenomena, particularly resonance theory, and, for orbits with high eccentricity, Einstein's theory of general relativity.[22]

An expert in the lunar theory, Brown became interested in asteroids as a means to subject gravitational theory to ever more exacting tests. He first studied the Trojans, a group of asteroids rigidly influenced by Jupiter's gravity, to see if their stability differed from that of any major planet or other well-known groups of asteroids. In 1923, he extended this study to the Kirkwood gaps, empty orbits in the asteroid belt discovered by the nineteenth-century American astronomer Daniel Kirkwood. Brown announced that a general resonance theory might account for these gaps, important for theoretical work on the three-body problem, but did not pursue the matter further.[23]

One astronomer intrigued by this problem was Brown's junior colleague, Brouwer. Better trained in astrophysics than Brown, Brouwer saw asteroid motions as a vehicle for integrating physical and mathematical styles of research in astronomy.

[21] Clemence (1951, 222; 1970) and Schlesinger and Brouwer (1940).

[22] H. N. Russell to W. D. Alter, Sept. 24, 1925, Box 1, Folder 60, HNR; Brouwer (1935a, 13–16), Clemence (1951, 219), Clerke (1896, 327), Doggett (in press), Herrmann (1984, 29–30), Kirkwood (1873, 10), Russell (1935b, 7), *Wolf* (1933).

[23] Brown (1923, 69), Kirkwood (1888), Watson (1941, 20), Schlesinger and Brouwer (1940, 250–1, 255).

His curiosity likely grew in the mid-1920s when he learned of new asteroid research by Kiyotsugu Hirayama, Professor of Astronomy at the Tokyo Imperial University. By tracing the proper orbital eccentricities and inclinations of asteroids back in time, Hirayama discovered that the longitudes of perihelion and nodes for many of them coincided. This prompted Hirayama to suggest that such asteroids had originated from a common parent asteroid that was later disrupted – an old idea but one presented in mathematical form for the first time by this Japanese astronomer. Hirayama argued for five asteroid "families," each containing fifteen to forty-four asteroids apiece. Fourteen percent of all asteroids then known were included in one of these five families. Brouwer found this result surprising because most astronomers then rejected the "exploded planet" hypothesis for the origin of asteroids; its popularity rose, however, after the dean of American astronomy, Henry Norris Russell, sanctioned it in the late 1920s.[24] In 1935, Brouwer began an asteroid observation program at Yale's Southern Station in Johannesburg, coordinating it with parallel studies at Leiden and the Allegheny Observatory in Pennsylvania. Although this project was partly intended to improve the accuracy of fundamental star charts, for which accurate observations of asteroid positions were crucial, Brouwer was aware that this work would also permit him to carry on work similar to Hirayama's on the origin of asteroids. At about the same time, Brouwer came into close contact with Wallace J. Eckert, a former Brown student who had organized the Thomas J. Watson Astronomical Computing Bureau at Columbia University. Jointly run by the AAS and International Business Machines (IBM), which supplied its Hollerith card-punch machines without cost, the Computing Bureau allowed Eckert and Brouwer to pioneer the use of these machines to refine theories of planetary motion. This undertaking further kindled Brouwer's interest in asteroids and their relationship to the solar system's larger members, topics to which he turned particularly after 1945.[25]

Asteroid research at Berkeley proceeded at an even faster pace. Led by Leuschner, founder and director of the Students' Observatory and chair of Berkeley's Department of Astronomy, the California program sought to create accurate orbits for all known asteroids.[26] Leuschner's fascination with this problem rose early in his career, when he was given responsibility for determining the orbits of twenty-two asteroids discovered by the wealthy amateur astronomer James Craig Watson, who bequeathed his estate for this purpose. While pursuing this task, Leuschner came to reject the classical or "pure" orbital computations that Brown and most mathematical astronomers routinely used to calculate planetary orbits. Believing this method too cumbersome and time consuming for asteroids, whose orbits were frequently perturbed by the major planets, Leuschner instead developed a statistical approach

[24] Brouwer (1935a, 17–18), Hirayama (1923, 18), and Russell, Dugan, and Stewart (1926–7, 357). Late-nineteenth-century opposition to this hypothesis is noted in Clerke (1896, 100).

[25] Bashe et al. (1986, 22–4), Brouwer (1935a,b), McPherson (1984/1940), and Schlesinger (1941).

[26] Herget (1980) and Osterbrock, Gustafson, and Shiloh Unruh (1988, 173–7).

that made use of general tables of perturbations. After demonstrating the promise of this approach with the Watson asteroids, Leuschner proposed extending this technique to the thousands of asteroids for which sufficient observations existed to compute orbits. In the 1920s, Leuschner began sifting through published records for all known asteroids to find the "elements of fundamental value" required by his method. This became the basis of his "Research Surveys" project, one of the most ambitious efforts in celestial mechanics undertaken in the early twentieth century.[27]

Leuschner's mathematical program was important in several ways. Because Berkeley offered the largest U.S. graduate program for astronomy in the early twentieth century, many American astronomers were exposed to mathematical astronomy and celestial mechanics. During the 1920s and 1930s, aided by its close association with Lick Observatory, Leuschner's department graduated on average three to six new Ph.D.'s per year, more than those of Princeton and Chicago combined. Moreover, Leuschner insisted that all graduate students, even those intent on careers in astrophysics, learn methods of orbital computation and mathematical analysis. In 1929, for example, Leuschner resisted pressure from Russell to waive the celestial mechanics requirement for visiting graduate students, declaring that mathematical astronomy remained an essential foundation for all astronomical work. He also encouraged graduate students to take part in the Research Surveys. Through the 1930s this undertaking, whose permanent staff included Holgar Thiele, a former Lick astronomer, was carried out in a large room in the campus library, making it easily accessible to departmental students.[28]

Moreover, Leuschner used asteroid studies to stress a distinct style of research at Berkeley, one that emphasized a merger of mathematical and astrophysical approaches. In his own studies of asteroids and comets, Leuschner avoided astrophysical techniques, preferring to concentrate instead on mathematical investigations of orbital motion. Nevertheless, he urged new astrophysical studies of these bodies to graduate students, and stressed the importance of undertaking orbital analyses the better to understand their physical properties. Increasingly drawn to the idea that asteroids are dormant comets bereft of volatile materials, Leuschner urged new photometric studies of variations in asteroid brightness, and called attention to Hirayama's proposed asteroid families as a pathway for understanding the physical evolution of asteroidal bodies. In 1929, for instance, Leuschner encouraged Nicholas T. Bobrovnikoff, a Russian-born, Chicago-trained astrophysicist interested in solar sys-

[27] Leuschner to W. W. Brierley, Oct. 31, 1927, series SI-1, Box 11, Folder 159, IEB; idem to H. N. Russell, May 22, 1928, Box 27, OAL; idem to F. B. Hanson, May 6, 1935, Box 26, OAL; idem to J. C. Merriam, Dec. 7, 1935, Box 2, OAL; and Paul Herget OHI (David H. DeVorkin, Apr. 19–20, 1977, SHMA at AIP); see also Herrick (1955), Leuschner (1935, i, iv, xi).

[28] Leuschner to F. C. Leonard, Apr. 15, 1930, Box 26, AOL; idem to Russell, Jan. 22, 1929, Box 48, Folder 15, HNR; see also S. A. Mitchell to E. Alderman, Feb. 16, 1926, Box 3, UVa; H. Shapley to W. Rose, Nov. 20, 1926, Series 1.1, Box 18, Folder 265, IEB.

tem bodies, to use an NRC fellowship to study the spectra of comet tails. After Leuschner introduced him to Berkeley physicist Raymond Birge and Lick astrophysicists Trumpler and C. Donald Shane, Bobrovnikoff analyzed physical observations of Halley's Comet during its close passage of 1910, producing a landmark reduction of these data. Leuschner's influence was greater still, however, on his own Ph.D. students. Products of his small but influential research school included Nicholson, W. Dinsmore Alter, Paul Herget, Samuel Herrick, and Fred L. Whipple, all of whom addressed solar system research later in their careers. Although some, like Herrick, focused largely on celestial mechanics, most employed a distinct mix of physico-mathematical approaches, allowing graduates like Whipple later to enter such hybrid fields as geophysics.[29]

More than has been realized, Leuschner's Research Surveys program also helped set international standards for accuracy in asteroid computations – an important point, given that the mathematical skills of U.S. scientists in the early twentieth century were often lower than their European colleagues.[30] Prior to World War I, the task of coordinating international efforts to observe asteroids and computing orbits for them had been handled by the Rechen-Institut of Berlin. Immediately following the war, however, this responsibility was transferred to the newly created International Astronomical Union (IAU). Since Axis scientists were then barred from participating in the IAU, its creation kept Rechen-Institut personnel from resuming their prewar role. Initially Leuschner, raised in Germany and fluent in German, was sympathetic to the request of August Kopff, the Rechen-Institut director, to prepare new asteroid ephemerides informally, and encouraged Lick and Yerkes astronomers to aid Kopff by observing asteroids whose orbits were poorly determined.[31] Later, however, learning of large inaccuracies in many Rechen-Institut ephemerides, and frustrated by Kopff's refusal to use his statistical methods, Leuschner, as president of the IAU's asteroids commission, sought computing reforms. In 1935, the same year that his Research Surveys published precise elements for 1,091 minor planets, Leuschner invited B. V. Numerov's Leningrad Astronomical Institute to become more active in asteroid research, and pressured Kopff to produce more accurate orbits. The outbreak of World War II, and Numerov's tragic arrest and death in the Great Purge of 1936–7, dashed these plans; yet Leuschner's energetic defense of his method helped make him a leader within the early IAU,

[29] Leuschner, "Report of the Committee on Asteroids, Comets, and Satellites," Jan. 2, 1925, Box 25, OAL; idem to H. N. Russell, May 22, 1928, Box 27, OAL; idem to J. C. Merriam, Dec. 7, 1935, Box 2, OAL; Bobrovnikoff (1927); see also Herget (1980, 135) and Osterbrock (1986). Further information on Leuschner's students appears in Appendix Table A.2.

[30] Servos (1986).

[31] Leuschner and Ernest Brown also served on a National Research Council committee devoted to asteroid research; see E. B. Frost, "Committee on Asteroids, Comets, and Satellites," Dec. 3, 1924, Box 25, OAL; Leuschner to Kopff, Dec. 20, 1924, OAL. On the early history of the Rechen-Institut, see Bauschinger (1897).

strengthening the international standing of American astronomers in solar system research.[32]

Finally, the Research Survey also served as the centerpiece for Leuschner's efforts to raise funds for Berkeley's Department of Astronomy. Extramural support played an important role in sustaining California's large graduate program in astronomy, particularly as the Great Depression worsened in the early 1930s. Leuschner was well positioned to seek such support: In addition to his authority as department chair, Leuschner administered science programs at the University of California, and later chaired the University's Board of Research. During the 1920s, through these responsibilities, Leuschner came to know leading figures at the major new patrons of American science, particularly Rose and CIW President John C. Merriam. Leuschner cultivated both men as potential supporters of the asteroid program, stressing its value for improving Newtonian theory and for better understanding the origin of minor planets. Ultimately Leuschner received large grants from the International Education Board, the NRC, and the James Craig Watson Trust, whose $25 thousand endowment made it the third-largest trust administered by the National Academy of Sciences (NAS). Sharing little of the anxiety of other leaders of the physical sciences that accepting government support would corrupt pure science, Leuschner scrambled to secure Roosevelt-era Works Progress Administration (WPA) grants for researchers in 1933. He used them to support numerous graduate students during the Great Depression, including Priscilla Bok (née Fairfield) and Nicholas U. Mayall, later director of the Kitt Peak National Observatory. His fund-raising kept the Berkeley astronomy department budget relatively stable throughout the 1930s, maintaining its production of new Ph.D.'s.[33]

Leuschner's entrepreneurial style was reflected in the careers of several of his students, much as they copied his mathematical-physical orientation. Although some showed little inclination to seek extramural patronage – among them Herrick and Nicholson, each in a stable, well-funded environment (the Naval Observatory and Mount Wilson) that discouraged outside aid – other Leuschner students became the most entrepreneurially minded U.S. astronomers of their generation. Whipple (discussed in the next section) shared Leuschner's enthusiasm for government funding and the long-running research programs they enabled. Leuschner's influence was equally evident in the career of Dinsmore Alter. After earning his Ph.D. in 1916 and

[32] Leuschner to A. J. Stratton, Apr. 18, 1932, Box 1, AOL; idem to J. C. Merriam, Dec. 7, 1935, Box 2, AOL; and idem to E. W. Brown, Dec. 9, 1935, Box 2, AOL; see also Leuschner and Thiele (1929). Faulty Rechen-Institut ephemerides in the 1930s vexed many American researchers; see, e.g., P. Herget to Kopff, Jan. 1, 1951, and D. Brouwer to G. P. Kuiper, Feb. 5, 1952, both Box 69, Folder 142, DB; see also Watson (1941, 13, 23–4). Numerov's career is analyzed in McCutcheon (1991).

[33] Leuschner to H. N. Russell, May 22, 1926, Box 27, and idem to G. Gray, Mar. 22, 1939, Box 26, both HNR; idem to F. Flugel, Feb. 9, 1934, Box 18, OAL; "Trust Funds of the National Academy of Science," attached to P. Brockett to members of the National Academy, May 17, 1932, Box 52, HNR; see also Leuschner (1935, i–ii). Departmental finances were assessed using "Budgets for Astronomy," Box 18, OAL; they averaged about $32,000 per year from 1931 to 1935.

receiving a Guggenheim postdoctoral fellowship in physics, Alter accepted a post at Kansas, soon making asteroid research a priority for the university's new 27-inch reflector. In the early 1930s, as the Great Depression weakened departmental support, Alter attempted to fund his own asteroid "Computing Center" by calculating long-term meteorological forecasts for regional farmers. For this Alter used statistical correlations that he believed linked solar activity to rainfall, long a topic of interest to him. Alter's program faltered, partly because Russell, Whipple, and Donald H. Menzel, discovering mathematical errors in theoretical cosmogonal models he circulated to them, muted their endorsements of his work. However, it is important to note that Alter, like Leuschner, saw asteroid research as a promising career option. Leuschner's research school in small-body studies clearly influenced the character of solar system astronomy in the United States well through the 1950s, when Whipple became Leuschner's most productive heir.[34]

Mathematical astronomy thus provided an important institutional and intellectual foundation for solar system research in America during the 1920s and 1930s. Yet astrophysics had then become the dominant branch of the discipline. Other solar system phenomena – including meteors, planetary atmospheres, and cosmogony – were also studied during this period, but only rarely were these investigations made, like Lowell's of Mars, to understand the planets per se. As astrophysics matured, its leading practitioners sought to reinterpret the principal phenomena of the solar system in terms of what was known about the universe at large. Astronomers moved easily among what only later came to be regarded as distinct areas of astronomy. Their research strongly shaped what became solar system astronomy's component fields.

THE HARVARD ARIZONA METEOR EXPEDITION

Between 1926 and 1938, the Harvard College Observatory (HCO) emerged as a major U.S. center for meteor research. In 1930, Harlow Shapley, director of the HCO, arranged for several staff members and assistants to embark on an "expedition" to the darkened skies of Flagstaff, Arizona, designed to determine the average velocity of meteors. Shapley hired two young astrophysicists, Ernst Öpik and Fred L. Whipple, to direct and facilitate this work. He also encouraged other staff members and graduate students to study the origins of meteors and other small bodies in the solar system. By contemporary standards, Shapley's program was large; as the British astronomer Bernard Lovell later commented, "the subject of meteor astronomy was

[34] Alter to H. N. Russell, Sept. 12, 1925, Box 32, Folder 15, and Russell to Alter, Nov. 6, 1928, Box 1, Folder 60, both HNR; Leuschner to Alter, Mar. 12, 1929, Box 1, AOL; Shapley to Alter, Sept. 25, 1925, Box 1, DHCOs; see also Levy (1991, 94). Alter eventually left Kansas to direct the Griffith Observatory and Planetarium in Los Angeles, where he later took up lunar studies; see Doel (1992, 246–7).

dominated by the great Arizona expedition for the study of meteors."[35] It indeed became one of the most influential transient institutions in solar system astronomy in the early twentieth century.

The Harvard–Arizona meteor expedition, and especially Shapley's involvement in it, may seem surprising. Virtually all historical accounts of the HCO during the 1920s and 1930s have focused on its contributions to stellar and galactic research, including Shapley's own studies of globular clusters, Cepheid variable stars, and the Magellanic Clouds. In 1920 Shapley squared off against Heber Curtis of Lick in a debate, sponsored by the NAS, over whether spiral nebulae are distant "island universes" or smaller objects within the Milky Way. This debate, as Robert W. Smith has justly noted, was "one of the most famous events in the history of twentieth century astronomy."[36] Many others at the observatory were similarly devoted to stellar research. The careers of two of Harvard's most well-known women "computers," Henrietta Swan Leavitt and Annie Jump Cannon, had been made by their painstaking analyses of stellar spectra recorded on Harvard's massive and unique collection of plates, a research program urged by former Harvard director Pickering. Another young staff member, Cecilia Payne, focused on the abundance of gaseous elements in the atmosphere of the Sun. By 1930, leading U.S. astronomers regarded Harvard as a prominent East Coast center for astrophysical research. This orientation was strengthened when, in the late 1920s, Shapley worked to create a formal graduate program in astronomy associated with the observatory, emphasizing stellar astrophysics in its course work. Against this background, Shapley's decision to promote meteor research at Harvard seems something of an anomaly.[37]

The appearance of conflict is illusory, however. Shapley regarded meteor studies as a branch of galactic research, for he shared the view of a majority of astronomers at the time that meteors had an interstellar origin, and thereby offered a promising means for studying the environment of remote regions of space. Justifying the project in 1932, Shapley declared that "[t]he nature of interstellar and intergalactic medium through which radiation, stars, clusters, and galaxies move is found to be of so much significance in our understanding of galactic distances and structure that fundamental research on the contents of space has become necessary."[38] But Shapley saw other justifications for supporting this project: The abundance of plates in Harvard's collection on which meteor trails and spectra appeared, he argued, gave HCO a competitive edge for initiating work in this field. Most important, Shapley believed that linking astronomy, meteorology, chemistry, geology, and geochemistry would create "a new epoch for [the] H.C.O." by revealing new insights into physical phenomena. Close to Rose, and well aware of the Rockefeller Foundation's

[35] Lovell (1954, 7).

[36] Smith (1982, 55).

[37] Smith (1982, 55–96) and Kidwell (1986, 157–72). Not even Bart Bok's (1978) obituary of Shapley mentions the meteor expedition, or his role in planning it.

[38] Shapley, Öpik, and Boothroyd (1932, 16).

sympathy for interdisciplinary studies, Shapley saw meteor astronomy as a way to build a broader methodological foundation for contemporary astrophysics.[39]

The connection between meteors and interstellar space, which members of the Harvard expedition sought to substantiate, had emerged from research on the relation of meteors to the solar system. Until the last decades of the nineteenth century, astronomers remained unsure whether meteors belonged to the solar system or originated beyond it – that is, whether they were a problem of cosmogony or cosmology. In the 1860s, American astronomer Hubert A. Newton and Italian astronomer Giovanni Schiaparelli had presented results indicating that comets and meteors shared a common origin. Their work convinced many researchers, as American astronomer Daniel Kirkwood wrote in 1873, that "meteor clouds are but the scattered fragments of comets." Since comets were suspected, but not demonstrated, to have an interstellar origin, the inference that many astronomers drew was that most meteors also derived from interstellar space.[40]

A direct way of resolving the question involved calculating the average velocity of meteors. By the 1860s astronomers argued that meteors traveling faster than forty-two kilometers per second (kps) when entering Earth's atmosphere were moving in hyperbolic orbits, that is, independent of the Sun's gravitational field; such meteors would thus be interstellar. By the turn of the century investigations aimed at achieving an average meteor velocity were underway at a number of observatories in the United States and Europe. These studies indicated a preponderance of interstellar meteors. The 1895 meteor catalogue of the German astronomer Gustav von Niessl, for instance, found a majority of meteors with velocities of nearly 60 kps, more than 25 percent faster than the critical hyperbolic threshold. Meteor data compiled by von Niessl's successor, Cuno Hoffmeister, in his 1925 *Katalog der Bestimmungsgrößen für 611 Bahnen großer Meteore*, indicated that 79 percent of all observed meteors had hyperbolic velocities. However, not all astronomers believed that the origins question was resolved, particularly following new results by Danish astronomer Elis Strömgren in 1914. Strömgren was surprised to find that the orbits of several well-observed comets were parabolic and not hyperbolic, which he had expected on the assumption that comets and meteors were linked. Thereafter most astronomers adopted von Niessl's proposal that two classes of meteors exist: fast meteors, which came directly from interstellar space, and slow meteors, which originated from comets.[41]

39 Meteor "expeditions" were not solely an American phenomenon: In 1930, the German astronomer Cuno Hoffmeister led an "exploring voyage" to the Caribbean Sea in order to study daily variations in meteor activity, and in 1934 joined another colleague at 35° south latitude to study other predicted periodicities in meteor behavior; see Hoffmeister (1931; 1937a, 50, 207). Expeditions were also launched by physicists during this period to resolve controversies over cosmic rays; see DeVorkin (1989b) and Kargon (1982, 153–60).

40 Quoted from Kirkwood (1873, 3, 49); see also Abetti (1975), Berendzen (1974), Clerke (1896, 115–39, 363–83), Olivier (1925, 59), and Russell et al. (1926–7, 459).

41 Hoffmeister (1922; 1937b) and von Niessl and Hoffmeister (1925). Another widely cited proof of hyperbolic meteor velocities was the well-observed meteorite shower in Pultusk, Poland, on January 30,

By the 1920s, meteor research was maturing as a distinct field of astronomy. Although it had no permanent institutional base, a number of monographs and textbooks were devoted to the subject, including the 1925 *Meteors* by Charles P. Olivier, an astronomer then at the University of Virginia. A half dozen individuals published one or more articles annually in this field for five years or more during the period 1927–38. Virtually all contributors believed that a majority of meteors originated outside the solar system. As Hoffmeister declared, the importance of meteors derived from the probability that interstellar meteors "might be nothing more than the particles of interstellar clouds." So confident were many German scientists of meteors belonging to the interstellar realm that Hoffmeister's own *Die Meteore* was issued as a volume in the *Probleme der Kosmischen Physik* series, edited by the Akademische Verlagsgesellschaft of Leipzig.[42]

Despite the perceived connections between meteor astronomy and stellar astrophysics, the meteor project represented a new undertaking for Shapley, as the HCO had engaged in little meteor research until that time. Shapley first toyed with the idea of meteor studies in 1924, and sketched plans for a major undertaking by the following year; but expedition preparations began in 1927, shortly after he called Willard J. Fisher to join the observatory staff. A native of Waterford, New York, with a Ph.D. in physics from Cornell in 1908, Fisher had worked in various areas of applied physics since leaving Ithaca, including the viscosity of gas and the behavior of the upper atmosphere. This work convinced Fisher of the value of meteors as "transient visitors from outer space."[43] He complained nevertheless that the von Niessl–Hoffmeister velocity catalogues, compiled from visual observations, were not sufficiently accurate to allow firm conclusions about such phenomena as interstellar meteor streams, whose discovery von Niessl claimed. To investigate further, Fisher applied to the Milton Fund of Harvard to make a photographic study of meteor velocities, a project similar to one that the Yale astronomer William L. Elkin had conducted about the turn of the century. In 1927 Shapley wrote Edwin Frost, then director of the Yale Observatory, requesting permission to borrow Elkin's equipment.[44]

A particular reason that the meteor project attracted Shapley was his deep interest in the interstellar medium. In 1930, for example, Shapley declared that dark-edged nebulosities such as the Orion Nebula were sources of comets and meteor streams, suggesting the value of studying local meteoric phenomena. Together with Payne, Shapley also suggested the possibility that the cyanogen absorption band in hot stars resulted from millions of iron and stone meteors "plunging with enormous velocity

Footnote 41 *(cont.)*
1868, whose velocity was calculated by the astronomer J. G. Galle as 56 kps, well above the interplanetary threshold; see Lovell (1954, 154) and Paneth (1940, 11–12).

[42] Hoffmeister (1937b, 209) and Olivier (1925, 20–1, 260).

[43] Fisher (1927, 511); see also Hoffleit (1988).

[44] Shapley to W. S. Adams, Nov. 23, 1927, Box 1, DHCOs.

down through the atmospheres of stars." For Shapley, meteor research seemed a promising pathway to study the physical characteristics of stars and the interstellar medium. Since Trumpler also presented evidence for abundant interstellar material in 1930, many astronomers felt that additional work on this problem was required.[45] Yet institutional issues may have influenced Shapley even more. As the astronomer Dorrit Hoffleit remarked of Yale circa 1900, "the unavailability of a properly equipped observatory made the field [of meteor astronomy] especially attractive just because specialized equipment was not yet essential" for its pursuit.[46] This point is important, for the situation of astronomy at Harvard by the late 1920s resembled that at Yale one generation before. Despite its reputation in astrophysics, Harvard's principal research instruments (including its reconditioned 60-inch reflector at the Harvard southern station in Bloemfontein, South Africa, and its 61-inch reflecting telescope planned for a nearby Massachusetts site) were no longer competitive with Mount Wilson's 100-inch telescope and those of other major western mountaintop observatories. This deficiency posed special problems for Shapley, anxious not only about attracting staff but in recruiting students to his new graduate program in astronomy, in 1927 one among seventeen similar programs. Moreover, in contrast to most American observatories, the HCO was financially separate from its parent institution, leaving it dependent on outside contributions for basic expenses. Shapley clearly saw meteor astronomy as a means to build the observatory's endowment and to compete for graduate students and staff.[47]

The final factor behind Shapley's decision, and perhaps the most important, was the enthusiasm of U.S. scientists in the 1920s for "cooperative" research spanning disciplinary boundaries. The cooperative ideal in American science had several roots: For leading national scientists like Hale, it signified the informal links between federal, academic, and industrial organizations initially fostered during World War I that avoided encumbering and politically distrusted permanent liaisons. Most of all, however, it reflected the efforts of Rose and other foundation managers to support cross-field collaborations as a promising way to advance knowledge. Although Shapley showed no particular interest in interdisciplinary research while working under Hale at Mount Wilson (1914–21), he soon promoted such programs after becoming HCO director. At the bequest of Harvard administrators, for example, Shapley successfully urged the General Education Board to give two million dollars

[45] Quoted from Shapley (1930, 68); this concept ws first suggested in Shapley and Payne (1928). Shapley's idea that meteors manufactured stellar spectral lines apparently came from a similar suggestion by his Harvard predecessor, E. C. Pickering (see W. Fisher, "Memo for Dr. Shapley," n.d. [1924], Box 6, DHCOs), but it is unclear whether Shapley also knew a similar suggestion by the late-nineteenth-century astrophysicist Norman Lockyer, described in Meadows (1972, 179–84, 207). Shapley had also ignored the effects of interstellar matter in estimating the size of the Milky Way earlier in his career, a factor that influenced colleagues to reject his "big galaxy" model; see Smith (1982, 58–9).

[46] Hoffleit (1988, 118).

[47] W. Rose, "Excerpt from Record of Doctor Rose's Interviews," June 16, 1927, Series 1.1, Box 18, Folder 264, RF; see also Kidwell (1986, 158).

toward a biology laboratory at Harvard in 1925, allowing increased interaction between isolated Harvard centers of zoology, genetics, and natural history.[48]

For Shapley, meteor research seemed a way to achieve a similar synthesis in the physical sciences. "The realm of the shooting stars," Shapley wrote, "is where knowledge will come of the origin and growth of planets, of the chemistry of the wandering meteoric bodies in space, and of the nature of the nebulae that may be the parents of stars."[49] In 1926 Shapley presented Rose an ambitious plan to create a Harvard-based "Institute of Cosmogony" to coordinate work in astronomy, geochemistry, geophysics, and meteorology. Such an institute, Shapley argued, could address such large-scale problems as Earth's evolution better than traditional academic departments. Tutored by his mentor Russell, who had helped secure a major gift from the General Education Board for cooperative research at Princeton the previous year, Shapley initially seemed confident that his plan would work. He had several points in his favor: By 1928 Shapley had received offers to lead the University of Chicago and to succeed the retiring Rose at the Rockefeller Foundation.[50]

Shapley's plans nevertheless suffered two setbacks. One problem was his speculative 1928 article with Payne that linked meteor infall to stellar absorption lines. Soon after it appeared, Shapley wrote Russell, seeking his former advisor's blessing for his concept. In a private letter, Russell bluntly criticized the idea. Urged by his former student to reconsider, Russell invested several weeks in a quantitative appraisal. He soon published this in the *Astrophysical Journal*, concluding that meteoric matter would fail to produce even narrow Frauenhofer lines in solar-type stars. While grudgingly conceding the argument, Shapley urged Russell to acknowledge that meteor research "will contribute more to cosmogony than all our preliminary suggestions. . . . I do not want you to scare me and other astronomers off the meteor problem." Yet Shapley himself grew cautious about the meteor expedition, advising a supporter that he would not publish general claims until results seemed beyond challenge.[51]

Another problem was securing funds for his proposed institute. Unwilling to see additional institutes proliferating on campus, largely because they reduced university leverage to require faculty teaching, Harvard president A. Lawrence Lowell refused to endorse Shapley's Institute of Cosmogony. Worse still for Shapley, his final, revised proposal reached Rose only in October 1929; although approved, Shapley found it impossible after the stock market crash to raise the matching funds that

[48] Doel (in press-a), Kargon (1977), and Kohler (1991, 149, 220–1).

[49] Quoted from Shapley, [revised institute proposal], Mar. 21, 1929, pp. 3–4, Record Group 1.1, Series 200D, Box 139, Folder 1718, RF; see also idem, [untitled memo], Mar. 31, 1927, Box 16, DHCOs.

[50] Rose, "Excerpts from record of Dr. Rose's interviews," Oct. 27, 1926, and Shapley to Rose, Apr. 4, 1927, both Series 1.1, Box 18, Folder 264, IEB; Shapley to Lowell, Feb. 27, 1928, and F. Schlesinger to A. L. Lowell, Mar. 14, 1929, both Box 11, DHCOs. See also Shapley to Lowell, Mar. 13, 1929, Box 11, and idem to F. Schlesinger, Apr. 14, 1931, Box 24, both DHCOs.

[51] Russell (1929); quoted from Shapley to Russell, Nov. 10, 1928, Box 61, Folder 29, HNR. See also Shapley to Russell, Nov. 10, 1928, Box 61, Folder 29, HNR; idem to D. M. Barringer, Apr. 29, 1929, Box 2, DHCOs; and Shapley (1930).

Rockefeller awards required. In 1930 Shapley did help bring to Harvard a major grant for experimental geophysics from the Rockefeller Foundation, partially fulfilling his ambition to create links between the physical sciences, and raised over a million dollars in new endowments for the HCO, half from the General Education Board; yet he made no further efforts to create a cosmogonal institute.[52] Narrowing his sights to the meteor expedition alone, Shapley raised smaller sums for it from the J. Lawrence Smith fund of the NAS, the Rumford Fund of the American Association for the Advancement of Science, and the Heckscher Fund of Cornell – secured when Samuel Boothroyd, a Cornell astronomer, was appointed a senior member of the expedition.

Despite financial pressures, the Harvard Arizona Meteor Expedition took place as planned, running from November 1930 through mid-1932. Although Shapley made many preliminary arrangements, and paid a personal visit to Lowell Observatory in March 1930 to make certain director V. M. Slipher would fulfill his promise to make available its grounds and facilities, actual operation of the expedition fell to Ernst Öpik. Öpik had worked at the Moscow Imperial Observatory and earned his Ph.D. at the Tartu University (Estonia) in the early 1920s; his research prior to this appointment had focused on galactic structure as well as on meteor astronomy. He also shared Shapley's view that meteors were an important tool for studying stellar phenomena. Under Öpik's leadership, the project became well staffed, not unlike solar eclipse expeditions of the day: Five astronomers and observers were included in it full-time, a number supplemented by part-time participants.

Published criticisms of Hoffmeister's visual method of calculating meteor velocities notwithstanding, Öpik did not follow Willard Fisher's lead in applying photographic techniques, probably worried that available photographic emulsions would record only a restricted number of bright meteors. Instead, Öpik opted for visual studies that adopted the baseline approach utilized by Elkin. Observing stations separated by twenty-four miles were set up near Flagstaff, so that simultaneous observation of any meteor would yield, through trigonometric analysis, its height and real path through the upper atmosphere. (To reduce errors in determining the beginning and ending point of meteors, observers worked inside specially constructed "reticule" houses, whose windows featured iron grids marked in degrees.) At the Flagstaff station, a "rocking mirror" velocity device was also employed to verify direct determinations of meteor velocities (i.e., the apparent path length of the meteor in degrees divided by the length of time it remained visible). An observer studying the sky through the rocking mirror instrument would see meteors crossing his or

[52] Shapley, "Observatory Report [to the President of Harvard University]," draft pp. 1–5, in "Harvard University – Astronomy," n.d. [1932], Box 139, Folder 1718, Record Group 1.1, Series 200D, RF; idem to Boothroyd, Apr. 20, 1932, Box 24, DHCOs; and Shapley et al. (1932, 17). The J. Lawrence Smith Fund was restricted by deed to meteor studies, much as the Watson endowment was limited to asteroids; see Lankford (1987, 273, 278–9). While A. Lawrence Lowell's opposition to Shapley's proposed institute followed from his general antipathy to university institutes, lingering resentment towards his flamboyant sibling Percival may also have played a role.

her field of view describe a pseudocycloidal trajectory. High-velocity meteors would trace open loops, whereas closed loops would result for those of slower velocity. Several hundred meteors were recorded before the expedition ended. By 1935, employing low-wage "computers" at Tartu's Computing Bureau, Öpik carefully reduced these observations, publishing thirteen articles on the expedition's results. He concluded that nearly 70 percent of the Arizona meteors were hyperbolic, and hence interstellar.[53]

Despite losing his chance to build an institute of cosmogony, Shapley used the expedition to enhance small-body research at Harvard. He considered calling Olivier, the Virginia meteor expert, then approached Bobrovnikoff, the comet expert and former Leuschner associate. After Bobrovnikoff, committed elsewhere, reluctantly declined, Shapley eventually aided Harlen T. Stetson, a Perkins Observatory astronomer interested in cosmic–terrestrial relationships, and worked to integrate Stetson's results into Harvard's informal Committee on Experimental Geology and Geophysics, largely funded by the Rockefeller Foundation.[54] More significantly, Shapley knitted colleagues at several neighboring universities into an informal version of his planned multidisciplinary institute. The most important result of these efforts was the Harvard Summer School in Astronomy, which Shapley created in 1935. By making MIT chemist William Urry and Tufts geologist A. C. Lane participants in the Summer School, and drawing them into the orbit of the Harvard geophysics committee, Shapley created a context for interdisciplinary discussions of cosmogony and planetary formation. Already Shapley was actively discussing lunar phenomena with his geology colleague and fellow geophysics committee member, Reginald Daly. Shapley used Daly to sharpen his ideas about the significance of meteorite impacts as a geological phenomenon, a view then shared by few individuals in either discipline. Indeed, in drafts of his institute proposals to Rose in 1926, Shapley had speculated that asteroid impact collisions might have shaped the early history of Mars, adding that comets possibly had struck Earth more frequently in the remote past than now. In the 1960s these concepts would become central to planetary geology; but at the time, still smarting over Russell's criticisms of his scheme for interpreting stellar spectra, Shapley made no effort to publish them.[55]

53 See Shapley to Russell, Mar. 25, 1931, Box 61, Folder 30, HNR; idem to Whipple, June 15, 1931, HUG 4876.806, FLW–HU; idem, "Observatory Report [to the President of Harvard University]," 1931–2: 3–4; see also Öpik (1930) and Shapley et al. (1932, 20). No adequate biography of Öpik exists; see Hoffmeister (1929, 160–1) and Smith (1982, 103). Astronomers later agreed that Elkin had failed to place adequate distance between his baseline stations, one reason for the limited success of the earlier Yale meteor program.

54 Shapley to Bobrovnikoff, May 9, 1930, Box 2, idem to Lowell, Feb. 18, 1929, Box 11, and idem to R. Daly, Sept. 20 and Nov. 15, 1933, Box 29, all DHCOs. Shapley seemed aware that Olivier received extremely modest support from Virginia; see Olivier to E. A. Alderman, Apr. 18, 1927, Box 3, UVa.

55 On the Harvard Summer School, see Shapley to Dean Black, Sept. 22, 1932, Box 36, DHCOs, and DeVorkin (1984, 50); on his ideas concerning impact, see Shapley to D. M. Barringer, May 8, 1929, Box 2, and idem to Daly, Feb. 14, 1925, Box 4, DHCOs; and idem, "The Harvard Observatory Development," Mar. 21, 1929, draft p. 12, Record Group 1.1, Series 200D, Box 139, Folder 1718, RF; see also Hoyt (1987, 259).

Other Harvard astronomers also turned to meteor research. Although unconnected with the Arizona expedition, possibly because of his opposition to visual measures of meteor velocities, Fisher published widely on meteors, the zodiacal light, and the terrestrial atmosphere through the early 1930s, and laid the groundwork for photographic studies of meteors. He also aided other graduate students working broadly on small bodies in the solar system. Dorrit Hoffleit, for instance, joined Fisher in searching the Harvard plate collection for meteor trails. She also examined the lunar dark side through several lunations for signs of meteorite impacts, and wrote two papers on applying photometric techniques to meteor light curves. Peter M. Millman, arriving at Harvard before the expedition began, wrote his 1932 dissertation on meteor spectra – the first systematic investigation of this subject. After gaining an appointment at Toronto, Millman kept meteor research his main field of study, publishing eight papers on meteor spectra between 1933 and 1935. This work represented a significant change from previous research, for it underscored new emphasis on quantitative and physical studies of small bodies. It also created at Harvard an active center of solar system research – one that, like Leuschner's program at Berkeley, turned young researchers to this field. By the mid-1930s, having produced over fifty papers in this area, Harvard was the most active of several U.S. centers for meteor astronomy. Olivier declared to the IAU in 1933, "never before have so many people been working in this branch of science."[56]

The interdisciplinary connections that Shapley attempted to forge at Harvard also influenced the research of his students. This was particularly so in the case of Fletcher Guard Watson, Jr., a Harvard Ph.D. in astronomy. Interested in meteors from his undergraduate days at Pomona College, where he had attempted meteor photography under Walter T. Whitney, Watson in the mid-1930s began a broad-based assessment of small bodies and their implications for the problem of Earth's origin. In his investigation, directed by Shapley as well as by astrophysicist Bart J. Bok, Watson assessed evidence concerning comets, asteroids, meteors, and the zodiacal light. He also made geochemical investigations of meteorite samples in Harvard's collections (which he undertook with Urry at MIT), and surveyed the growing number of terrestrial features that certain geologists suspected had impact origins, in part to compute the number of asteroids whose orbits crossed that of Earth. He defended this work in language similar to Shapley's, declaring that small-body research made important contributions to such problems as "the structure of interstellar matter, the structure of Earth's upper atmosphere, the frequency of meteor-crater formation."[57] His 1938 dissertation was a pioneering effort at interdisciplinary integration. Watson employed his geochemical analyses of meteorites to support Hiraya-

[56] Shapley to W. S. Adams, May 9, 1927, Box 1, DHCOs; Hoffleit to F. Whipple, Nov. 28, 1955, Box 11, GPK; Dorrit Hoffleit OHI (David DeVorkin, Aug. 4, 1979, SHMA at AIP, pp. 13–18); see also Jarrell (1988, 155); quoted from Olivier (1933).

[57] Watson (1938, vi); see also Watson (1935, 1936a,b).

ma's argument that asteroids resulted from a disrupted planet, and claimed that similarities between asteroid and lunar polarizations suggested that asteroids were themselves cratered. He also gave the first modern estimate of the number of Earth-crossing asteroids, declaring that major impacts occurred roughly every hundred thousand years. Watson's thesis – which appeared as the semipopular *Between the Planets* in 1941 – was a product of the Harvard Committee on Experimental Geology and Geophysics no less than the HCO.[58]

Meteor research at Harvard was further stimulated by an internal debate between staff members over the expedition's chief conclusion, that most meteors are interstellar. One of the first astronomers to challenge Öpik's high-velocity estimates was none other than Fisher. Well before the Harvard–Arizona expedition began, Fisher had criticized von Niessl's and Hoffmeister's work, noting that estimates of meteor durations made without the benefit of precise timepieces – as many in the von Niessl–Hoffmeister catalogues had been – rendered the velocities derived from them questionable. Fisher also criticized their assumptions about the amount of drag expected on meteoric particles entering the upper atmosphere, which they resolved with a compensating factor; to Fisher, too little was known about the density of the upper atmosphere to make such judgments. "The exact paths and speeds, deceleration in the air-path, and the spectral evidence of the physical nature of the process, with all their implications as to the nature of the atmosphere and the cosmic relations of the meteors themselves," Fisher had argued in 1927, "can be dealt with exactly only by photography." Although Fisher refrained from criticizing Öpik's interpretations in print, his concerns were quickly communicated to other Harvard astronomers. Already in the early 1930s, when the expedition was underway, Fisher developed a small program of meteor photography near the Harvard campus.[59]

The man who played a key role in resolving this controversy was Fred L. Whipple, then newly appointed to the HCO. Whipple had an unusually broad background for an American astronomer educated in the 1920s: He had studied celestial mechanics under Leuschner at Berkeley, but he had also studied radiation and atomic structure with Birge, and had written his dissertation on the physical properties of cepheid variable stars under the informal supervision of Menzel. In 1931, as he completed his Ph.D., Whipple was recruited by Shapley to aid the meteor project; he was likely seen as a replacement for Öpik, who left the United States for Tartu in 1932. During the time he and Öpik overlapped in Cambridge, Whipple developed a considerable interest in the problem of meteors, accepted Öpik's arguments for their interstellar origin, and came to regard Öpik as a mentor figure.[60]

[58] Watson (1938, ii, 70–1; 1940), Shapley to R. Daly, Jan. 13, 1936, Box 29, DHCOs; H. N. Russell to Watson, Jan. 2, 1937, Box 29, Folder 63, HNR; Watson (1941), Fletcher G. Watson, Jr., OHI (Ronald E. Doel, Nov. 20, 1990, AIP).

[59] Fisher (1927 [quote at 512]), Hoffleit (1988), and Watson (1938, 82, 90, 92, 110).

[60] "Programme of the Final Ph.D. Examination of Fred L. Whipple," 1931, and Leuschner to G. D. Birkhoff, June 3, 1939, both Box 6, AOL; Shapley, "Observatory Report [to the President of Harvard University]" 1931–2: 1–4 and 1932–3: 1–3; Whipple to C. C. Wylie, Feb. 21, 1938, HUG 4876.806, FLW–HU; idem to Leonard Carmichael, June 20, 1958, Box 16, LC[si]; see also DeVorkin (1992).

Not long thereafter, however, apparently through Fisher, Whipple came to regard visual evidence for interstellar meteors as uncertain. Between 1934 and 1936 Whipple ran the photographic meteor project that Fisher had initiated. Using the Harvard sky survey cameras – one at Cambridge, the second twenty-six miles to the west at the HCO's new Oak Ridge facility – Whipple measured the paths of meteors simultaneously recorded by both stations. He placed considerable value in these data for two reasons. One was improvements in the quality and speed of photographic emulsions, which made the recording of meteors more efficient and accurate. The second involved improvements in the reliability of 60-cycle AC motors, used to drive a rotating shutter that interrupted the paths of meteors, thus providing precise indications of their duration. His calculations suggested that the majority of meteors exhibited not hyperbolic but rather parabolic velocities, considerably under the threshold of 42 kps. This result challenged the faster visual determinations from the Harvard expedition.[61]

The major difficulty that Whipple faced was not the lack of evidence for interstellar meteors, but finding a way to reconcile his results with those of Öpik, whose work he continued to hold in high regard. He found a solution to this dilemma in the work of a young astronomer at the University of Iowa, Charles C. Wylie. Wylie had turned to meteor astronomy for much the same reason that Shapley had embraced it at Harvard: Although trained as a photoelectric astronomer under Illinois's Joel Stebbins, the American pioneer of this field, Wylie discovered that the observatory facilities at Iowa did not allow him to continue in this arena. Instead, intrigued by reports of bright meteors and meteorite showers over Iowa and Kansas throughout the first third of the century, Wylie sought to investigate the possible interstellar origin of these bodies.[62] By the mid-1930s, Wylie became convinced that the velocities determined by visual methods were too high, as Fisher and Watson had argued. Through interviews with meteor observers, Wylie came to believe that while most observers accurately stated meteor durations, there was "a conspicuous psychological tendency to lengthen the path of a meteor," often by a factor of 2. Applying this correction to the von Niessl–Hoffmeister catalogue, and integrating his own research on air resistance, Wylie recalculated an average velocity at 34 kps, suggesting that "the percentage of extra-solar meteors is negligible."[63] Whipple, who began corresponding with Wylie in the mid-1930s, valued Wylie's experimental evidence as a confirmation of the solar origin of meteors and as a way to challenge Öpik's interpretations of the Arizona data without directly attacking Öpik himself.[64]

[61] Shapley to O. Struve, Jan. 10, 1936, Box 51, DHCOs; idem to P. Brockett, Mar. 16, 1938, HUG 4876.806, FLW–HU; see also Whipple (1938, 20).

[62] Wylie (1922; 1933); see also *Contributions of the University of Iowa Observatory* 1 (1931–40), nos. 1–9.

[63] Wylie (1937 [quotes at 210, 214]); see also Wylie to F. Watson, Jr., June 18, 1935, FGW.

[64] Wylie to Whipple, Jan. 20 and Feb. 11, 1938, and Whipple to Wylie, Jan. 14, 1938, both HUG 4876.806, FLW–HU.

Wylie's results, linked with Whipple's photographic meteor data, soon convinced numerous American astronomers that most meteors are not interstellar. International acceptance of this conclusion came more slowly: Not until Whipple extended his photographic meteor program after 1945, and the British scientist Bernard Lovell reached similar results with his meteor study at the Jodrell Bank radio observatory of Manchester University, did Western European astronomers accept the lack of evidence for hyperbolic velocities. This delay likely owed to the relative youth of Whipple and Wylie in the 1930s and the disruption of normal scientific communications during World War II, which kept word of American acceptance of their results from traveling overseas.[65] The Harvard meteor program, however, was already an important development in the 1930s. It created an informal community of earth scientists and astronomers commonly interested in the evolution of the planets and stellar systems, provided a training ground for astronomers in this field, and removed meteor astronomy from the realm of stellar and interstellar research, linking it to solar system studies and geophysics. It also demonstrated the essential role that foundation grants played in launching this interdisciplinary field.

COSMOGONY AS A GOAL: HENRY NORRIS RUSSELL

Meteor astronomy was not the only field of astrophysics that stimulated new solar system research in the early twentieth century. A similar role was played by studies of natural abundances. By the end of the nineteenth century, the distribution of chemical elements in Earth, the planets, the Sun, stars, and the universe at large had become a major question for scientists in several fields. For astronomers, the relationship of Earth's composition with that of the Sun formed the basis of astronomical spectroscopy. Through the early 1920s, virtually all astronomers agreed with the 1890 declaration by Henry Rowland, the Johns Hopkins spectroscopist, that "were the whole Earth heated to the temperature of the Sun, its spectrum would probably resemble that of the Sun very closely."[66] Geochemists (who studied a parallel relationship, that of abundances of Earth and meteorites) found abundances of equal importance, since theories about Earth's geochemical evolution rested on this foundation. While many American astronomers in the early twentieth century focused on stellar spectroscopy, a number of them also investigated the chemical abundances of planetary atmospheres. They believed that such information would increase knowledge about the Sun's composition, assuming a chemical connection between the Sun and its planetary family.[67]

[65] Lovell (1954, 13) and Whipple (1938); see also Lovell (1990). On American acceptance of the Whipple–Wylie interpretation, see, e.g., Watson (1941, 104).

[66] DeVorkin and Kenat (1983b, 180).

[67] Paneth (1940).

Of American astrophysicists who turned to this issue, none was more influential than Russell. Throughout the 1920s and 1930s, Russell encouraged colleagues to investigate the compositions of planetary atmosphere and, in 1935, he synthesized these results in a widely read monograph, *The Solar System and Its Origin*. He encouraged several graduate students, including Menzel, Theodore Dunham, Jr., and Lyman Spitzer, Jr., to take up research problems involving the planets. Each of these astronomers, like Russell, occasionally returned to planetary studies later in their careers. In addition, Russell urged astronomers at other American observatories to take on solar system research; among those he influenced were Arthur Adel and V. M. Slipher at Lowell and Walter S. Adams and Dunham at Mount Wilson. Finally, Russell's own investigations of the natural abundances of the elements were later regarded as a foundation for studies of the chemical structure of planets. More than has been recognized, Russell's studies of cosmogony helped lead to the downfall of close-encounter theories (involving interactions between the Sun and a passing star), stimulating fresh interest in one of astronomy's most intractable problems.[68]

What makes Russell's contributions to solar system research particularly interesting is his high standing in astronomy. A theoretical astrophysicist, Russell became the most influential American astronomer in the first half of the twentieth century. A native of Oyster Bay, New York, who studied at Princeton, Russell returned to Princeton for good in 1905 after a postdoctoral fellowship at Cambridge University. Early in his career, Russell worked on eclipsing binary stars, showing how analyses of their motions revealed information on the internal structure of these stars. He then demonstrated a relationship between stellar spectra and luminosity, later referred to the Hertzsprung–Russell diagram, and developed the first modern theory of stellar evolution. In 1929, Russell became the first astronomer to argue forcibly that hydrogen is the most abundant element in the Sun and hence, by inference, in other stars and throughout the universe. Russell was also affiliated with Mount Wilson and a leading figure in the American Astronomical Society and the American Philosophical Society. By the time that Russell turned to planetary atmospheres and the origin of the solar system in the 1930s, he was at the pinnacle of his career.[69] This is significant in two respects. Like many of his colleagues, Russell did not see planetary research as an end in itself, or as a distinct field, but rather as a set of problems within stellar astrophysics, broadly defined, that awaited solution. Moreover, because of Russell's wide-ranging influence, solar system research in America came to reflect his particular interests and methodological approaches.

Several factors stimulated Russell's interest in the solar system. One was his role as a textbook editor. Early in the 1920s, Russell and his Princeton colleagues, Raymond S. Dugan and John Q. Stewart, began writing a revised, enlarged edition of

[68] See H. C. Urey, "[Report on the Conference on the Abundance of the Elements]," n.d. [ca. Nov. 1952], Box 24, HCU.

[69] Brush (1978a, 87–91; 1981, 88–90), DeVorkin (1993), DeVorkin and Kenat (1983b).

Charles Young's *Astronomy*, a standard text in astronomy. Perhaps because he had done a number of studies involving the solar system early in his early career (his 1900 dissertation, for instance, had addressed the orbit of the minor planet Eros), Russell devoted considerable energy to editing the first volume of the Russell–Dugan–Stewart *Astronomy*, wholly dedicated to the solar system. Russell quickly sensed that updating the volume would allow him to request new observations and studies. Before the revised edition appeared in 1926, Russell persuaded the Mount Wilson astronomer Edison Pettit to make new spectroscopic plates of Venus, intended to reveal its atmospheric composition and rotation rate. He also collaborated with Stewart to measure more accurately the diameters of satellites of Saturn and Uranus using photoelectric observations of their eclipses. This work heightened Russell's already broad interest in solar system research, and produced the first modern U.S. textbook in this field.[70]

A second factor that steered Russell toward solar system studies was his appointment as Research Associate at Mount Wilson in 1921. Engineered by Hale, who wanted to bring to Mount Wilson a theoretical astronomer able to exploit the rich observational data produced by staff astronomers, Russell's appointment stimulated new research and put him in the role, as one historian noted, of a "fox raiding the hedgehogs."[71] While Russell used his annual two-month visits to Pasadena primarily to study stellar and solar spectra, he also perceived an opportunity to turn this facility's peerless resources to unsolved questions in solar system astronomy. Appointed by CIW President Merriam to an interdisciplinary committee to study the Moon's surface (see Chapter 5), Russell requested new investigations of the thermal conductivity of the lunar surface, a matter "of very considerable theoretical interest."[72] He also encouraged spectroscopic and radiometric work on the planets by Adams, Pettit, Nicholson, and Clair E. St. John of Mount Wilson. Between 1921 and 1930, ten separate studies of this kind were made at this facility, nearly as many as those produced by Lowell astronomers at this time. Russell valued these observations not merely for revealing planetary characteristics but for what they suggested about the Sun's composition, assuming a chemical connection between the Sun and its planetary family.[73]

Russell's interest in planetary phenomena also grew as he became better acquainted with staff members at Lowell. Russell frequently stopped at Flagstaff during his annual pilgrimage to Mount Wilson. During one visit, Russell made a side trip to nearby Meteor Crater, which convinced him of its meteoritic origin. He also advised Lowell astronomers about research opportunities, and sought to limit fallout

[70] Russell to Pettit, Jan. 25, 1923, and Dec. 6, 1926, both Box 21, Folder 3, and idem to G. E. Hale, Jan. 31, 1920, Box 11, Folder 28, all HNR; see also Pettit and Nicholson (1924a,b).

[71] DeVorkin (1993, 103).

[72] Russell to Pettit, Dec. 6, 1926, Box 21, Folder 3, and see Merriam to Russell, Oct. 9, 1926, Box 50, Folder 8, both HNR; see also Wright (1938).

[73] See Russell (1935a). Mount Wilson and Lowell figures are drawn from the *Astronomischer Jahresbericht.* Minor notices about particular planetary observations are omitted.

from the Martian canal controversy. Russell enthusiastically backed its planned trans-Neptunian search for Planet X, and helped it to secure extramural funding from the American Philosophical Society. He also advised Roger Lowell Putnam, the observatory's sole trustee, in 1927 that a draft book manuscript by V. M. Slipher that defended Percival Lowell's interpretation of Mars would damage the observatory's reputation. Putnam killed the book project, forestalling a divisive reawakening of the canal controversy.[74]

Although Russell did not engage directly in planetary research during the 1920s, he steered several Princeton graduate students toward these problems, and aided their efforts to find solutions. A revealing example of this occurred in 1923. Intrigued by the radiometric studies of planetary temperatures then underway in Flagstaff by Carl Otto Lampland of Lowell and the physicist William W. Coblentz of the National Bureau of Standards, and aware that his graduate student Menzel had taken an interest in Coblentz's work, Russell requested Menzel to use these raw radiometric measurements to determine the temperature of Jupiter. The value of this problem for Russell was apparently the long-unresolved question of whether Jupiter more closely resembles other planets or the Sun. Throughout the early twentieth century, most astronomers accepted arguments developed in the 1880s that Jupiter was a solarlike body heated to the point of incandescence, either as a result of Helmholtz–Kelvin gravitational contractions or residual stores of primordial heat. Such arguments, however, rested on qualitative analogies with the Sun. To facilitate this work, Russell developed a theory for the dependency of planetary heat on observed temperature, adopting his studies of stellar temperatures to the cooler infrared spectrum where planetary atmospheres show their strongest emissions. Armed with this theory, Menzel calculated a temperature for the Jovian cloud tops at only −110° C, about that expected for a blackbody in radiative equilibrium at Jupiter's distance from the Sun. High internal temperatures seemed excluded by this evidence, a noteworthy finding for many astronomers.[75]

Russell's involvement proved significant in several ways. Coblentz used Russell's theory as a key to translate his extensive series of radiometric measurements into actual planetary temperatures, including his widely reported 1924 values for Mars (which indicated summertime temperatures reaching 60° F along the equator). In addition, Menzel and Russell argued that Jupiter's cold temperature provided experimental evidence for a new physical theory of the outer planets simultaneously published by Harold Jeffreys, the Cambridge University geophysicist, which held that these planets consisted of rock and metal cores surrounded by vast shells of ice

[74] Russell to O. Veblen, May 25, 1928, Box 19, HNR; V. M. Slipher to Putnam, Feb. 6, 1928, LOW. For Russell's advice to Putnam, see Putnam to Shapley, June 17, 1927, and Shapley to Putnam, Mar. 10, 1930, both Box 15, DHCOs.

[75] Russell to E. Pettit, Jan. 25, 1923, Box 21, Folder 3, HNR; Coblentz and Lampland (1924a,b); see also Hufbauer (1981, 279) and Menzel (1923, 65–74; 1939). On Russell's interest in radiative transfer theory, see DeVorkin and Kenat (1983a, 106).

and their enormous atmospheres. Russell successfully urged Pettit to begin new radiometric studies of the nighttime hemisphere of Venus with the 100-inch telescope, hopeful that the thickness of its atmosphere could be found.[76] Finally, Menzel himself began additional investigations of planetary spectra, convinced now that the spectra of the outer planets, unlike those of stars, showed the lines of cold gases rather than those characteristic of high temperatures. Little progress was made in identifying the numerous unexplained absorption bands they contained, however, largely because laboratory data on the properties of substances at temperatures below zero centigrade were limited. Although Menzel, like Russell, turned to planetary spectroscopy only on occasion, this work illustrates Russell's role as a mentor, as well as the willingness of astrophysicists to apply familiar instrumental techniques to planetary research when problems of interest seemed within range of solution. Telescopic and instrumental limits were soon reached, but these investigations disclosed for the first time quantitative values for the temperatures of Mars and the nonstellar character of Jupiter and the other giant planets.[77]

Until 1929 these excursions into solar system studies had for Russell the character of puzzle solving, a kind of activity in which he delighted. By the early 1930s, however, Russell began venturing into this field more seriously. What caused this turn of events was his continuing work on natural abundances of elements in the solar atmosphere. In particular, it owed to growing doubt after 1925 among European and American astrophysicists, including Payne, Charlotte Moore, Antonie Pannekoek, and Albrecht Unsöld, that Earth's composition resembled that of the Sun and stars; they believed instead that evidence pointed to far greater abundances of hydrogen and helium in stellar bodies. This was a poignant conflict for Russell, who, more than many of his colleagues in astronomy, had invested considerable effort to comprehend the geochemistry of Earth as well as stellar bodies. In 1914 Russell had declared his acceptance of a close correlation between solar and terrestrial abundances.[78] Seven years later, in a widely cited 1921 paper, Russell employed estimates of uranium still present in Earth's crust (compared with abundances of radioactive daughter elements, such as lead) to calculate a maximum age for Earth; he had come up with an age of two to eight (American) billion years. Moreover, Russell had continued to follow geochemical studies closely throughout the 1920s, all of which assumed close solar–terrestrial abundances. In 1927 he served on an NRC committee that reviewed the internal constitution of Earth, affirming again the close relation of solar and terrestrial abundances.[79]

Within two years, however, Russell revised his views about the general chemical makeup of the Sun, stars, and planets. Hydrogen, Russell now declared, "is far more

[76] Coblentz (1925, 472, 474), Jeffreys (1923; 1924), Menzel, Coblentz, and Lampland (1925), and Russell (1935a, 157, 162).

[77] Russell to T. Dunham, Jr., Apr. 20, 1931, Box 7, Folder 61, HNR; see also Menzel (1923; 1924, 226; 1925), Phillips (1929).

[78] DeVorkin and Kenat (1983b, 181, 205–17).

[79] Russell (1921, 86); see also Brush (1978a, 89).

abundant than the other elements." This discovery marked a watershed in his thinking on solar atmospheres and stellar structure, and Russell invested much effort in the early 1930s refining his new interpretation of cosmic abundances, as well as convincing other astronomers of its veracity. For Russell, however, it is important to stress that the conceptual shift demanded by the high cosmic abundance for hydrogen was not confined to theories of stellar structure, but to his ideas about the structure and evolution of Earth as well. As a result, Russell turned to the planetary system and its origin.[80]

A number of astrophysicists began to investigate planetary atmospheres after Russell presented his arguments for the high abundance of solar hydrogen in 1929. Although improvements in photographic emulsions and high-dispersion spectrographic instruments were partially responsible for this upsurge of interest, a greater motivating factor was the sharp contrast now accepted between solar and terrestrial abundances, suggesting that similar variations in composition existed among the planets. Menzel, by 1930 at Lick, reanalyzed the spectra of the outer planets to evaluate the hydrogen content of these bodies.[81] At Mount Wilson, Adams joined Dunham, a former Russell student who had earned his doctorate in 1927, to make new visual- and infrared-light spectra of planetary atmospheres. Dunham employed a high-dispersion camera of his own design, new infrared plates secured from Eastman Kodak's Research Laboratory, and a pressurized absorption tube to study absorption lines produced by gases at cold temperatures and high pressures. In 1932, Adams and Dunham found that several unexplained absorption bands on Jupiter and Saturn were caused by methane and ammonia, confirming work by the young Göttingen physical chemist and astronomer, Rupert Wildt. That same year, they used infrared plates for a difficult four-hour spectrograph of Venus at the 100-inch telescope, revealing absorption bands at $\lambda 7820$, $\lambda 7883$, and $\lambda 8689$. Dunham identified these as caused by carbon dioxide, discovering the major constituent of this planet's atmosphere.[82]

Significant new work on planetary atmospheres also began at Lowell in the early 1930s. In 1933, Putnam – anxious to develop infrared astronomy and, through Russell, aware of Mount Wilson advances in this field – sought to add a spectroscopist at Lowell. He soon hired Adel, a newly minted Ph.D. in physics from Michigan, the leading U.S. school for infrared spectroscopy, to analyze the spectra of oxygen, carbon dioxide, methane, and ammonia under simulated atmospheric conditions. Staying at Michigan for his initial two-year contract (1933–5), Adel experimented with a Hilgar spectrograph-equipped, high-pressure absorption tube of 22.5 m, then calculated absorption line displacements for these gases. Comparing his results with V. M. Slipher's spectroscopic studies of planetary atmospheres (for which, in part,

[80] DeVorkin and Kenat (1983b, 209) and Hufbauer (1991, 103–6).

[81] Menzel (1930).

[82] Theodore Dunham, Jr., OHI (David H. DeVorkin, Apr. 30, 1977, SHMA at AIP); see also Adams and Dunham (1932, 243), Dunham (1938, 268; 1956), Russell (1935a, 157–9), and Wildt (1932).

Slipher had been awarded the Gold Medal of the Royal Astronomical Society in 1933), Adel found that most absorption bands for Jupiter and Saturn were caused by methane and ammonia. A further confirmation of Wildt's findings, Adel's work identified other unknown bands as belonging to methane, solving an outstanding puzzle in planetary spectra.[83]

Russell was himself drawn to this topic, although he focused initially on what Earth and its atmosphere could tell about the natural distribution of elements. He began corresponding with Wildt and Viktor Goldschmidt, the eminent European geochemist and natural-abundance expert, seeking to better correlate astronomical and terrestrial abundance estimates.[84] Russell also began new research in this area. In 1933, Russell teamed with Menzel to compare the abundance of gases found in Earth's atmosphere with what they termed the "permanent" gases in stars and gaseous nebula. Although they found "general agreement" between the terrestrial and cosmical abundances of most elements, a number of variations were apparent. Neon (which Menzel and other astronomers had discovered in several stars earlier that same year) provided a striking example, since the cosmic abundance of this element is five hundred times that within the atmosphere of Earth. This difference led Russell and Menzel to conclude that Earth had lost virtually all of its primitive atmosphere, for along with neon, hydrogen and helium are similarly depleted compared to the Sun. Russell, the paper's senior author, seemed even more intrigued by the low terrestrial abundance of nitrogen. Molecular nitrogen, Russell observed, is a gas of moderate atomic weight, and gravitational separation did not seem sufficient to explain its absence on Earth. He therefore leaned toward the possibility that "both nitrogen and neon may be less abundant in the sun than elsewhere," and used this result to support the idea that certain elements had a nonuniform cosmic distribution. This was one of the first instances where Russell employed his knowledge of geochemistry to infer inhomogeneities among stellar objects, and also caused him to think more about planetary evolution. These chemical factors suggested to Russell a molten origin for Earth.[85]

In 1934, for the first time, Russell turned wholeheartedly to problems of the solar system. By then he had been thinking about planetary geochemistry for some years, and accepted an invitation to deliver the Page–Barbour Lectures at the University of Virginia with the intention of using this forum to review existing theories on the origin of the solar system.[86] Like most astronomers, Russell through the early

[83] Russell to R. Wildt, June 18, 1934, Box 30, Folder 12, HNR; Arthur Adel OHI (Robert W. Smith, Aug. 13, 1987, at AIP); see also Adel and Slipher (1934a, 240–1; 1934b, 902–4) and Hoyt (1980b, esp. 432). Slipher's Darwin lecture of 1933 was titled "The Studies of the Spectra of Planets"; see Slipher to R. Putnam, Mar. 25, 1933, LOW.

[84] Russell to Wildt, June 18, 1934, Box 30, Folder 12, and Goldschmidt to Russell, Apr. 15, 1933, Box 42, Folder 22, both HNR.

[85] Quoted from Russell and Menzel (1933, 997–8); see also Russell to Menzel, Mar. 1, 1932, and Oct. 2 and 11, 1933, all Box 17, HNR; Russell (1935b, 163),

[86] Russell to Jeffreys, Mar. 19, 1934, Box 13, Folder 57, HNR; see also Brush (1978a, 90).

1930s had accepted a dualistic origin for the solar system, which held as its fundamental assumption that the planets had been created by a chance close approach of another star to the Sun, casting solar-type material into neighboring space. Russell was especially attracted to encounter theories argued more than a decade earlier by the Cambridge scientists Jeffreys and James Jeans. Jeffreys's theory required that two stars actually collide, rather than pass at extremely narrow distance, in order to produce a planetary system. Russell had reviewed Jeffreys's theory sympathetically in the 1926 edition of *Astronomy*, as it provided for him a satisfactory account of many of the solar system's features. (It explained, for instance, the concentration of the solar system's angular momentum in the planets, rather than the Sun, a puzzle that had caused astronomers to reject monistic, or nebular, theories for its origin.) Nevertheless, as he wrote Jeffreys early in 1934, he doubted that it could account for the great distances that separated the Sun from the orbits of the major planets. In addition, owing to the newly discovered high cosmic abundance of hydrogen, as well as the prediction by the British astrophysicist Arthur Eddington that subsurface regions of the Sun had temperatures of one million degrees, he could not see how material torn from the Sun had coalesced into planets rather than thermally diffusing through space. Interest in these dynamical problems in part inspired Russell to prepare his 1935 *The Solar System and Its Origin*, one of the most influential books about the solar family written between the world wars.[87]

Another factor that heightened Russell's interest in the solar system's origin was concern by astronomers about the ages of Earth and the universe resulting from new geochemical studies of atomic abundances. Through the early 1930s, Russell continued to believe that his 1921 estimate of the age of Earth's crust (2–8 billion years) rested on a firm foundation. A 1931 NRC committee report on the age of Earth did much to enhance his confidence: In it, based on new analyses of radioactive isotopes, the eminent geologists Adolph Knopf and Charles Schuchert provisionally estimated Earth's age as 1.6–3.0 billion years. Astronomers followed these results with interest. Until about 1930, Russell, Shapley, and other American astronomers accepted arguments based on calculations of stellar cluster movements and the expected longevity of stars that placed the age of the universe at ten billion years or more, or between three and five times that of Earth. During the early 1930s, however, astronomers became far less certain that the universe was indeed that old. Flaws in mathematical analyses of motions in stellar clusters and new theories of stellar energy production accounted for some of these doubts. Even more important was new observational evidence, reported by the Mount Wilson astronomer Edwin P. Hubble in 1929, that the redshifting of light from distant island universes (galaxies) revealed the universe was itself expanding. Hubble's result, one of the most significant discoveries in twentieth-century astronomy, made astronomers aware as never before of the time scale of the universe. Fitting Hubble's data to relativistic models of the universe,

[87] On its influence see, e.g, Kuiper (1949, v), Urey (1952c, 4–7), and Watson (1938, 302).

mathematical astronomers and physicists such as Willem de Sitter and Lemaître calculated its rate of expansion, finding an age for the universe of two to three billion years at most.[88]

This decrease in the accepted age of the universe, and simultaneous increase in the age of Earth accepted by geochemists, placed both ages in the same ballpark for the first time, a situation not unnoticed by American astronomers. Buoyed by the Arizona Meteor Expedition, Öpik, for instance, declared that geochemical determinations of meteorite ages (which roughly matched those for Earth) indicated that "our stellar universe itself . . . may be not very much older than the solar system." Öpik urged greater cooperation between astronomers and meteorite geochemists to address the problem of cosmology, arguing that geochemical evolution could have been influenced by the birth of the universe.[89] Russell was equally intrigued by these developments. In 1933 he strongly urged a younger colleague working on stellar evolution to pay attention to "the question of the time scale" emerging from the theories of Lemaître, de Sitter, R. C. Tolman, and E. A. Milne, for their conclusions set limits on the amount of stellar evolution that could have occurred.[90] In taking on the problem of cosmogony, Russell thus saw himself addressing an important episode in the early history of the universe. Cosmogony, for him, had become conceptually linked to cosmology and stellar astrophysics.

In the published text of his Virginia lecture, Russell approached this connection cautiously but directly. By the time he prepared his publication draft, Russell had become convinced that Jeffreys's theory was fatally flawed. His concluding section therefore sought to illuminate dynamical errors in Jeffreys's argument that a two-star collision could draw planetary material outward to the orbital distance of the outer planets, while reaffirming the value of interdisciplinary approaches to such problems. In a much-quoted assessment of dualistic cosmogonies, Russell observed that "[p]ast studies of the origin of our system appear . . . to have ended in an impasse." At the same time, however, Russell stressed the then-accepted hyperbolic velocities of meteors calculated from the Harvard meteor expedition, the geochemical age determinations of meteorites, the composition of planetary atmospheres, and the new age determinations of Earth as matters of considerable astrophysical significance. "One of the most striking results of modern investigation," he wrote, "has been the way in which several different and quite independent lines of evidence indicate that a very great event occurred about two thousand million years ago." This date was fundamental not merely for the solar system, but for the material universe as a whole. The history of our planet, he argued, "is thus tied up with that of the whole relativistic 'world.'" His particular solution involved combining cosmogony with cosmology, hence linking it to geochemistry. Planetary systems, he argued, may have formed when stars emerging from Lemaître's initial "atom" collided with one

88 Bok (1946, 61–2), Burchfield (1975, 201–5), and Smith (1982, 191–3).

89 Quoted from Öpik (1933, 71).

90 Russell to W. Markowitz, Feb. 20, 1933, HNR (microfilm ed.).

another in large numbers, thus overcoming dynamical objections to dualistic cosmogonies and assuring the production of large numbers of planetary systems.[91]

Russell's particular cosmogonal scheme attracted little support from American astronomers, who viewed it as ad hoc, and after 1934 Russell continued to concentrate largely on stellar spectra. Nevertheless, *The Solar System and Its Origin* stimulated further work in solar system astronomy. Russell himself remained an influential supporter of research into planetary atmospheres and cosmogony. Intrigued by Wildt's work on planetary atmospheres, Russell helped arrange a visiting appointment for him at Mount Wilson in 1935; two years later, aware that Wildt refused to return to Nazi-dominated Göttingen, he appointed Wildt Research Associate at Princeton. While at Princeton, Wildt continued working on planetary physics as well as stellar astrophysics, developing his model of a rock-ice core within Jupiter. In 1940 Wildt argued that a "considerable greenhouse effect" likely existed in Venus's carbon-dioxide-rich atmosphere – later regarded as one of his most important planetary contributions. Russell championed Wildt's papers, successfully urging their prompt inclusion in the *Astrophysical Journal*.[92]

It was another close associate of Russell, Lyman Spitzer, Jr., who produced at this same time one of the most important results in early-twentieth-century cosmogony. In 1939, a year after writing a dissertation on theoretical astrophysics under Russell, Spitzer considered the behavior of material condensation in extremely hot environments, such as a filament of solar material. Although Russell had not believed it possible that material at such high temperature could coalesce into small solid bodies, he had only raised a qualitative objection to this crucial plank in Jeffreys's collisional cosmogony while lecturing in Virginia four years before. Spitzer demonstrated quantitatively that material at these temperatures would diffuse rather than coalesce. His refutation of the collisional condensation process was soon viewed by American astronomers as the deathblow to encounter theories for the solar system.[93]

Russell's and Spitzer's condemnation of close-encounter cosmogonies marked an important juncture in the history of twentieth-century cosmogony. It awakened interest in the origin of the planetary system and stimulated new discussions about the significance of geochemistry for understanding planetary formation. It also caused astronomers to think of planetary systems – and life – once again as possibly frequent phenomena in the galaxy, rather than the extremely rare events predicted by close-encounter theories. In the 1940s, this factor would encourage renewed interest in cosmogonal research.

Solar system research in America prior to 1940 shared two general characteristics. Scientists who investigated solar system phenomena did not see themselves contribut-

[91] Russell (1935b, 133–4, 137).

[92] Struve to Russell, Oct. 8, 1937, Box 64, Folder 40, HNR; see also Whipple (1941, 163–4) and Wildt (1940, 266; 1942).

[93] Lyman Spitzer, Jr., OHI (David DeVorkin, Apr. 8, 1977, SHMA at AIP); see also Brush (1978a, 90–1) and Spitzer (1939).

ing to a distinct *field* of astronomy, but rather to problems of general disciplinary interest. The bearing of meteorite ages upon the age of the universe, and the relationship of terrestrial chemical abundances to those of the Sun and other stars, were questions that astronomers found important. Shapley and Russell began their respective studies of meteors and planetary compositions hoping to learn more about interstellar and stellar phenomena. Although astrophysical techniques were difficult to apply to the comparatively cold members of the planetary family, astronomers showed no hesitation to use them whenever solutions to outstanding problems, from the origin of asteroids to the structure of the giant planets, seemed within reach. American astronomy remained a unified, inclusive discipline, with common training and traditions, research instruments, and disciplinary journals. Astronomers readily crossed boundaries to contribute to what only later came to be perceived as distinct fields.[94]

The most vigorous programs in solar system astronomy during this period were interdisciplinary. Sustained by centers for mathematical astronomy at Berkeley and Yale, through such temporary agencies as the Harvard meteor expedition, and in the research programs at Harvard, Lowell, and Mount Wilson, solar system investigations reflected not only the particular interests of its leaders but their success in convincing extramural patrons to support work outside astronomy's core research programs. Leuschner successfully attracted private and federal support for his large asteroid program, while Shapley pitched his ambitious plan to blend astronomy, geochemistry, physics, and geology in a new disciplinary matrix to a willing audience at the Rockefeller philanthropic boards. That Shapley's bid to create a Harvard Institute of Cosmogony failed is less important than his success in creating an informal institutional context for scientists interested in problems on astronomy's border with geochemistry and geology. Meteor research by Watson and Whipple was shaped by Shapley's ideal of "cooperative" study as well as the problems each attempted to solve. When Russell declared in 1935 that carbon dioxide on Venus and ammonia on Jupiter were found only through the combined contributions of "astronomy, physics, chemistry, geology, biology, and technology," he endorsed a cooperative style of research he had long emphasized to colleagues and his students.[95]

Interdisciplinary studies of the solar system resumed after the disruptions of World War II, particularly as new federal patrons multiplied. While work on small-body problems continued at many U.S. centers for astronomy, military interest in atomic abundances, the mechanics of high-energy explosions, and other topics helped to transform solar system astronomy from a diverse set of astrophysical problems to a distinct, intellectually coherent field. In the years immediately after 1945, however, it was another branch of solar system research that attracted greatest attention. This field was planetary meteorology.

[94] Similar issues affected twentieth-century physics; see Weart (1992b).

[95] Russell (1935a, 168).

2

Planetary Atmospheres and Military Patrons, 1945–1955

World War II marked an important watershed for American astronomy. The war disrupted research programs and dispersed American astronomers to wartime projects far from their home facilities. At the same time, it spurred development of new instruments, exposed astronomers to research outside their own specialties, and increased federal and military patronage to this community. The needs of military patrons after World War II proved particularly beneficial to certain fields of astronomy, including solar astrophysics and solar system research. More important, wartime scientific and technical triumphs left many astronomers optimistic that large, unsolved problems in their discipline would yield to renewed attack. Although American astronomers were slower than their colleagues in physics to clamor for federal assistance, most believed that a new era of research was at hand.[1]

Among the first fields of astronomy to profit from these developments was the study of planetary atmospheres. The tremendous significance of accurate forecasts for modern military operations was a major lesson that military planners drew from World War II. After 1945, these patrons gave generous support for meteorological research in such fields as climate modification and global circulation. Other military planners, anticipating that the German V-2 foreshadowed the development of guided missiles, provided ample funding for investigations of the properties of the upper atmosphere.[2]

Aware of military interest in atmospheric research, several astronomers launched new interdisciplinary programs along astronomy's border with meteorology. This chapter explores the emergence of planetary atmospheres programs at three major centers of American astronomy: the Yerkes–McDonald observatories (Chicago), the Lowell Observatory, and the Harvard College Observatory. These projects, although short-lived, were the first created by astronomers interested in the solar system after 1945. More important, they helped sustain the vitality of these pivotal U.S. centers of astronomy in the critical postwar years.

[1] Hufbauer (1991, 119–59), Needell (1991), and Smith (1989, 22–5).

[2] DeVorkin (1992) and Nebeker (1995, 111–32). Solar astrophysics also expanded rapidly after 1945 because of military interest in long-range communications; see Hufbauer (1991, 119–209).

INFRARED ATMOSPHERIC STUDIES

As we saw in Chapter 1, American astronomers had addressed the problem of planetary atmospheres in the 1920s and 1930s. Skilled spectroscopists such as Adel had applied molecular theory to discover constituents in planetary atmospheres. Others, including Adams, Campbell, Dunham, Menzel, Russell, and Wildt, had turned to planetary atmospheres as a means to explore broad problems in astrophysics, including the physical structure of the giant planets and the evolution of the solar system.

Although foundations and local institutional support had sustained this work in the interwar years – and would continue to play an important role after 1945 – federal and military patronage greatly influenced postwar research in this field. Nowhere was this better illustrated than in the study of planetary atmospheres begun at the Yerkes–McDonald observatories of the University of Chicago. Led by Gerard P. Kuiper, who directed these observatories between 1947 and 1949, this program sought to address such problems as the composition of planetary atmospheres and their long-term stability. Kuiper saw this project as a means to develop new links between the astronomical and meteorological communities, declaring in 1948 that wartime advances in the study of global climatology would provide astronomers "new tools for the interpretation of planetary phenomena." Moreover, Kuiper did not work alone on this project. Nearly a third of the Yerkes–McDonald staff became involved in aspects of atmospheric research in the immediate postwar years, seeing them as closely related to observatory programs involving stellar atmospheres and interstellar matter.[3]

Kuiper influenced this project in many ways, but his influence on solar system astronomy in America was greater still. Born in 1905 in Harenscarspel, a small town in northern Holland, Kuiper had studied at the University of Leiden, a leading center of European astronomy. He had worked principally under Jan Woltjer and Willem de Sitter, two of Europe's leading mathematical astronomers at the time; his interest in celestial mechanics and origin of the solar system was heightened by these men. At the same time, he had studied astrophysics under Ejnar Hertzsprung, a prominent astrophysicist who, independently of Russell, had discovered the stellar mass–luminosity relationship. Armed with a Ph.D. on the structure of binary stars, a good command of English, and extraordinary stamina for observational work, Kuiper had traveled to the Lick Observatory in 1933. Finding permanent employment there blocked by resentment of foreigners, Kuiper had accepted a temporary post at Harvard in 1935. Two years later Otto Struve, the Yerkes director, had invited him to join the newly founded McDonald Observatory in Texas, operated jointly with Yerkes. While Kuiper made his home in Williams Bay, Wisconsin, where Yerkes is located, he commuted to Texas for his annual four weeks of telescope time at McDonald, whose 82-inch spectroscopic reflector was the second largest telescope in

[3] Quoted from Kuiper (1949, v). Earlier accounts stressed the isolation of Kuiper's research; see, e.g., Sagan (1974, 117–18) and Tatarewicz (1990a, 20–1).

the United States. By then Kuiper had gained a reputation as a bright, brash researcher zealously devoted to observational work, with a self-confidence at times bordering on arrogance.[4]

Although Kuiper had, while at Leiden, published a synthesis of Martian atmospheric studies and developed an interest in cosmogony, he nevertheless earned his early reputation in stellar astrophysics. His discovery of binary stars of extremely short periods, which won him the admiration of Russell, led him to study stellar motions within several local clusters. This provided an effective foundation for calibrating the stellar temperature scale. Kuiper also studied low-mass stars such as white dwarfs and faint blue stars, his interest in them kindled by studies on degenerate matter by his friend Subramanyan Chandrasekhar, the eminent Yerkes theoretical astrophysicist. By 1942 Kuiper had discovered twenty-one of the thirty-odd white dwarf stars then known. He confided to Struve his hopes of publishing a monograph on dwarf stars, followed by one on double stars, intending each to provide a basis for discriminating between theories of stellar energy production. It was for him the start of a fruitful research program.[5]

Kuiper's turn toward solar system astronomy – a fateful move for his career, and for the future of the field – resulted from a specific discovery that he made in January 1944. At that time Kuiper returned to McDonald from wartime work at the Radio Research Laboratory in Cambridge, Massachusetts, for his scheduled month of telescope time. Kuiper's trip testified to his personal tenacity; few fellow researchers on wartime assignment received similar privileges.[6] After completing routine spectroscopic binary work, Kuiper turned the 82-inch telescope on Jupiter, Saturn, Uranus, and Neptune to expose spectra of their atmospheres. Possibly he did so recalling Russell's 1935 interpretation of planetary spectra. Kuiper believed that the resolution of the McDonald spectroscope was sufficient to determine abundances of constituent atmospheric gases with better precision than that possible in the mid-1930s, allowing him to secure important refinements.

Instead, Kuiper found a surprise. His spectra revealed that Titan, Saturn's largest moon, had strong absorption bands at 6,190 and 7,260 angstroms, indicating a methane atmosphere surrounding the satellite. In his influential 1935 review, Russell had concluded, on the basis of mass, density, and absolute surface temperatures, that "[l]arge planets have atmospheres containing hydrogen compounds; middle-sized planets, atmospheres containing oxygen compounds; and small planets no atmo-

[4] Kuiper, "Data for the University Records," Feb. 10, 1933, and idem to S. Chandrasekhar, May 11, 1936, both Box 28, GPK; A. O. Leuschner to P. Peters, Sept. 23, 1937, Box 23, AOL; Struve to R. T. Birge, Oct. 1, 1940, OS[mic]; idem to H. Shapley, Sept. 30, 1935, Box 51, DHCOs; see also DeVorkin (1982).

[5] Russell to A. O. Leuschner, Feb. 25, 1935, Box 15, Folder 59, HNR; Struve to H. Shapley, Aug. 10, 1940, Box 50, DHCOs; Kuiper to Struve, Apr. 4, 1942, Box 31, GPK; idem, "Appraisal of the Work of Kuiper," undated MS [ca. 1945], Box 29, GPK; Cruikshank (1993), DeVorkin and Kenat (in press), Gehrels (1988, 95), Kuiper (1931), and Tatarewicz (1990a, 19–22). On Kuiper's relation with Chandrasekhar, see, e.g., Kuiper to Chandrasekhar, Jan. 16, 1941, Box 28, GPK.

[6] Few American astronomers felt able to continue astronomical research with their wartime responsibilities; see Oort to N. U. Mayall, Aug. 30, 1945, Box 31, JHO, and also DeVorkin (1989a, 3–4).

spheres at all." By possessing a measurable atmosphere, Titan challenged Russell's widely accepted arguments concerning the origin and evolution of planets, as well as assumptions concerning the stability of planetary atmospheres. Kuiper thus believed that Titan's atmosphere had important implications for cosmogony, a subject in turmoil following the refutation of encounter theories by Russell and Spitzer.[7]

For many astronomers, the significance of Kuiper's discovery was in providing evidence about the thermal history of planets after the solar system began to form. Through letters exchanged with Russell, Kuiper came to believe that Titan's methane atmosphere meant that Titan had never been as hot as the Moon for any appreciable period of time. In his own report on the discovery, which he quickly placed in the *Astrophysical Journal,* Kuiper extended this conclusion. Titan's current atmosphere, he declared, would become unstable with even a twofold increase in temperature. Since initially high surface temperatures were "commonly assumed for all bodies in the solar system," this implied that Titan's atmosphere had formed after the satellite's initial high temperature phase. Moreover, because Titan is at a considerably greater distance from the Sun than the terrestrial planets, Kuiper argued that the initial atmospheres of Mars and Venus, as well as that of Earth, "must have escaped from the crust after the crust was essentially cooled off." His commitment to a molten origin for planets wavered only occasionally through the 1950s, and became a focus of his subsequent controversy with Harold C. Urey.[8]

Kuiper believed that his discovery pointed to fruitful research pathways in solar system astronomy. In the early 1940s, the formation of satellite systems about planets remained a stumbling block for general theories of cosmogony. While the discovery did not immediately incline Kuiper toward cosmogonal research, he did argue in his 1944 discovery account that similarities in the compositions of the atmospheres of Saturn and Titan demonstrated that both worlds shared a common genesis. Furthermore, the presence of this atmosphere (given the thermal constraints already established) seemed to rule out the possibility that Titan had been captured from an elliptical orbit extending into the inner region of the solar system. Because the existence of satellite systems seemed difficult for astronomers to explain on dynamical grounds, Russell proclaimed the significance of Kuiper's evidence that planetary satellites are geochemically linked to their parents. The discovery of Titan's atmosphere thus suggested to Kuiper new research programs with potentially high dividends. The mass and density of both Pluto and Triton (a large moon of Neptune) seemed higher than Titan, suggesting these bodies also possessed atmospheres. In addition, as Kuiper wrote his Yale colleague Dirk Brouwer, the discovery required modern redeterminations of such fundamental constants as the diameters and densi-

[7] Quoted from Russell (1935b, 51–2); cf. idem (1935a, 16). For Kuiper's perceptions of this discovery, see Kuiper to Struve, Jan. 7 and 9, 1944, and Struve to Kuiper, Jan. 14, 1944, all Box 28, GPK; see also Russell to Kuiper, Jan. 17, 1944, Box 47, HNR; and Kuiper to Russell, Jan. 18, 1944, Box 31, GPK.

[8] Quoted from Kuiper (1944, 381, 382); see also Russell to Kuiper, Jan. 17, 1944, Box 14, HNR; idem to Kuiper, Jan. 25, 1944, Box 31, and Kuiper to Brouwer, Jan. 30, 1944, Box 29, both GPK.

ties of the outer planet satellites, for the question of atmospheric stability depended on these critical values.[9]

It was the war, however, that placed these studies in new context and made available to Kuiper unanticipated resources with which to pursue them. One influence was the wartime Radio Research Laboratory. There Kuiper came in frequent contact with Fred L. Whipple, the Harvard meteor astronomer and fellow RRL colleague whom Kuiper had befriended in the 1930s. As we shall see, Whipple was then seeking military support to apply his meteor investigations to study the properties of the upper atmosphere. Both men now found a common interest in planetary atmospheres.[10]

A more decisive step was a military assignment that Kuiper undertook in early 1945. Owing to his fluency in Dutch, German, and French, and his familiarity with scientists and research facilities in Western Europe, Kuiper was recruited for the secret American ALSOS mission. Administered by the U.S. Army, ALSOS teams were deployed behind advancing Allied troops to interview Axis scientists about their wartime weapons research, including the rumored German atomic bomb.[11] He used leave time to visit European astronomers, among them Bernard Lyot of the Meudon Observatory, widely noted for his invention of the coronagraph and measurements of polarized light from the Moon and planets. Kuiper gained firsthand information from Lyot about his polarization studies of planetary atmospheres, and arranged an instrumentation exchange that provided him a double-image micrometer particularly suited for measuring the diameter of planetary satellites;[12] however, this wartime assignment also made Kuiper particularly aware of heightened military interest in atmospheric phenomena, and ways in which astronomers could contribute to their study. On an ALSOS assignment in Germany, Kuiper interviewed Erich Regener, the Austrian-born physicist, becoming one of the first U.S. scientists to discover Regener's plans to place scientific instruments on board a Peenemünde V-2 launch to study the upper atmosphere. Kuiper later declared Regener's work and studies of solar–ionospheric relationships by the German astronomer Karl-Otto Kiepenheuer as the most interesting wartime German astronomical work. Kuiper became sufficiently identified with studies of this kind to receive an invitation to join the U.S. V-2 Rocket Research panel formed after the war, a privilege extended to few astronomers.[13]

[9] Kuiper to Brouwer, Jan. 30, 1944, and idem to Russell, Jan. 31, 1944, both Box 29, GPK; see also Kuiper (1944, 381).

[10] Kuiper to Chandrasekhar, Aug. 3, 1943, Box 28, GPK; Whipple to Kuiper, June 13, 1946, and Kuiper to Whipple, June 17, 1946, Box 32, GPK; see also F. Terman to O. Struve, Dec. 4, 1945, OS[mic].

[11] On ALSOS, see Doel (1992, 249) and Walker (1989, 153–60).

[12] Audouin Dollfus OHI (Ronald E. Doel, Sept. 2, 1986 and Jan. 14, 1987, AIP); Kuiper to Struve, Jan. 1, 1945, Box 28, and idem to W. Adams, Feb. 6, 1945, Box 29, both GPK; Bernard Lyot, "Correspondence file on I.A.U.," n.d. [ca. 1946], BL; see also Hufbauer (1994), Kuiper (1952b, 307), and Tatarewicz (1990a, 1–12).

[13] Kuiper to Struve, Nov. 12, 1945, Box 28, GPK; see also DeVorkin (1992, 49–51), Kuiper (1946). Kuiper's increased involvement in atmospheric research is suggested in Kuiper to J. A. Hynek, June 13, 1946, Box 29, GPK; idem to Whipple, June 17, 1946, Box 32, GPK.

Still, returning to Yerkes in late 1945, Kuiper faced a critical decision. He remained convinced that a study of planetary atmospheres could yield important results applicable to global atmospheric circulation, the stability of planetary atmospheres, and the origin of the solar system. In letters to colleagues, he argued that Titan's atmosphere, combined with wartime measurements of unexpectedly high temperatures in Earth's upper atmosphere, invalidated key assumptions that James Jeans had employed in his classic *Dynamical Theory of Gases,* still the standard source on the stability of planetary atmospheres. Moreover, Kuiper anticipated that other observatories would also take up new research in this field, increasing its vitality. He noted that a Harvard College Observatory symposium planned for December 1946 would include discussions of global climatology and upper-atmospheric research, and he expected that Mount Wilson astronomers would turn again to high-dispersion spectroscopy of planetary atmospheres. Finally, Kuiper believed that the military might become a patron for such work. Solar system astronomy therefore appeared to Kuiper a way to maintain the competitiveness of Yerkes–McDonald among American astronomical centers. Yet Kuiper also felt attached to his prewar program of stellar research. Typing a hurried memorandum on his future research plans, Kuiper mentioned the white dwarf and blue star research he had eagerly conceived in 1942, but failed to mention solar system astronomy at all.[14]

What ultimately convinced Kuiper to move toward planetary research was a combination of instrumental, intellectual, and institutional factors. A Schmidt telescope under construction at McDonald, essential for his faint-blue-star survey, remained unfinished in late 1945. His inability to resolve spectral difficulties inherent in interpreting white dwarf structure also frustrated him; Kuiper seemed to realize that his work on this topic had temporarily at least reached a dead end.[15] Most of all, through contacts in the RRL and ALSOS, Kuiper gained inside knowledge of wartime advances in infrared detectors, a technology especially suited for solar system research. In November 1945, Kuiper discovered that German scientists had made rapid progress in building detectors using lead sulfide cells. Sifting through still-classified records, he found that American efforts to build similar detectors had been led by Robert J. Cashman, a Northwestern University physicist whose Evanston, Illinois, laboratory was just hours from Yerkes. Cashman's cells recorded infrared radiation out to 40,000 and 50,000 angstroms, far beyond the 10,000-Å limit of infrared films. Seeing "enormous possibilities" in applying Cashman cells to astronomy, and convinced that no other American astronomers knew of Cashman's device,

[14] Kuiper, "Outline of Research Programs," Dec. 11, 1945, Box 29, and idem to P. Swings, May 17, 1947, Box 31, both GPK; and Kuiper to O. Struve, Jan. 9, 1944, Dec. 18, 1946, and Jan. 13, 1947, all Box 28, GPK. On the 1946 climate conference see Harlow Shapley, *ARHCO,* 1946–7: 1. Jeans's 1904 *Dynamical Theory of Gases* was last reissued in 1925. As it turned out, the Mount Wilson–Palomar observatories, under the postwar direction of Ira S. Bowen, did little research in planetary atmospheres through the 1950s.

[15] DeVorkin and Kenat (in press).

Kuiper persuaded Struve to provide $500 to build a far-infrared astronomical spectrograph at Yerkes–McDonald, the first of its kind.[16]

Patronage was a final motivation behind Kuiper's turn to solar system research. His new infrared detector was partially suited for stellar research as well as solar system astronomy; indeed, an early test of the detector using the McDonald 82-inch telescope successfully recorded the first far-infrared spectrum of Betelgeuse, a cool red giant star. Yet by 1946 Kuiper was aware that potential military patrons for this work favored infrared astronomy aimed at exploring atmospheric phenomena. In a proposal to the Navy's Office of Research and Invention, Kuiper stressed his new instrument's value for studying the transmission properties of Earth's atmosphere, using the Sun and the planets as extraterrestrial sources; he downplayed a possible high-resolution study, advising a colleague that "for obvious reasons, Navy communication systems could not make use of it." In 1946, Kuiper received this grant. Arguing that planetary atmospheres research could help maintain the competitive advantage of Yerkes–McDonald, Kuiper never again returned full time to stellar studies.[17]

One astronomer who reached a similar conclusion was Struve, leader of the Yerkes–McDonald observatories and Kuiper's superior at Chicago. Struve's support was significant, for it illustrated how wartime scientific and technical developments, combined with new sources of federal patronage, came to provide a new institutional home for solar system studies at Chicago. Yerkes director since 1932, a stellar spectroscopist widely respected by colleagues, Struve was by the early 1940s a prominent leader of the American astronomical community. As an administrator, Struve had earned much praise for linking the newly constructed McDonald Observatory (willed to the University of Texas, which then lacked a Department of Astronomy) with Chicago, and for recruiting to McDonald eminent astrophysicists such as Kuiper, Chandrasekhar, Bengt Strömgren, and William W. Morgan. More than a dozen years younger than Russell, who had reached the pinnacle of his career in the 1930s, Struve was gaining stature as a leading figure in American astronomy at the outbreak of World War II, and determined to maintain Yerkes and McDonald as distinguished scientific institutions.[18]

Struve's decision to incorporate solar system astronomy into the research mission of these observatories was motivated by three factors. World War II had been a trying time for Struve. Compared to physics at Chicago, which prospered during the war through the Manhattan Project, Yerkes–McDonald suffered from limited

[16] S. Goudsmit to Scientific Members of ALSOS, July 3, 1945, and Kuiper to Struve, Nov. 23, 1945, both Box 28, GPK; Kuiper, "Final Report: IR Spectra," memorandum, June 20, 1961, Box 33, GPK; see also Kuiper, Wilson, and Cashman (1947, 243).

[17] Quoted from Kuiper to A. Adel, June 26, 1946, Box 29; see also Kuiper, "Infra-Red Project," n.d. [1946], Box 29, and idem to R. Cashman, June 3, 1946, Box 35, both GPK. An effort to recruit Adel to assist Kuiper on this Navy grant failed; see Adel to Kuiper, June 29, 1946, Box 29, GPK; Arthur Adel OHI (Robert W. Smith, Aug. 13, 1987, at AIP); see also Struve (1947b, 146).

[18] DeVorkin (1982, 597), Skolovskaya (1976), and Struve (1947a, 227–44).

funds and attrition as staff members departed for wartime research projects far from the rural environs of these facilities. Struve also worried that planned expansions at competing institutions such as Harvard and Mount Wilson would make these facilities more attractive than Yerkes–McDonald once the war ended, causing him to lose staff painstakingly recruited less than ten years before. In autumn 1944, Struve warned University of Chicago President Robert Maynard Hutchins that a new, million-dollar, 120-inch reflector planned for Lick and a proposed 94-inch telescope at Michigan might induce Kuiper to leave, since he and Kuiper both viewed the McDonald spectroscopes as seriously outdated. Although nearly all Yerkes–McDonald staff astronomers on wartime leave did return after 1945, the painful process of contemplating drastic policy options and his conviction that astronomy had no military utility caused Struve to reconsider what kinds of astronomical research were best suited for American observatories.[19]

These wartime anxieties convinced Struve to advocate a new, flexible research policy for Yerkes–McDonald, one that stressed competitiveness and openness to extramural patronage over consistency and long-term goals. Rejecting what he later characterized as "the accumulation and slow digestion of a broad basis of observational and theoretical facts" as the optimum pathway to discovery, Struve declared that many research programs begun by his predecessor at Yerkes, Edwin Frost, had "absorbed perhaps twenty years of telescope time, yielded little new information not immediately surpassed by results from other observatories," and thus were "sterile and dull." This made explicit what senior American astronomers had come to recognize: Long-running projects like the international "Carte du Ciel," once central to astronomical research, were no longer seen as competitive or prestigious to young astrophysicists entering the discipline.[20] Struve instead embraced simultaneous work on "several smaller research programs" that utilized telescope time efficiently "and promised quick results." Meeting with Hutchins early in 1947, Struve declared that "[t]he scope of astronomy has increased enormously since the end of the war," owing to new research possibilities involving rockets, infrared spectroscopy, microwave radar, cosmic rays, and nuclear processes in stars, requiring that the department of astronomy "be re-oriented with regard to these changes." This was good news for Jesse Greenstein, a young staff member whose postwar interest in rocket-based spectroscopy had earned him an appointment on the V-2 rocket panel. It was equally good for Kuiper, whose planned study of planetary atmospheres also fit the bill.[21]

[19] Struve's concerns led to his innovative 1944 proposal to finance a large new research telescope through a consortium of midwestern universities, an idea that foreshadowed the organization of the Kitt Peak National Observatory; see Struve to Hutchins, July 3, 1944, and Aug. 17, 1945, and idem to Herzberg, Jan. 30, 1947, all OS[mic]; see also DeVorkin (1982, 599–600, 607).

[20] Struve and Zebergs (1962, 31); similar statements first appeared in Struve (1942, 470). Harvard's own photographic sky surveys, in operation since the late nineteenth century, were also sharply scaled back at this time; see Bok to P. Buck, Oct. 16, 1952, Box 7, FLW–SI.

[21] Quoted from Struve, "Discussed with Dr. Hutchins," Jan. 10, 1947, OS[mic]; see also DeVorkin (1989a).

Struve did not base his decision to support Kuiper's new interest in solar system astronomy solely on its perceived utility. Like Shapley and Russell in the 1930s, Struve saw Kuiper's work as a means of extending research in stellar astrophysics. He was particularly intrigued by the implications that Titan's atmosphere had for cosmogonal evolution, seeing this as closely related to the problem of spectroscopic binary stars that he and Kuiper had collaboratively pursued before World War II. In 1945, Struve declared that Kuiper's work would "profoundly affect the theory of the origin of the solar system, and the question of the origin of double stars."[22] Furthermore, Struve believed planetary atmospheres offered a means of investigating the natural distribution of atomic elements, a research area that remained one of astronomy's most fundamental challenges. During the 1920s and 1930s, determining the bulk composition of the Sun and stars had preoccupied Russell, Menzel, and other American astronomers, who saw this problem as vital to studies of the origin and evolution of stars. Natural abundances had also intrigued Struve as well as the Danish-born Strömgren, one of Struve's first appointments to McDonald. Struve had wanted Strömgren to lead astrophysical research in this area at Yerkes–McDonald; but after Strömgren resigned in 1937, Struve sought a replacement. The person he ultimately chose was Gerhard Herzberg, a German-born physicist then at the University of Saskatchewan.[23]

Struve's decision to recruit Herzberg illustrates the degree to which Struve, like Kuiper himself, saw an advantage in reaching beyond the discipline of astronomy to stimulate new fields of research. Although Herzberg in the 1930s had suggested a method of detecting molecular nitrogen in planetary atmospheres and faint, cool stars – which most astronomers until then had believed a theoretical impossibility – and had participated in a Yerkes symposium on interstellar lines in 1938, he was principally known as an infrared spectroscopist whose research was often inapplicable to astronomy. Yet Struve believed Herzberg could aid Yerkes–McDonald programs involving dwarf binary stars, the interstellar medium, and planetary atmospheres, and strengthen ties to Chicago's burgeoning physics department, where Herzberg would hold a joint appointment. For his part, Herzberg welcomed the opportunity to become part of a larger academic environment than that available to him in provincial Saskatchewan, and the chance to return to problems involving planetary and stellar atmospheres that he had postponed during the war. Although his appointment to Chicago was delayed by Canadian and American authorities because of his German citizenship and subsequent classification as an enemy alien, Herzberg arrived in the United States in July 1945. To prepare, Struve ordered the refurbishing of the spectroscopic laboratory at Yerkes Observatory designed and built by George Ellery Hale, the founder of Yerkes. He added a 21-ft grating spectrograph, an infrared spectrograph, and a 75-ft absorption tube capable of produc-

[22] Struve to Hutchins, Aug. 17, 1945 (quoted), and idem to Bok, Jan. 21, 1944, both OS[mic].
[23] DeVorkin (1982, 602).

ing pathlengths of 5,500 ft; the latter was built principally to study the pressures and constitutions of planetary atmospheres. It represented a considerable investment for the observatory, both financially and in terms of setting future research priorities.[24]

This new infrastructure for solar system research soon yielded significant dividends. By autumn 1946, occasionally joined by other Yerkes–McDonald staff, Kuiper and Herzberg began concentrated work in this field. Initially both researchers focused on the composition and distribution of gases in planetary atmospheres. Kuiper turned first to Mars, interested in the problem of atmospheric stability that his discovery of Titan's atmosphere had raised. On the basis of thermal considerations (after such factors as planetary mass and mean solar distance were considered), he calculated that the atmospheres of Titan and Mars "are nearly on the verge of instability."[25] Kuiper nevertheless achieved his most important results as an observer and spectroscopist. Using the Cashman cell spectrometer on the 82-inch McDonald reflector, he identified carbon dioxide in the Martian atmosphere and declared that frozen water was present at the Martian poles, challenging earlier arguments that they comprised carbon dioxide; colleagues soon accepted this latter result as indirect evidence of the existence of water vapor on this planet. (Kuiper himself used these findings to speculate that lichenlike plants were possibly present on Mars.) Although Kuiper's specific discoveries did not permit him to reach general conclusions regarding the stability of planetary atmospheres, he used the spectroscope to study physical conditions elsewhere in the solar system that influenced the retention or evolution of planetary atmospheres. For instance, he declared in 1948 that particles within the rings of Saturn are composed of ice or covered with hoar frost, rather than consisting of bare rock as he and other astronomers had expected. This research program was significant in several ways. It increased Kuiper's contacts with members of neighboring disciplines whose work also touched on planetary problems, among them the Harvard high-pressure physicist Percy Bridgman and the Chicago geologist Rollin Chamberlin. Moreover, insights that Kuiper gained about the thermal history of individual planets through these observations helped guide his subsequent research into the origin of the solar system.[26]

Herzberg, by contrast, took the atmosphere of Venus and of the outer planets of the solar system as his primary planetary topics. His research on the outer planets allowed him to address the problem of detecting molecular hydrogen in planetary

[24] H. N. Russell to Herzberg, memorandum, n.d. [ca. Feb. 1937], Box 11, HNR; Struve to Herzberg, Oct. 30, 1943, and idem to Hutchins, May 2, 1945, both OS[mic]; idem to Hutchins, Aug. 17, 1945, Box 2, UChi; Gerhard Herzberg OHI (Brenda P. Winnewisser, Feb. 28 and Mar. 2, 1989, at AIP, pp. 61–74); Herzberg (1938, 428; 1952, 406). On Yerkes–McDonald finances, see Struve to W. W. Morgan, Feb. 2, 1947, YO.

[25] Kuiper (1952a, 11, 13).

[26] Kuiper also made spectroscopic observations of Pluto and Triton to find evidence of atmospheres, but his observations detected none; he nevertheless discovered a second satellite of Neptune, Nereid; see H. N. Russell to Kuiper, Mar. 8, 1948, Box 47, GPK; see also Kuiper (1952b, 361, 364). On Kuiper's speculations about Martian life, see Dick (1996).

atmospheres, work he had sought to do under Russell's guidance ten years earlier. His studies of Venus, on the other hand, seemed inspired by investigations Wildt and Russell had made of this planet's atmosphere at Mount Wilson in 1939, which confirmed the large abundance of carbon dioxide reported by Adams and Dunham in 1932 and the absence of detectable water vapor or molecular oxygen. Most of this work was done at the laboratory bench, but in 1946 Herzberg sought access to the coronagraph at the Harvard-operated High Altitude observatory in Climax, Colorado to detect a possible lunar atmosphere, noting that the "problem of observing the fluorescence of the lunar atmosphere is of course entirely similar to the problem of photographing the solar corona." He apparently decided against attempting the observation after learning that the coronagraph's drive mechanism could not be adjusted to follow the Moon's motion through the sky, making the required long exposures difficult to make. The proposal nevertheless demonstrated Herzberg's intent to study planetary atmospheres as general phenomena. Although he and Kuiper also used the infrared spectrograph to record spectra for white dwarf stars and giant red stars, both strong infrared emitters, most of their efforts were devoted to solar system research.[27]

Yet another group of Yerkes–McDonald astronomers took on atmospheric problems of a different kind, including the distribution of small particles within planetary atmospheres and the behavior of particles and gases within Earth's ionosphere. In 1946, the Dutch astronomer Hendrik van de Hulst, an expert in interstellar dust and small particles within the solar system, used a two-year Yerkes appointment to study the behavior of particles within planetary atmospheres, employing data from Kuiper's observational studies to test a new theory of scattering developed by Chandrasekhar. Pol Swings, a former Yerkes astronomer who also became a research associate in 1947, made spectroscopic studies of night-sky emissions, investigating their relation to magnetic and auroral disturbances. At the same time, Greenstein maintained his involvement with the V-2 panel; while his rocket-based research centered on ultraviolet solar spectroscopy, he gained considerable acquaintance with upper-atmospheric studies in the process. As a result, nearly a third of the Yerkes–McDonald staff became involved in aspects of planetary atmospheric research, part- or full-time, between 1946 and 1947.[28]

Struve seemed pleased with the growth of these programs. Military patronage provided a large proportion of the funds required for this atmospheric research: In addition to Kuiper's Navy funds, Swings's spectral emissions work was funded by a grant from the Johns Hopkins University's Applied Physics Laboratory, which judged this study relevant to its applied work on ballistic missiles. Greenstein's work was also funded by the APL. Moreover, Struve was relieved to find that these patrons

[27] Russell to Dunham, Dec. 16, 1939, Box 7, HNR; Herzberg, "Proposal of an Attempt to Detect a Lunar Atmosphere," n.d. [ca. Jan. or Feb. 1946] (quoted); and Shapley to Herzberg, Feb. 12, 1946, both DHCOm; and Struve (1947b, 149); see also Herzberg (1952) and Jarrell (1988, 136).

[28] Kuiper (1948, 72).

imposed no special requirements on research work. Atmospheric studies and solar system astronomy thus fit well into Struve's plans to develop flexible research strategies for the observatories.[29]

A high point for the Yerkes program on planetary atmospheres came in September 1947. To commemorate the observatory's fiftieth anniversary, Kuiper organized a major conference on "The Atmospheres of Earth and Other Planets." Kuiper saw the meeting as a vehicle for bringing together researchers from various disciplines and to advertise growing Yerkes–McDonald competence in this field, particularly to potential patrons. At Whipple's urging, Kuiper invited such authorities as Harry Wexler of the U.S. Weather Bureau and Raymond J. Seeger, then with the Office of Naval Research (ONR); Kuiper also recruited leading researchers connected to the V-2 rocket program, including Iowa physicist James A. Van Allen. The list of symposium speakers illuminated the number of American astronomers interested in atmospheric phenomena after World War II. In addition to van de Hulst, Swings, Adel, Greenstein, and Kuiper from Chicago, presenters included Whipple, Dunham, and Spitzer, who had become a consultant on the upper atmosphere with the RAND Corporation following the war.[30]

Because the symposium largely served as a forum to report wartime advances in atmospheric science, participants reached consensus on few substantial issues. Its lasting contribution, however, was in forging contacts among members of these fields. The meeting sparked a long and fruitful association between Whipple and nuclear physicist Harrison Brown of Chicago, who was just beginning his study of meteorite abundances (see Chapter 3); it also introduced Kuiper to researchers working on the physics of Earth's upper atmosphere and geologists interested in Earth's early atmosphere. Moreover, Kuiper's edited conference volume, *The Atmospheres of the Earth and Planets,* quickly sold out its first printing. The importance of the conference lay not only in the research that participants reported, but in setting out questions to be answered and in defining the field as a coherent whole.[31]

As it turned out, the symposium was a watershed for participation by Yerkes–McDonald astronomers in atmospheric research, rather than its point of departure. Two factors – one involving instruments, the other changes in personnel – came to cripple the efforts of Chicago astronomers to expand work in this field. One problem involved the new infrared spectrometer itself. In testing this instrument in late 1946, Kuiper had become aware of its two potentially serious drawbacks: limited sensitivity and low spectral resolution. He had hoped Cashman would construct a second cell with a larger surface area and an analyzing slit, but Cashman did not do so. Furthermore, Kuiper found that the spectrometer's tightly compartmentalized design did not permit cooling of the cell with dry ice, an improvement he calculated

[29] Oort to Struve, Nov. 13, 1945, Box 31, JHO; Swings to Oort, Dec. 21, 1949, Box 32, JHO; see also DeVorkin (1989b, 3–4), Kuiper (1948, 72), and Struve (1947a, 290–1).

[30] Whipple to Kuiper, July 1, 1947, Box 3, FLW–HU; Kuiper (1949); see also DeVorkin (1989a; 1992).

[31] Rupert Wildt, "[Review of Kuiper, ed., *Atmospheres of Earth and Planets*]" copy in Box 29, GPK.

would increase its signal–noise ratio by a factor of 3 or better. Since abundance determinations within planetary atmospheres required precise bandwidths, the limited resolution of the McDonald spectrometer restricted the kinds of atmospheric measurements possible even at these previously unexplored wavelengths. Kuiper found, for example, that his detection of water vapor in the Martian polar caps was just within the resolving limit of the 82-inch reflector, owing to the extremely small angular extent of the caps (about two square seconds of arc) and the low intensity of radiation they emitted. Additional atmospheric observations seemed unlikely to reveal new spectral features. Kuiper's far-infrared detector remained unique, but Struve and Kuiper hesitated to invest additional telescope time in using it.[32]

A more serious limitation was the loss of skilled spectroscopists from Yerkes–McDonald, caused by the scarcity of such individuals at a time of greatly expanded postwar opportunities for infrared spectroscopists in industry and government. Interpreting the constitution and structure of planetary atmospheres at infrared wavelengths required a more extensive knowledge of molecular spectroscopy than that needed for studying the spectral features of most stars at visual wavelengths. Assessing the relative strength of particular bands seen against the heavy molecular absorptions characteristic of planetary atmospheres was rendered difficult, for example, by the apparent similarity of extremely weak and strong abundances; solving such problems meant spectroscopists had to investigate radiative transfer within planetary atmospheres. Few astrophysicists of Kuiper's generation had the requisite spectroscopic skills to undertake such analyses, and it was for this reason that Struve had felt grateful that Herzberg, with his extensive background in infrared spectroscopy, had joined the Yerkes–McDonald staff. Yet although Herzberg enjoyed the challenge of astronomical puzzles, he felt distant from fellow spectroscopists and isolated in rural Williams Bay. When the Canadian National Research Council in Ottawa offered him the directorship of a major national laboratory for fundamental spectroscopic research in late 1947, Herzberg quickly accepted. This left Kuiper's program at a severe disadvantage: Although he discovered several new absorption bands in the infrared spectra of Uranus and Neptune in 1948, for example, he could not make laboratory identifications of these features on his own.[33]

Kuiper soon cut back his studies of planetary atmospheres, but did not end them. Despite Struve's anxieties regarding growing popular interest in Kuiper's findings about Mars – he sharply criticized Kuiper for allowing McDonald support staff to advise reporters that his infrared studies indicated the likelihood of Martian life, straining relations between these strong-willed men – Struve encouraged Kuiper to take on new editorial work about the solar system on the whole. The ultimate result,

[32] Russell to Kuiper, Mar. 8, 1948, Box 31, GPK; Kuiper to Whipple, Mar. 22, 1948, Box 32, and idem to Bridgman, Mar. 23, 1948, Box 29, GPK; see also Kuiper (1952b, 351, 353, 361, 364) and Kuiper et al. (1947, 245).

[33] Herzberg to Struve, June 24, 1947, OS[mi]; see also Kuiper (1948; 1952b, 351) and Rabkin (1987). Herzberg nonetheless continued work on planetary atmospheric spectra in Canada; see Jarrell (1988, 166).

as we shall see, was Kuiper's influential, multivolume *Solar System* series, which helped define this increasingly distinct, interdisciplinary community.[34] The planetary atmospheres program at Yerkes–McDonald, while in the tradition of astrophysical investigations of the solar system initiated in the 1920s and 1930s, showed how new instruments and patrons could reconstitute the field.

GLOBAL CIRCULATION AND THE SOLAR CONSTANT: JOINT VENTURES

A separate program involving planetary atmospheres began shortly after World War II at Lowell Observatory. In contrast to postwar work at Chicago, however, astronomers joined meteorologists at Lowell to concentrate on the problem of global atmospheric circulation, defined by the eminent Swedish-American meteorologist Carl-Gustav Rossby as "the quantitative analysis of the state and motion of the atmosphere based upon the laws of physics."[35] Of interest to commercial airlines and military agencies as well as to meteorologists and climatologists, global circulation emerged as a central concern for participants at the 1947 Yerkes atmospheres conference. Kuiper wanted researchers to discover whether Mars has a general circulation similar to Earth's, given the physical similarities between the two planets. Rollin Chamberlin, son of the eminent geologist Thomas C. Chamberlin, similarly observed that "the atmospheres of the other planets, closely related to the Earth and yet differing in many respects, can give comparative data bearing significantly on certain problems of the terrestrial atmosphere."[36]

The Lowell project was initiated in 1947 by Harry Wexler, a meteorologist and Chief of the Scientific Services Division of the U.S Weather Bureau. The "Project on Planetary Atmospheres," Wexler confidently declared, would "reveal much of value that can be applied to the general circulation of the earth's atmosphere and to dynamic meteorology."[37] It was not the first time that astronomers and meteorologists had attempted to collaborate with one another: Adolphe Quetelet, Urbain Le Verrier, Clevelend Abbe, Samuel Langeley, and Charles G. Abbot were the most illustrious of those who had attempted such joint studies after 1800. Wexler's project was significant for two reasons, however. It marked a new, intense level of cooperation, spurred by the confidence of meteorologists after 1945 that the problem of global circulation seemed capable of solution;[38] and it channeled significant feder-

[34] Kuiper to W. Bartky, Feb. 6, 1948, and Chandrasekhar to Kuiper, Feb. 24, 1948, both Box 28, GPK; Struve to R. Hemens, Mar. 28, 1946, Box 31, GPK; and "Press Release," Mar. 4, 1948, Kuiper 1948 folder, YO.

[35] Rossby (1959, 9).

[36] Quoted from Chamberlin (1949, 257); see also Kuiper (1952a, 3).

[37] Wexler to Reichelderfer, Oct. 31, 1947, Box 3, HW.

[38] On Quetelet and Le Verrier, see Fleming (1990, 47–8, 165–6); on Abbot, see DeVorkin (1990).

al funding to Lowell. Like many American centers of astronomy, Lowell found itself with reduced staff and limited resources after World War II. The atmospheres project thus illuminates how federal and military contracts sustained postwar American astronomy, preferentially aiding solar system studies.

Wexler's concern with the atmospheres of other planets rose from his fascination in the global dynamics of Earth's atmosphere, an interest kindled early in his career. A Harvard undergraduate in mathematics, Wexler had initially begun graduate studies in astronomy. He had soon transferred to meteorology, however, earning a Ph.D. in meteorology in 1938 from the Massachusetts Institute of Technology, the first American university to offer graduate training in this branch of geophysics. While a student there, working under Rossby's guidance, he had published several papers on the upper atmosphere, including on the radiative cooling of the air and the patterns of polar storms. During the war he had worked on defense problems at the U.S. Weather Bureau, including the application of radar to weather observations, and in 1944 had flown on board a reconnaissance plane into the Great Atlantic Hurricane, becoming the first meteorologist to penetrate a storm of this magnitude by air. His training under Rossby, his role as an advisor to the Army Air Force's Weather Service, and his awareness of wartime increases in meteorological data all convinced him that general circulation was a particularly promising research topic for postwar meteorology, both at the U.S. Weather Bureau in specific and within university departments of meteorology in general.[39]

Wexler first began to think about studying the atmospheres of other planets to gain insight into global circulation early in 1947. That spring, Wexler attended a Weather Bureau seminar in Washington presented by James B. Edson of the White Sands Proving Grounds. Edson, a former Lowell Observatory staff member in the 1930s who had turned to guided missile research during the war, proposed the possibility, as Wexler jotted in his notebook, of "throwing light on the general circulation problem by comparing that of the earth with other planets such as Mars and Jupiter." Edson's superiors at White Sands, sensing the importance of atmospheric circulation to the problem of guiding ballistic missiles, had encouraged him to pursue the topic. Since Lowell's collection of planetary photographs was "greater than that of all other observatories in the world combined," Edson argued that Flagstaff was the best location for such a project. Wexler found Edson's presentation convincing and, more important, within the postwar mandate of the Weather Bureau. With the approval of bureau chief F. W. Reichelderfer, himself eager to increase the bureau's scientific competence, Wexler set about to initiate such a project.[40]

[39] "Biographical Sketch [of Harry Wexler]," n.d., "Biographical Sketch of Dr. Harry Wexler, U.S. Weather Bureau," Oct. 25, 1955, and "Note on the Chicago Conference on Problems of Meteorological Research, Dec. 9–13, 1946," all Box 37, HW; Whipple to Kuiper, July 1, 1947, Box 3, FLW–HU; Harmon Craig OHI (Ronald E. Doel, Nov. 27, 1990, at AIP, p. 72); and Athelstan Spilhaus OHI (Ronald E. Doel, Nov. 10, 1989, at AIP, pp. 21–5).

[40] Wexler to Reichelderfer, Oct. 31, 1947, Box 3, HW; on Reichelderfer, see Bates and Fuller (1986, 142–3, 272).

Two specific factors also drew Wexler to planetary atmospheric research. One was the military's growing interest in weather prediction and modification. By 1946 Wexler had become involved in a collaborative project with John von Neumann, the eminent Hungarian-born physicist at the Institute for Advanced Studies in Princeton, New Jersey, involving the general global circulation of Earth. Shortly after the war, Vannevar Bush, the MIT engineer who had led the massive wartime Office of Scientific Research and Development (OSRD), confidentially urged leading meteorologists to investigate if weather control was possible. While discussing this proposed project, von Neumann had posed a fundamental but disturbing question: How could one know if weather modification had succeeded? It was this question that led von Neumann to employ his new Princeton-based electronic computer to compute solutions to problems in dynamic meteorology and hydrology. Members of what became known as the Meteorological Computing Project, including (in addition to Wexler) meteorologists Bernhard Haurwitz, Chaim L. Pekeris, Hans Panofsky, and Jule Charney, hoped to calculate numerical forecasts of weather activity based on theoretical models of atmospheric circulation, a goal which had eluded meteorologists in part because earlier computers were too slow to compute them.[41] Wexler perceived great value in von Neumann's numerical forecast work. The Meteorological Computing Project, he advised a colleague in 1947, would likely improve the foundations of meteorology. In this context, Wexler saw the possibility of securing global circulation data from other planets a rewarding means for meteorologists to test the universality of global circulations models derived from terrestrial data.[42]

In addition, Wexler's interest in planetary atmospheres grew out of his own postwar research on the upper atmosphere, particularly the effects of variations in solar energy upon Earth's climate and global circulation. One of the more difficult questions meteorologists faced in studying climatic variations and weather anomalies in the atmosphere of Earth was whether they are principally controlled by factors internal to the atmosphere, such as cloud cover, or external to it, perhaps through variations in the Sun's radiative output. Like a large number of theoretical meteorologists of his time, Wexler was intrigued with this particular solar–terrestrial relationship, which he termed the "heating from above" problem. In 1946, Wexler was deeply impressed by a paper written by Haurwitz, who argued that ultraviolet

[41] Wexler declared that "[d]evelopment of these electronic computers is a wartime phenomenon which represents an abrupt discontinuity in speed over the prewar mechanical and electrical computers," making it possible for a theoretical meteorologist "to set up a difficult problem and solve it in a few days instead of several months that hand or early machine methods would have required" (Wexler to Reichelderfer, Sept. 12, 1946, Box 2, HW). On von Neumann's meteorological computing project, see Aspray (1990, 152), Nebeker (1995, 135–51) and especially Athelstan Spilhaus OHI (Ronald E. Doel, Nov. 10, 1989, at AIP, pp. 49–52).

[42] Wexler to George E. Forsythe, July 22, 1946, Box 2, HW; see also Chamberlin (1949, 257) and Hess (1958, 809). Wexler was apparently unaware of an earlier suggestion by Harold Jeffreys that Jupiter's atmosphere could test general circulation theories, perhaps because many meteorologists regarded them as too poorly established to make such research rewarding; see Jeffreys (1935). I am grateful to Rik Nebeker for calling this item to my attention.

energy generated by solar flares could cause heating in the ozonosphere and hence influence the lower atmosphere. Wexler's own calculations, begun in response to Haurwitz's argument, suggested that such events could create an upward bulge at the top of the atmosphere, causing air masses to bob up and down at regular and predictable intervals.[43] This was a testable hypothesis, one that could best be addressed, Wexler realized, by marshaling forces with astronomers and other researchers familiar with the behavior of the upper atmosphere.

What further kindled his interest in these problems was his regular contacts with scientists in neighboring disciplines. Required to oversee research grants funded by the Weather Bureau, Wexler was frequently on the road to research centers throughout the country, an activity like that performed by Rose and Warren Weaver for the Rockefeller Foundation prior to World War II. This effort familiarized him with research groups interested in the upper atmosphere, and made him an important conduit of information. In late 1947, for instance, Wexler arranged a Weather Bureau grant for ozonosphere research by the New Mexico physicist Victor Regener, who had investigated high-altitude ozone and the upper atmosphere in Germany with his father, Erich Regener, before the war.[44] In addition, Wexler became a member of the Upper Atmospheric Advisory Subcommittee of the National Advisory Committee on Aeronautics, and began attending meetings of the V-2 rocket panel. In so doing he became intrigued with the work of fellow panel member Whipple, whose own postwar investigations of the upper atmosphere indicated seasonal periodicities of the kind Wexler was investigating.[45] The planetary atmospheres project that Edson proposed thus interested Wexler because it also promised to provide evidence of the Sun's influence upon long-period weather anomalies on planets other than Earth, which he expected if solar variations caused such changes. The "year-by-year variation in the area of the ice caps of Mars," Wexler wrote in 1947, "offers a possible index which can be compared with our record of terrestrial weather." [46]

Important as Wexler perceived the cooperative project to be, a number of obstacles nearly forestalled it. Chief among them was the reluctance of Vesto M. Slipher, Lowell director since 1926, to host meteorologists at Flagstaff or to accept government contracts for research. In part this stemmed from Slipher's advancing age (73), his distracting involvement in real estate investments, and his reluctance to be challenged by new research techniques. It also stemmed, however, from Flagstaff's geographic isolation from centers of wartime research and development, Lowell's lack of university affiliation, and the increasing isolation of Lowell's senior astronomers, including Slipher and Carl Otto Lampland, from the American

[43] Wexler to Reichelderfer, Sept. 17 and Oct. 31, 1947, both Box 3, HW; idem to Rossby, Dec. 16, 1948, Box 4, HW; see also Haurwitz (1946).

[44] Wexler to Reichelderfer, Oct. 31, 1947, Box 3, HW; see also DeVorkin (1987; 1989a,b).

[45] Wexler to Reichelderfer, Sept. 17, 1947, Box 3, and idem, notes on V-2 rocket panel meeting, Sept. 5, 1946, Box 34, both HW.

[46] Wexler to Reichelderfer, Oct. 31, 1947, Box 3, HW.

astronomical community. Slipher's staff experienced none of the changes brought about by wartime contract research at major observatories on the East and West coasts.[47] Moreover, younger staff members who had left Lowell to take part in military research programs, including Edson and his brother-in-law, Clyde W. Tombaugh, did not – or were not invited to – return to Flagstaff after 1945. Finally, while extreme in Lowell's case, American astronomers were more reluctant than colleagues in physics to seek federal funds for their discipline, fearing government patronage would lead to loss of autonomy. Aware from Slipher's courteous but noncommittal response that he wanted no part of the Weather Bureau project, Wexler turned to Whipple for assistance.[48] Whipple introduced him to Roger L. Putnam, still the sole trustee of Lowell Observatory. Initially Putnam admitted doubts about the contract as well. He changed his mind after Harlow Shapley assured him that Whipple's meteor work, under Navy contract, enjoyed the same academic freedom as under private patronage. Putnam then pressured Slipher to accept the contract, declaring that it would attract young people to solar system astronomy, provide needed clerical support and new research equipment, and increase the observatory's competitiveness. As during the 1930s, prominent East Coast astronomers successfully steered Lowell's resources toward interdisciplinary problems they deemed of interest.[49]

Getting the project approved was only the first of Wexler's problems; staffing it was another. As Wexler had already discovered in developing von Neumann's Meteorological Computing Project, experienced meteorologists and climatologists not already reabsorbed by universities or engaged in government-funded meteorological studies were hard to come by.[50] Initially Wexler looked to Edson to head the project; Edson, however, decided to remain involved in guided missile research at the White Sands Proving Grounds, consenting only to consult part-time. Similarly, Wexler's choice for chief meteorologist, Hans Panofsky of New York University, also refused a full-time commitment. Problems in hiring staff pushed the project's beginning months beyond its projected January 1948 startup. Wexler finally resolved the impasse by hiring graduate students in meteorology and appointing Lowell Observatory staff members on the project part-time.[51] The only full-time me-

[47] On postwar perceptions of Lowell Observatory by American astronomers, see, e.g., Kuiper to G. de Vaucouleurs, Sept. 6, 1950, Box 32, and J. L. Greenstein to Members of Astronomy Consultant Panel, NSF, Dec. 2, 1953, Box 13, both GPK; on Slipher, see Slipher to Russell, Feb. 8, 1939, Box 61, HNR, and idem to Kuiper, n.d. [summer 1947], Box 31, GPK.

[48] A previous Weather Bureau–Lowell Observatory project, involving a study of the atmospheric transmission of solar radiation, had been abrogated "at the signature point" in 1939 by Slipher who, according to Wexler, stated "he did not think Percival Lowell would have approved using government funds." Wexler found this particularly frustrating, since the Blue Hill Meteorological Observatory in Massachusetts, founded by A. Lawrence Rotch, Lowell's cousin, "will resort to almost any means to gather funds – Government or otherwise" (Wexler to Reichelderfer, memo, Oct. 31, 1947, Box 3, HW).

[49] Wexler to Reichelderfer, memo, Oct. 31, 1947, Box 3, HW; Putnam to Slipher, Aug. 19, 1947, Slipher to Putnam, Aug. 21, 1947, and idem to Putnam, Oct. 5, 1947, all LOW.

[50] Nebeker (1995, 140).

[51] Putnam to Slipher, Dec. 19, 1947, and Mar. 9, 1948, both LOW.

teorologist he hired was a young instructor on leave from the University of Florida, Seymour L. Hess.

Hess quickly became the project's dominant presence. A native of Brooklyn, New York, Hess had entered the new graduate program in meteorology at Chicago in 1941, then thriving under Rossby, who had left MIT to take charge of this rapidly expanding effort. Like many Ph.D. candidates who entered meteorology during the early 1940s, Hess had quickly become involved in research projects for the Navy and Army-Air Force. Under Rossby's influence, Hess, like Wexler, developed a fascination with global circulation, studying the magnitude of poleward decrease of temperature in the troposphere.[52] His interest in global transport, then shared by relatively few theoretical meteorologists, was likely what encouraged Wexler to select Hess to run the Lowell program. In the summer of 1948 Hess arrived at Lowell, where he soon busied himself in recruiting colleagues and assistants, reviewing existing data, and studying astronomical techniques.

Within a year, Hess realized that the project could not fulfill the interdisciplinary ambitions Wexler had foreseen. One reason was that available planetary data were at too coarse a resolution to permit detailed investigations of either global circulation or temperature variations across their globes. Photographic data for Mars, for instance, were too fragmentary and on too large a scale to reveal clear global patterns. In a 1949 article draft, Hess optimistically noted that a catalogue of apparent velocities of Martian clouds that he and his colleagues were preparing "should give a more quantitative picture of the atmospheric circulation of Mars than is now available." He nevertheless felt obliged to add, "[s]uch cases are rare, however, and it may be that the number of such observations will be so small as to constitute too infirm a basis for a climatology of Martian winds."[53] In addition, Hess came to realize that even rudimentary astronomical data on Mars – including surface temperature variations, the amount of water vapor present in its clouds and polar caps, and the total mass of its atmosphere – all remained in dispute. Meteorologists viewed the amount of water vapor in the Martian atmosphere as a crucial measure, since the latent energy of water vapor largely determined the distribution of cloud cover and the intensity of storms on Earth. Because astronomers were no less interested in this value on account of its bearing on possible Martian life, numerous attempts to measure it had been made; yet the study most widely cited by astronomers in the late 1940s, a 1925 spectroscopic measurement by Adams and Saint-John at Mount Wilson, gave only an upper limit for this amount, rather than an actual value. Finally, Hess found that senior Lowell staff, including Earl C. Slipher and Lampland, were reluctant to share their unpublished data. Lacking leverage to force collaboration, he and fellow project meteorologists made do with whatever data they could find.[54]

[52] Bates (1989), Baum (1985), Phillips (1956, 124), and Rossby (1949, 19).

[53] S. Hess, "A Meteorological Approach to the Question of Water Vapor on Mars and the Mass of the Martian Atmosphere," 1949, draft MS, p. 14, Box 54, Folder 46, HNR; see Hess (1948).

[54] G. de Vaucouleurs OHI (Ronald E. Doel, Nov. 20 and 23, 1991, at AIP), and DeVorkin (1977, 48).

Despite these difficulties, Hess planned a two-pronged strategy to investigate the *meteorology* of Mars and the *climatology* of Jupiter, two of the most consistently studied planets. Hess made Martian meteorology the focus of his own study. He may have done so in part because the Weather Bureau had in 1946 participated in a major study of thunderstorms, funded by the Navy, Air Force, the National Advisory Council for Aeronautics, and the University of Chicago;[55] a section of this Thunderstorm Project focused on the environment of clouds in the rarified heights of the terrestrial atmosphere, a problem analogous to the problem of cloud formation in the thinner atmosphere of Mars. Using Thunderstorm Project data, which indicated that high-level convective cumulus clouds required only 0.6 mm of precipitable water vapor for formation, Hess estimated that 0.4 mm of precipitable water would suffice to produce the diffuse clouds observed on Mars. Since this volume of water vapor was below that which astronomers could successfully detect, Hess assumed this amount was present. Integrating additional data, such as Kuiper's measurement of carbon dioxide, Hess calculated a surface pressure for Mars of eighty millibars, less than one hundredth that for Earth at sea level.[56] The calculation was significant, for it marked the first time that atmospheric parameters for Mars were derived from meteorological principles, rather than from astronomical measurements alone.

The project's second branch, focused on the global circulation of Jupiter, was directed by Hans Panofsky. The most interdisciplinary scientist on the Lowell project, Panofsky had earned a Ph.D. in astronomy at Berkeley in 1941 but soon taken up "emergency" meteorological teaching and research at UCLA directed by Joseph Kaplan – adjusting his career path but leaving him with solid contacts among American astronomers and meteorologists.[57] To make sense of the large variety of cloud formations apparent in Jupiter's atmosphere, Panofsky concentrated on large-scale, long-lived features. He interpreted the Great Red Spot, the immense oval viewed on the planet since at least the nineteenth century, as an entrenched center of low pressure, and with Hess further determined the probable concentration of ammonia in Jupiter's atmosphere. Both researchers used this value to argue that its cloud-top temperature seemed 45° C warmer than that expected from solar heating alone, indicating either considerable internal heating or an appreciable greenhouse effect. Hess and Panofsky succeeded in publishing these results in several widely read sources, including the *Journal of Meteorology* and the massive *Compendium of Meteorology* of 1951, which summarized meteorological advances made since the war.[58]

55 Bates and Fuller (1986, 140).

56 Lowell Observatory (1952); see also Hess (1948; 1950).

57 Panofsky's thesis, directed by C. D. Shane, addressed absorption line profiles of atmospheric oxygen, a topic well suited to meteorological research; see "Programme of the Final Examination for the Degree of Philosophy for Hans Arnold Panofsky," June 26, 1941, Box 5, AOL, and Shane to R. W. Shaw, Jan. 5, 1942, Box 5, AOL.

58 Lowell Observatory (1952, 87); see also Hess (1950) and Hess and Panofsky (1951, esp. 394–5).

In the early 1950s, the Project on Planetary Atmospheres shifted direction in two ways. One change involved sponsorship. Beginning in 1950, Wexler handed over the project to the Geophysics Directorate of the Air Force Cambridge Research Center (AFCRC), responsible for Air Force research on the upper atmosphere. He did so chiefly because the Weather Bureau's negligible extramural research budget (limited to just $90,000 per year) made it impossible to continue Bureau support, a stark reminder of the differential between military and civilian funding for basic science in the early cold-war years.[59] Second, the focus of the Lowell project staff shifted from meteorology and global circulation studies to what Wexler called the "extra-terrestrial control of long-period weather anomalies" – that is, to investigations of possible changes in weather patterns caused by variations in the Sun's energy. Detecting such patterns soon became a major research goal at Lowell.

Assessing solar–terrestrial relationships and the Sun's influence upon meteorological phenomena had interested astronomers since the late eighteenth century, when William Herschel attempted to determine the constancy of solar radiation. By the early twentieth century, determining a precise value for the solar constant (defined as the total amount of solar energy received by Earth per unit area at mean distance) had become the chief research mission of the Smithsonian Astrophysical Observatory, directed by Charles G. Abbot. In the 1920s, Abbot had announced that the solar constant varied within a range of about 3 percent, arguing that these variations caused climatic changes on Earth. During the 1930s and 1940s, Abbot had divided his time between making additional measurements of the solar constant and calculating what he claimed were correlations between his solar data and observed weather patterns. He urged meteorologists to make use of them in exploring the relationship of weather to solar variability.[60]

American meteorologists by 1950 were of two minds about Abbot's proposed solar–terrestrial link. Many, like Wexler, were intrigued by the likelihood of solar influence on global circulation. (Wexler had initially justified the Lowell project in the hope it would throw light "on the age-old problem of whether climatic variations and weather anomalies in the earth's atmosphere are the result of external or internal controls.")[61] Project meteorologists shared Wexler's interest. Hess sought to study variations in the polar ice caps of Mars, intending to correlate them "with such quantities as the deviations from normal of the solar constant, relative sun spot numbers, and temperatures on earth." Yale Mintz, a UCLA meteorologist who joined the project part time, similarly tried to correlate the times of sunspot maxima and the outbreak of disturbances in the atmosphere of Jupiter; by 1952 he became intrigued by evidence suggesting that this was so. Since such correlations were easier

[59] Wexler advised Rossby that Reichelderfer felt the Weather Bureau was "somewhat vulnerable by supporting research into the atmospheres of other planets when we haven't quite solved the problem of the terrestrial atmosphere" (Wexler to Rossby, Dec. 16, 1948, HW); see also Whitnah (1961, 212).

[60] DeVorkin (1990), Doel (1990b), and Hufbauer (1991, 82–96).

[61] Wexler to Reichelderfer, Oct. 31, 1947, Box 3, HW.

to derive from Lowell data than from wind velocities or temperature gradients, project members were increasingly drawn to them.[62]

At the same time, meteorologists and astronomers disputed Abbot's confident claim that significant variations in the solar constant were established beyond a doubt. In 1946, for example, Haurwitz, whose solar–terrestrial studies had inspired Wexler's own subsequent research, asserted that "terrestrial observations of the solar constant show only slight variations of doubtful reality." Other meteorologists, however, such as MIT's Hurd C. Willett, countered that large-scale changes in global circulation were too consistent to be explained either by internal dynamic or thermodynamic instabilities, pointing to solar variability as a probable cause.[63]

Aware of these differences, and convinced of the significance of additional studies of solar–terrestrial studies for climatological studies, the project's new Air Force sponsor hosted two interdisciplinary conferences on planetary atmospheres. The first, convened at Lowell in March 1950, brought together a broad mix of specialists, including astronomers Edison Pettit, Franklin E. Roach, and Kuiper, as well as Hess and Edson. Although those attending the 1950 conference reached no consensus over Abbot's claim that the solar constant varied, participants did reaffirm their desire for additional investigations of solar radiation. Hess summarized a dilemma faced by meteorologists: If the variations were real, then theoretical meteorologists would tackle the problem of how these changes could affect the weather. "[U]ntil the variability is satisfactorily proved, however," he declared, "the theorists will not seriously attack so difficult a problem. This is potentially one of the most important meteorological questions."[64]

One way out of this dilemma – ultimately with important ramifications for Lowell – was suggested by a young staff member, Henry Giclas. Giclas proposed a new method for monitoring the solar constant independent of the Smithsonian pyrheliometer studies on which Abbot had based his claims: using reflected planetary light. Photoelectric measurements of sunlight reflected from Uranus would indicate the steadiness or variability of the Sun's output; simultaneous measurements of stars in the same region of the sky would allow for a subtraction of atmospheric effects. The method itself was not new: Joel Stebbins, the noted Wisconsin pioneer of photoelectric photometry, had studied the four brighter satellites of Jupiter at Lick Observatory photometrically for six weeks in 1926, concluding that no variations were present;[65] but Giclas believed this interval too short to be meaningful, and proposed an expanded program employing modern photoelectric equipment. After a second Air Force–sponsored atmospheres conference in 1951, Giclas and Harold Johnson, another junior staff member with a talent for building photoelectric detectors, began

62 Quoted from Lowell Observatory (1952, 289); see Hess (1948).

63 Quoted from Haurwitz (1946, 161); see Willett (1948, 803; 1949). For American astronomers's reactions to Abbot's claims, see DeVorkin (1990) and Doel (1990b).

64 Lowell Observatory (1952, 92). Similar issues were raised at a two-day Harvard conference on climatic change in 1952; see Shapley (1953).

65 Lowell Observatory (1952, 91).

a new study of the solar constant at Lowell. They monitored the brightness of Uranus and Neptune with a Johnson-designed photoelectric photometer, using new and upgraded equipment Lowell received through a new Air Force contract.[66] The project thus established photoelectric photometry as a new specialty at Lowell, challenging the hegemony of Abbot's Smithsonian solar constant program in the process.[67]

Project members such as Hess and Panofsky wanted to continue work on possible correlations between planetary atmospheric phenomena and variations in the solar constant. In 1952, however, AFCRC officials decided to continue funding only for the Giclas–Johnson solar constant studies, canceling all other branches of the project. Two factors led to this decision. One was that project meteorologists had reached the limits of their data. Aside from Hess's important early discussion of Martian meteorology, and Panofsky's contributions to the global circulation of Jupiter, staff members had produced few publishable results in the project's four-year history. Large-scale meteorological phenomena from Hadley cells to jet streams could not be resolved using existing data, and many project members concluded individual studies simply with pleas for additional observations.[68] Even Hess found it difficult to generalize about planetary atmospheres. When he attempted to test the generality of Rossby's conclusions about global circulation through his own investigation of Jupiter, studying variations of methane, he found that the theory seemed limited to Earth – a potentially important result but one meteorologists were reluctant to accept because of limited knowledge of the actual structure and thermal environment of Jupiter's atmosphere. Although the Lowell study had been designed to enhance knowledge of general circulation theory and related phenomena, it instead disintegrated into a profusion of narrow studies, each focused on particular planets.[69]

A second and equally important factor was a shift in research emphasis within the project's new sponsor, the AFCRC's Geophysical Research Directorate. In 1951 many theoretical meteorologists were attracted to new experimental results produced by Dave Fultz, a former Rossby student who had developed Chicago's meteorological research facility, the Hydrodynamics Laboratory. Fultz's rotating-pan experiments, which pivoted a pool of water heated at the pan's periphery and cooled at its center (simulating Earth's own circulation), derived from Rossby's use of the technique as a classroom demonstration; yet leading meteorologists became con-

[66] The Weather Bureau–Air Force contracts paid, e.g., for realuminizing the 42-inch reflector mirror, used in part for the photometric study; see Wexler to Reichelderfer, Oct. 31, 1947, Box 3, HW; J. Hall to Putnam, Jan. 6, 1957, and idem, "Present Policy of the Lowell Observatory," July 10, 1959, both LOW.

[67] DeVorkin (1990); for a fuller account of Lowell's photoelectric program, see Doel (1990b).

[68] See, e.g., Ralph Shapiro, "A Quantitative Study of Bright and Dark Spots on Jupiter's Surface," in Lowell Observatory (1952, 97–109). The limited output of the Lowell project is also apparent in review articles about global circulation; see Charney (1959, 178).

[69] Lowell Observatory (1952, 285) and Hess (1958); Rossby's arguments were presented in Rossby (1949, 47).

vinced that Fultz had also succeeded in demonstrating the physical forces underlying the water's flow, thus making the rotating pan an important tool for laboratory-based studies of atmospheric circulation. Fultz's results encouraged patrons of global circulation research to invest heavily in experimental hydrodynamics research, marginalizing the study of other planetary atmospheres as a means for testing global circulation theories for Earth.[70]

Cancellation of the Lowell atmospheres project left a mixed legacy. Since experimental hydrodynamic facilities developed largely within departments of meteorology, opportunities for collaborative research between astronomers and theoretical meteorologists declined after the early 1950s. Meteorologists remained interested in the Lowell results, however, and requests for the final project report exhausted the Air Force's supply by 1953.[71] More important, the project established Lowell as a center for planetary meteorology. In 1953 E. C. Slipher joined with Harold C. Urey and other scientists to create the International Mars Committee, which sponsored comprehensive observations of the planet in 1954 and 1956.[72] Hess returned to Lowell briefly as part of an ONR-sponsored study of theoretical climatology; he remained interested in planetary meteorology problems, including Wildt's 1940 prediction that the greenhouse effect would heat the lower atmosphere of Venus. Frank Gifford, one of the young U.S. Weather Bureau meteorologists formerly attached to the project, similarly visited Lowell to continue research involving the surface temperature of Mars; as with Hess, it led to new, published results.[73] In 1955, Albert G. Wilson, who had replaced V. M. Slipher as Lowell director the previous year, considered appointing Gifford a full member of the staff.

Moreover, the planetary atmospheres project proved important to Lowell Observatory, and to the growth of solar system astronomy. Despite its shortcomings, the Project on Planetary Atmospheres provided the first sustained transient home for meteorologists and astronomers. It exposed astronomical records to meteorological scrutiny, produced the first meteorological assessments of weather and climate on other planets, and alerted meteorologists to the promises and limitations of astronomical data for addressing the problem of global circulation. Like Kuiper's

[70] Athelstan Spilhaus OHI (Ronald E. Doel, Nov. 10, 1989, at AIP, p. 25); see also Hess (1959, 336), Lorenz (1970, 7), Phillips (1956, 123), and Rossby (1959, 23). In 1953, Robert M. White, chief of the Large-Scale Processes Section of the AFCRC's Atmospheric Analysis Laboratory, declared that the center would henceforth emphasize experimental hydrodynamics and numerical forecasting by computers, since these techniques made possible "a recasting of our theoretical and physical approaches to the problems of the atmosphere in terms of observable quantities"; see White (1953).

[71] R. Shapiro to R. Wildt, Oct. 14, 1953, Box 1, RW.

[72] The Mars Committee secured operating funds from the NSF and the National Geographic Society, producing an extensive archive of photographic and photometric measurements; see Tatarewicz (1990a, 22). Its meetings stimulated astronomers to consider using aircraft for spectroscopic measurements of Mars; see E. P. Martz to Kuiper, Nov. 20, 1953, Box 30, and H. C. Urey, "Supplemental Report: International Mars Committee," Box 11, both GPK.

[73] Hess to A. G. Wilson, Dec. 5 and 12, 1955, both LOW, and Hess to H. C. Urey, May 5, 1958, Box 42, HCU; on Gifford, who completed his Ph.D. in meteorology at Pennsylvania State University in 1955, see, e.g., A. G. Wilson to Gifford, Feb. 21, 1955, and Gifford to Wilson, Mar. 14, 1955, both LOW. Gifford completed analysis of Lampland's radiometric measurements of Mars after Lampland's death.

efforts at Yerkes–McDonald, the project heightened the interest of American astronomers in planetary atmospheres and strengthened the institutional foundations of the field.

METEORS AS ATMOSPHERIC PROBES

Interdisciplinary work involving atmospheric phenomena also began at the Harvard College Observatory (HCO) after World War II. In contrast to the Yerkes–McDonald and Lowell efforts, however, this was not a new undertaking, but rather an expansion of Harlow Shapley's 1931 meteor expedition. Moreover, the Harvard program focused not on the atmospheres of other planets, but on Earth's own. By the early 1950s this project became a stable part of the observatory's postwar research program, and reinforced Harvard's standing as a significant American center for solar system research.

The force behind this effort was Fred Lawrence Whipple, the Berkeley-trained astronomer who had ably extended the Harvard meteor expedition after Öpik's departure. Leaders of the expedition, it is important to recall, had argued that determining the velocity of meteors depended on the physical properties both of meteoritic particles and of the upper atmosphere through which they traveled, thus providing an indirect means for studying this region. For instance, Whipple had declared in 1938 that his baseline meteor photography program "should eventually yield . . . more direct information concerning the physics of the upper atmosphere"; but he had also perceived that exacting measurements of velocity and meteoric particle characteristics were necessary before further conclusions could be drawn. Finally, Whipple recognized the importance of finding stable, reliable patrons for work requiring costly, specialized instruments. Mentored by both Leuschner and Shapley, leading entrepreneurs of U.S. astronomy, Whipple saw value in promoting his project to extramural patrons for enlarged support.[74]

The crucible wherein Whipple's growing interest in the upper atmosphere was fused with the military patrons who ultimately came to fund it was World War II. Since 1936, Whipple's Harvard-funded photographic meteor program had operated without interruption; but in 1940, Whipple joined the newly formed Radio Research Laboratory on the Harvard campus to work on radar countermeasures.[75] Whipple soon grew eager to alert military officials to the significance of his now-halted meteor research, aware through RRL director Frank E. Terman of rising military interest in Earth's atmosphere. Working at a feverish pace through early 1942, Whipple nearly completed a major review of this problem for a planned Chi-

[74] Quoted from Whipple (1938, 499); see also C. Wylie to Whipple, Feb. 11, 1938, series HUG 4876.806, FLW–HU, and Whipple (1939).

[75] Kuiper to S. Chandrasekhar, Aug. 8, 1943, Box 28, GPK; see also DeVorkin (1992, 96) and Kidwell (1992).

cago symposium, then collapsed from exhaustion.[76] He finished it nevertheless, and in 1943 published "Meteors and the Earth's Upper Atmosphere" in the *Review of Modern Physics.* Whipple reviewed existing estimates of the density and temperature of the atmosphere above 30 km, the limiting elevation for which direct radiosonde data was available. Noting that various methods for determining the middle atmosphere's density–temperature profile (including the famous studies of anomalous reflections of gunfire sound in the 1920s and theoretical predictions based on free-period oscillations in the lower atmosphere) were all in general agreement, Whipple emphasized that meteor studies could redetermine this profile, and moreover extend it to an unprecedented height of roughly 110 km.[77]

Military patrons did take notice. That Whipple's review was published in the open scientific literature in 1943 reveals that upper-atmospheric research was not yet considered a sensitive topic; by contrast, research on solar–ionospheric relationships by Walter Orr Roberts, Whipple's junior colleague at Harvard, was "born classified" because of the perceived importance of ionospheric conditions for long-distance radio communications.[78] Between 1943 and the war's end, however, as Allied forces received intelligence reports about German rocket developments at Peenemünde, American military planners became increasingly interested in the upper atmosphere. By 1945, the Navy began funding the MIT Center for Analysis to reduce photographic meteors recorded by Harvard's two baseline camera stations, hoping to gain new density profiles for this region. Two former Harvard astronomers already at the Center for Analysis, Luigi Jacchia and Zdeněk Kopal, joined Whipple on this expanding project. In late 1945, Whipple proposed to Naval Ordnance Laboratory officials a system of new, faster meteor cameras to study larger numbers of meteors. Within months Whipple received a contract of $60,000 from the Bureau of Ordnance for this work, quite large by astronomical standards of the time.[79] In contrast to Kuiper, who was uncertain about pursuing stellar or solar system research upon returning from wartime duty, Whipple found that the war had given his prewar research new purpose and direction.

Major uncertainties that Whipple wanted to address were the size, structure, and composition of meteoric particles. From his meteor research in the 1930s, Whipple became convinced that these characteristics influenced how meteors travel through the upper atmosphere. Working with Öpik in the 1930s to calculate whether meteors were interstellar or "solar," Whipple had made numerous assumptions about these values. Most astronomers assumed that extremely fragile meteors would not withstand deceleration forces after encountering the atmosphere, while those of

[76] Whipple to Otto Struve, Mar. 27 and Apr. 6, 1942, and Struve to R. S. Mulliken, June 27, 1942, all Roll 13, Section 6, OS[mic]. I thank David DeVorkin for copies of these documents.

[77] Whipple (1943, 246–7); see also DeVorkin (1992, 273–82).

[78] Hufbauer (1991, 125–6) and Kidwell (1992).

[79] E. V. Appleton to Whipple, Feb. 19, 1947, Box 15, and Peter Millman, "Visit Report: MIT and Harvard University," Mar. 20, 1946, Box 3, both FLW–HU; see also DeVorkin (1992, 278) and Kidwell (1992).

stronger internal constitution would reach lower elevations; the challenge was to know how much to adjust for these differences in the general equations used to calculate meteor velocities and elevations. Whipple believed the solution would come from photographically investigating the paths of many hundreds of meteors, allowing him to distinguish random, sporadic meteors from those associated with meteor showers. Since astronomers generally accepted that meteoric particles, even if of solar origin, could provide important information on the size, structure, composition, and distribution of interstellar particles, Whipple had high hopes that his work would yield new information about the evolution and distribution of dust in the solar system. He optimistically advised Kuiper in June 1946 that the program would likely answer "the most fundamental problems concerning meteors" within a few years.[80]

At the same time, Whipple perceived that pursuing this work would require him to develop new contacts with meteorologists and upper-atmospheric specialists. Determining the characteristics of meteoric particles, he became convinced, required understanding such processes as heat transfer at high velocity and the distribution and composition of high-altitude gases, topics in which upper-air researchers were more expert than astronomers. Rather than hiring additional specialists for the project, Whipple sought to educate himself about upper-atmospheric physics, drawing on various experts in this field. Through Kuiper's ALSOS interview with Regener, Whipple secured information about Regener's study of atmospheric temperatures above 37 km. More significantly, Whipple joined the V-2 Upper Atmosphere Rocket Research Panel, perceiving at least two advantages in associating with the panel's scientists. One was that rocket-based measurements of the properties of the upper atmosphere were of obvious importance to his meteor program. The other was that he hoped to secure their backing to persuade military officials to fund experimental work on the problem of ballistic heat transfer, convinced that new laboratory studies were essential for advances in this field. After the failure of Caltech physicist Fritz Zwicky's 1946 attempt to create artificial meteors through the release of high-velocity pellets during a night launch of a V-2, Whipple's desire to develop certifiable laboratory methods grew stronger still.[81]

The importance of maintaining contacts with V-2 experimenters was soon reinforced by worrisome differences between Whipple's density–temperature profile of the upper atmosphere and that emerging from rocket studies. Shortly after the war, meteor reductions by Whipple, Jacchia, and Kopal suggested that seasonal fluctuations were occurring at the roof of the atmosphere. Previous studies of the upper atmosphere had given no indication of these fluctuations, which amounted to nearly 9 km in elevation; reports that such fluctuations existed therefore surprised and intrigued many meteorologists.[82]

80 Whipple (1943, 246, 248); quoted from Whipple to Kuiper, June 19, 1946, Box 32, GPK.

81 Whipple to Kuiper, June 3, 1946, Box 32, GPK, and R. Thomas to Whipple, Apr. 13, 1950, DHCOm; see also DeVorkin (1987).

82 Kuiper (1952a, 7), Whipple (1943, 264), and Whipple, Jacchia, and Kopal (1952, 155).

A second but more serious difference emerged in late 1946. Continued reductions of Harvard photographic meteors by Jacchia and Kopal indicated a higher density–temperature profile above 70 km than those deduced from the first V-2 rockets launched from the White Sands Proving Grounds in New Mexico. At 90 km, Whipple's results differed from those of the V-2 experimenters by a factor of 3. He felt certain that meteor determinations of the temperature–pressure curve were more accurate than those from rocket-based instruments, reflecting his confidence that the Harvard meteor program had placed meteor velocities and elevations on a firm foundation. In June 1948 Whipple advised William G. Dow, a University of Michigan engineer and V-2 experimenter who had helped devise high-altitude recording instruments, that he felt "extremely suspicious" of low pressure measurements Dow had obtained during a November 1946 V-2 flight. Although certain that systematic error (or the seasonal variations he, Jacchia, and Kopal had discerned) had adversely affected Dow's measurements, the discrepancy posed a considerable problem for Whipple. The difference between their respective density–temperature curves for the upper atmosphere was hardly negligible, and Whipple was well aware that the shape of this curve was of critical interest to military patrons involved in ballistic missile research. Moreover, if genuine variations in the density–temperature profile existed, they would affect his calculations of meteoric particle size and structure. This, in turn, would complicate his plans to map the distribution and origin of small particles in the solar system.[83]

To help resolve this puzzling discrepancy, Whipple introduced several changes into the Harvard program. Frustrated by the high percentage of cloudy nights that had long plagued operation of the Massachusetts-based meteor cameras, Whipple used his Bureau of Ordnance contract to establish a network of new stations. Where to locate them occupied Whipple during much of 1946 and 1947. Initially he considered sites in the vicinity of the Harvard-affiliated High Altitude Observatory in Climax, Colorado, then directed by Roberts; but the severity of Climax winters soon convinced him to abandon this option. Subsequently he considered locating them adjoining the McDonald Observatory in Texas or at Lowell, site of the original Harvard expedition, so that experienced technicians and film changers could be housed nearby; yet Whipple rejected these possibilities as well, instead establishing his new baseline stations in New Mexico. One reason he did so was because many astronomers judged the Las Cruces region to have a higher percentage of *clear* nights than the Colorado, Texas, and Arizona locations, which excelled in their quality of atmospheric seeing. Another factor that favored the New Mexico site, however, was its proximity to the V-2 test range at White Sands. In practice, this meant Whipple's cameras and the V-2 rocket instruments would essentially observe the same parcel of air. Since simultaneous observations would eliminate seasonal variations and latitude effects as possible causes of the discrepant density–tempera-

[83] Whipple to Spencer, May 19, 1948, and idem to Dow, June 9, 1948, both Spencer Papers, provided courtesy of David H. DeVorkin.

ture profiles determined by the two techniques, Whipple intended to use the New Mexico stations to force a resolution of this problem.[84]

If locating the camera stations was influenced by interdisciplinary considerations, the new meteor cameras, by contrast, represented an evolution of conventional astronomical optics. In negotiating his Navy contract, Whipple had successfully pressed for funds to develop new patrol cameras, pointing out that existing cameras recorded only extremely bright meteors and required several thousand hours to achieve exposures suitable for measurement. Moreover, Whipple had become convinced by Harvard astronomer James G. Baker that modifying existing camera designs would not yield a satisfactory solution. A skilled optical designer who had run the HCO's wartime Optical Research Laboratory, Baker designed a "Super-Schmidt" camera in 1946, based on the wide-field telescopes first produced by the Soviet astronomer Bernhardt Schmidt in 1931. Baker's Super-Schmidt cameras had a diameter of 12.5 inches, an extremely short focal ratio (0.85), and an effective field of 52 degrees; they were designed to increase fortyfold the number of recorded photographic meteors. Difficulties in manufacturing their optics delayed deployment of these cameras until the early 1950s. Unfazed, Whipple operated the New Mexico stations until then with surplus aerial cameras formerly installed at Oak Ridge.[85] Through the Navy Ordnance contract, Whipple effectively transformed the 1931 meteor expedition into a long-running institution.

As at Yerkes and Lowell, postwar contracts for meteorological research at the HCO helped sustain its astronomical programs during the immediate postwar years. Both Jacchia and Kopal remained full-time members of the meteor project, their research addressing atmospheric phenomena as well as the distribution of meteoric particles; Whipple recruited graduate students to the field of meteor astronomy, among them S. E. Hamid and Carl Bauer. Solar system research was not the sole field at the HCO to benefit from postwar military contracts, however (in contrast to the case at Lowell). By the late 1940s, Menzel, Russell's former student who had worked on solar–ionospheric relationships and radio communications during the war, launched extensive new research programs in solar astrophysics at Harvard funded by Navy and Air Force patrons. Menzel's rapidly expanding solar empire, including a new solar observatory on Sacramento Peak, New Mexico, soon dwarfed all other university-based astronomy projects in the United States. Yet the scale of Menzel's operations should not overshadow Whipple's efforts. By 1952, Whipple's meteor program operated with an annual budget of roughly $100,000 a year, about

[84] Whipple to G. P. Kuiper, June 19, 1946, and May 21 and July 30, 1947, as well as Kuiper to Whipple, May 12, 1947, all Box 32, GPK; see also Kellogg (1951, 757) and Jacchia and Whipple (1949b, 91; 1956, 986). Whipple also worked with the Canadian astronomer and former Harvard Ph.D. Peter Millman to develop a northern station. Funded by the Canadian government and placed near the Dominion Observatory's Meanook Magnetic Observatory in Alberta, these stations were expected to yield further information on latitude effects; see Jarrell (1988, 155) and Millman and McKinley (1967, 281–3).

[85] Baker (1945), Jacchia and Whipple (1956, 988), and Whipple (1949b).

twenty times the size of typical ONR research grants in astronomy and second only to Menzel's in terms of size and cost. Both provided stable funding to the observatory. (The HCO's budget, excluding extramural contracts, was only slightly above $215,000 in 1951.)[86]

The rapid growth of these postwar programs gave Whipple and Menzel considerable authority at the HCO, and shifted the general orientation of its research. By 1952 the number of publications by HCO staff in solar physics and meteoric–upper-atmospheric research exceeded that for stellar and galactic research for the first time.[87] Several factors were responsible for this shift, which, given Harvard's former leadership in these fields, is a matter of considerable importance. One was that these new programs, while well funded, were simultaneously restrictive in focus. Although the Baker Super-Schmidt meteor cameras were among the most optically advanced research instruments produced at the HCO after the war, they were suited for few stellar or galactic research programs. Solar system astronomy thus became the primary beneficiary of this instrumental advance.[88]

A more important factor, however, was that Harvard staff members found few funds available to them for work in stellar and galactic research. The irony of this situation was not lost on Shapley, who in 1945 completed his twenty-fifth year as HCO director. In 1942 a buoyant Shapley had challenged Struve regarding what he considered Struve's excessive pessimism as to the limited utility of astronomers for military research, noting that contracts for research were helping sustain his observatory despite the dispersal of staff astronomers to various local wartime projects. By 1946, however, Shapley discovered that he could secure virtually no funds for what he termed "pure astronomical research" involving variable stars, stellar spectra, and galaxies, which he called the "neglected" fields of astronomy. Shapley's situation at Harvard was then complicated by several additional factors. James B. Conant, Harvard's president, opposed Shapley's plans to launch a large endowment campaign in 1946, desiring to consolidate fund-raising behind a new university-wide science center. Shapley had little immediate success in raising funds for stellar and galactic astronomy from the Office of Naval Research, leaving little money for new work in these fields. Most important, Shapley's outspokenly liberal views collided with increasingly conservative cold-war attitudes, irritating Conant and chilling relations between them.[89] Shapley's concerns about military patronage of science included

[86] Figures are derived from "Departmental Summary – Income," undated 1952 memo, and Harvard College Observatory Council to Shapley, confidential memo, Oct. 3, 1952, both DHM; on Menzel, see Hufbauer (1991, 136–44).

[87] *ARHCO,* 1948–9 and 1951–2.

[88] Only one Super-Schmidt camera was to be installed at Harvard's Bloemenfontein station in South Africa, for example, presumably to aid stellar programs; see "Rearrangement of Budget Items on Harvard Observatory's Post-war Development Plans," Oct. 12, 1946, Box 7, FLW–SI.

[89] Harvard College Observatory Council minutes, Dec. 10, 1946, and Shapley, "Memorandum on the Support of Astronomical Research by the Office of Naval Research," n.d. [ca. 1946], both Box 7, FLW–SI; Shapley, *ARHCO,* 1946–7: 2; see also DeVorkin (1982, 607), Hershberg (1993, 617–19), and Shapley (1969, 150).

how military contracts concentrated research in particular fields. Former stellar astronomers such as Baker, Jacchia, and Kopal were supported solely by contract funds for meteor research, not by university appointments offering freedom of research. Angered by the new emphasis on utility, Shapley declared that the Navy was interested in meteors "because the shooting stars perform in the same level of the Earth's atmosphere where the shooting rockets and the rocket ships of the future are planning to operate."[90]

Such charges did not perturb Whipple, who in the early 1950s worried more about the continued divergence of his density–temperature profiles of the upper atmosphere from those of V-2 experimenters. In 1948, Whipple published a summary of Harvard results for the atmosphere above 70 km, criticizing V-2 results as based on indirect measurements and flawed by "several assumptions." His confidence in the meteor data was bolstered by several contingent issues. By then Whipple felt certain, perhaps through his contacts with Wexler, that seasonal variations in the density of the upper atmosphere existed and could be accounted for theoretically. Since these variations were not indicated by rocket data, this seemed to indict the V-2 method. After employing his meteor results in 1950 to support a new theory of cometary structure, which gained widespread acceptance by American astronomers (see Chapter 4), Whipple felt even more convinced of their validity. His confidence soared when early Baker Super-Schmidt results became available. In 1952, after reducing measurements of Geminid shower meteors that coincided with a V-2 rocket flight – again producing discrepant results – Whipple concluded that "relatively small numbers of independent rocket observations" was likely causing this persistent and troubling order-of-magnitude divergence.[91]

Whipple nevertheless remained frustrated by his futile efforts to convince his Navy patrons to construct a laboratory that would allow study of the cosmic aerodynamics of meteoric particles, and hence limit nagging uncertainties about their sizes and shapes. Harvard resources were insufficient to develop an interdisciplinary facility of this kind, yet Whipple believed that ground-based experimental ballistics were necessary to defend against charges that meteoric properties were too unreliably known to validate his approach.[92] In 1950, however, he pursued an alternative approach suggested by his former student, Richard N. Thomas, at Utah. A 1948 Ph.D. who had also worked at the Aberdeen Proving Ground, Thomas was well acquainted with Whipple's ambitions for meteor astronomy. While attending lectures on explosive chemical reactions given by Henry Eyring, the noted Utah chemist, Thomas realized that the problem of meteor flight seemed physically analogous. (The reactions Eyring studied took place when only the outer layer of an object was heated.) Thomas considered his insight valuable for two reasons. First, it indicated

[90] Shapley, "10th informal memorandum from Harlow Shapley," Aug. 1946, Box 4, FLW–SI.

[91] Shapley, *ARHCO,* 1951–2: 4, and Whipple (1949b, 93; 1952a, 18–19).

[92] Thomas to Whipple, Apr. 13, 1950, DHM, and B. Bok to Whipple, June 17, 1955, Box 15, FLW–HU.

that quantitative studies of the elusive reaction-rate problem were indeed possible using relatively low-cost techniques. Second, he hoped Whipple could use Eyring's results to convince Navy or Air Force agencies to support new experimental ballistic studies involving "pseudo-meteors." As Thomas reminded him, this work involved the crucial heat-transfer problem that confronted Air Force engineers in designing guided missiles, and therefore of crucial interest to military patrons.[93]

Thomas's suggestions led to renewed expansion of the Harvard meteor program. By late 1950, Whipple's former student began working with Eyring and his associates on new theoretical studies of chemical reaction rates. At the same time, Whipple used Thomas's arguments to convince Bureau of Ordnance officials to initiate high-velocity-projectile experiments with steel and aluminum pellets at the Ordnance Laboratory's Test Range Facility in Inyokern, California. In 1951, Whipple and Thomas published preliminary findings in the *Astrophysical Journal.* The velocities of 6 km per second (kps) achieved by Navy researchers, he noted, fell far short of the 11–72-kps velocities common for meteors. Despite the method's promise, Whipple labeled these studies "inconclusive."[94]

Not long after the Inyokern tests, however, Whipple began having second thoughts about the superiority of meteor results for calculating atmospheric densities and pressures. His doubts arose not from his experimental collaboration with Thomas but through the rich data on faint meteors produced when Baker's first Super-Schmidt cameras began operating in New Mexico in 1951. Until then, Whipple had seen no reason to doubt that all meteors have roughly the same density, even though he believed, as did most astronomers active in solar system research, that meteors derived from both cometary and asteroidal bodies. By 1952, however, Whipple came to believe that meteors of suspected cometary origin were more fragile than those of suspected asteroidal parentage. Whipple welcomed this new evidence, for it supported his "dirty snowball" model of cometary structure, developed in 1950 during a heated debate over the origin of comets. At the same time, he realized that apparent structural variations among different classes of meteors invalidated conclusions that he, Jacchia, and Kopal had reached regarding upper-atmospheric densities on the assumption that meteors were uniform in character. Whipple first disclosed this problem in a note added in proof to a 1954 article. The inconsistency, he admitted, "reduces the significance of the seasonal effects in upper-atmospheric densities as derived from photographic meteor data." With these words, Whipple effectively ended the seven-year-long controversy over the properties of the upper atmosphere.[95]

Whipple's admission did not indicate a failed research program; indeed, the early Super-Schmidt results deluged Harvard researchers with new information about

[93] Thomas to Whipple, Apr. 13, 1950, DHM; see also Whipple to G. Kuiper, July 1, 1947, Box 3, FLW–HU.

[94] Jacchia and Whipple (1956, 983) and Thomas and Whipple (1951, 465).

[95] Whipple (1952b; 1954, 513; 1955b).

the taxonomy, structure, and origin of small particles in the solar system. Rather, the problem that Whipple now faced was finding new patronage to continue this research. Since the Navy had supported Whipple's project solely to gain information about the upper atmosphere, Navy managers found little reason to sustain the project once its utility for meteorology was compromised. Aware that fellow American astronomers were loath to endorse extremely expensive projects that might endanger general funds for the discipline – Kuiper promptly alerted officials at the National Science Foundation that Whipple's million dollar program "has failed" – Whipple turned quickly to Air Force officials, negotiating a new mission for the specialized Super-Schmidt instruments.[96] In the summer of 1954, the White Sands cameras were transferred to the Air Force's new Sacramento Peak Solar Observatory in northern New Mexico, where they were employed to study winds in the high upper atmosphere through photography of persistent meteor "trains." Whipple had first suggested this possibility to military patrons in the late 1940s; it allowed him to continue operating the Super-Schmidts and to gather data of clearly astronomical character.[97]

The shift from Navy to Air Force patronage did little to slow meteor research at Harvard. With Thomas's aid, Whipple emphasized the value of using meteors to study heat transfer within the upper atmosphere, a subject he later termed "meteor-astroballistics" and one of increasing concern for designers of intercontinental ballistic missiles (ICBMs). New Air Force patronage also allowed Whipple and his associates to investigate related phenomena in neighboring branches of astrophysics, including stellar atmospheres and the interstellar medium.[98] More than Kuiper or the "astro-meteorological" researchers at Lowell, Whipple remained active in studying the physics of the upper atmosphere, and he later became active in the International Geophysical Year (1957–8). As at Yerkes and Lowell, however, subsequent research brought meteorologists and astronomers in contact with one another less frequently than immediately following the war. Despite their influence on the character and direction of solar system research after 1945, the dreams of cooperation between these communities remained largely unfulfilled.

It is perhaps tempting to view the efforts of American astronomers to develop programs in planetary atmospheres after 1945 as unsuccessful. After all, none survived in their original form for more than a few years. Despite the postwar optimism that new instruments and patronage could quickly solve stubborn problems in planetary

[96] Quoted from Kuiper to P. van de Kamp, Jan. 22, 1955, Box 31, GPK. Whipple subsequently declared that the meteor technique had outlived its usefulness "in view of the wide variety in the density and physical structure of meteoroids and the subsequent low precision of results"; see Whipple (1959, 605).

[97] *ARHCO,* 1958–9: 5; see also p. 201 of the present volume.

[98] Minutes of the HCO Council, July 8, 1955, Box 7, and Whipple to Carmichael, May 25, 1956, Box 4, both FLW. Whipple published sixteen research papers involving meteors between 1945 and 1954; half of these papers focused on upper-atmospheric results derived from meteors, half on meteor characteristics and their implications for solar system astronomy.

meteorology and atmospheric chemistry, Kuiper's program at Yerkes–McDonald lasted only until the late 1940s, the Lowell project until 1953, and the Harvard meteor study – under Navy sponsorship – to 1954. Each was dogged by instrumental limitations and staffing problems. None resulted in a permanent institution devoted to this field. Indeed, by the early 1950s, global atmospheric studies became a province of meteorology, whereas investigations of the upper atmosphere were increasingly undertaken by physicists and geophysicists directly involved in experimental rocket research.[99] As Kuiper and Whipple would repeatedly rediscover, such hybrid research programs were difficult to sustain.

Seeing these programs as failures, however, would overlook the significance they had for solar system research and American astronomy in general. By providing new instruments and detectors to astronomers at Yerkes–McDonald, Lowell, and Harvard, federal patrons gave solar system astronomy a stronger presence at these facilities than it had had before the war. All these observatories remained important centers for this field through the 1950s. Moreover, the transient structures created by these projects produced fruitful if brief collaborations between meteorologists and astronomers on problems of mutual interest, leading to significant new work on planetary atmospheres.

Finally, these programs introduced astronomers to military agencies eager to fund astronomical research. Although U.S. astronomers continued to rely far less on military funds for research through the 1950s than did members of other physical science disciplines, military patronage helped maintain the viability of American observatories in the lean years of 1946 and 1947. However, military contract funding also encouraged researchers to design proposals with short-term solutions. Those projects that did not achieve these promised ends faced heightened risk of disruption or discontinuance, regardless of their scientific merit; this bound researchers more closely with military missions as the cold war deepened.[100]

Although planetary atmospheres research diminished at American observatories in the early 1950s, fundamental questions raised by these studies – including the origin and evolution of the solar system – remained of intense interest to a number of astronomers. Many found that astronomical instruments remained better suited for studying solid bodies such as asteroids, meteors, comets, and the Moon (as is discussed in Chapter 4); but these studies of solid bodies were greatly influenced by postwar geochemical studies of meteorites. How planetary geochemistry so swiftly influenced solar system astronomy in America after 1945 is the subject of Chapter 3.

[99] Charney (1959, 178–80), Lorenz (1970, 7), Phillips (1956, 123), and Rossby (1959, 23–4); see also DeVorkin (1987) and Newell (1980, 33–44).

[100] Shapley, "Memorandum on the Support of Astronomical Research by the Office of Naval Research," Box 7, FLW–SI.

3

Astronomers, Geochemists, and Astro-chemistry, 1945–1955

A fundamental challenge that faced astronomers interested in the solar family was assessing the composition of planetary bodies. Astronomers believed that understanding what materials made up meteors, meteorites, and comets would reveal relationships among these bodies. Researchers also believed that the composition of planets would provide important clues about their internal structure and chemical evolution. Since the cold temperatures of planets (compared to stars) limited spectrographic data, astronomers turned to studies of planetary diameters and meteorites. Meteorites, Whipple declared, "provide a key to our knowledge about the formation of the asteroids and therefore to the solar system in general."[1]

Geochemists were equally concerned with meteorites, sharing with astronomers a deep interest in the natural abundance of the elements. As the astrophysicist William Fowler wrote in 1956, "the abundance curve is the product [of the history of Earth, the stars, and galaxies] and was shaped by cosmic events. From this curve we can learn much about the evolution of stars, about cosmology and about all the grand-scale subjects of modern science." Geochemists saw similar advantages in using abundance estimates to tackle fundamental questions about Earth. "It now appears likely," the geochemist Harrison Brown wrote in 1949, "that an intense study of the distribution of elements in meteorites will enable one to draw important conclusions concerning the structure of planets, the origin of our solar system, and the relative 'cosmic' abundances of chemical elements."[2] By the late 1940s, astronomers, geochemists, geophysicists, and geologists began collaborating with one another on these issues more than at any previous time in the twentieth century; planetary geochemistry became a central theme at the Rancho Santa Fe conference of 1950 and the Williams Bay conference two years later. Although several of these cooperative programs proved difficult to sustain (as we shall see in Chapter 4), geochemistry had greater impact on solar system astronomy after World War II than any other outside discipline.

[1] Whipple (1959a, 1653).
[2] Quotes from Fowler (1956, 83) and Brown and Goldberg (1949, 347).

Brown, Urey, and their associates, who addressed the structure and chemical evolution of the terrestrial planets (including the Moon and Earth) and the age of Earth, had unparalleled influence on the development of solar system astronomy in America. Their work is important for several reasons. They introduced astronomers to wide-ranging debates among earth scientists after 1945 as to the nature of Earth's core and Earth's thermal history. They stimulated astronomers to review the calculated age of the universe – an issue of particular interest, since geochemists belonged to an intellectual tradition that many astronomers regarded as inferior.[3] They also made Chicago and Caltech major academic centers of planetary geochemistry.

Most significantly, Brown and Urey pioneered the application of nuclear reactors and especially high-precision mass spectrographs to the problem of atomic abundance determinations. Mass spectrometers caused a shift from qualitative to quantitative geochemistry in the way that high-dispersion spectroscopy transformed astrophysical studies of atomic abundances; the rapid growth of geochemistry after World War II reflected attempts by such former atomic scientists as Brown and Urey to apply quantitative techniques to what until then had been a largely qualitative field. Their arguments about the evolution and thermal history of Earth influenced the views of American geologists and astronomers about these fundamental problems. Moreover, new postwar patrons, including the Atomic Energy Commission, also began to affect the growth and methodological orientation of academic geology in the United States. The geochemical revolution that began at Chicago, which reshaped solar system astronomy in America, sent shock waves across the intellectual and institutional landscape of the earth sciences.

PLANETARY GEOCHEMISTRY AT CHICAGO

American astronomers, as we first saw in Chapter 1, had turned to the problem of terrestrial and cosmic abundances early in the twentieth century, seeking to understand how the composition of Earth related to that of the Sun and the universe at large. Among those who remained interested in this issue at the close of World War II were Menzel, Russell, and particularly Wildt, the former Göttingen astronomer who left his temporary home at Princeton for a permanent post at Yale in 1946. Intrigued by the atmospheres and internal structures of planets as well as stars, Wildt published a lengthy, influential review of planetary geochemistry the following year. Wildt used the article to explore possible physical analogues between stars and giant planets like Jupiter, believing that this approach would yield insight into the internal structure of such planets; but he also used it to call attention to the limited accuracy of past abundance estimates, as well as inconclusive research into the properties of materials at high pressures and temperatures. "As yet," Wildt declared, "there is no sign of our having accomplished even an exhaustive inventory of the physical-

[3] Greenstein (1986, 18) and Paneth (1940, 9).

chemical principles relevant to planetary constitution and cosmogony." He urged geochemists to join with astronomers in studying planetary geochemistry, arguing that such efforts would yield a rational "astro-chemical" model of solar system formation and evolution.[4]

Harrison Scott Brown, a nuclear chemist at Chicago, was among the first to take up this challenge. Brown's investigations, begun in 1946, were significant for several reasons. His work in meteorite chemistry made him an important link between researchers in geochemistry and astronomy interested in the asteroid origins and planetary evolution. More important, Brown applied mass spectrographs, greatly improved during World War II, to studies of meteorite geochemistry. The accuracy of his determinations of atomic abundances in meteorites soon surpassed those made at natural history museums and traditional departments of geology using petrographic techniques. His work thus helped to shift the center of gravity of meteorite research to academic centers of nuclear chemistry, and created new links with astronomers interested in borderland research in "astro-chemistry."[5]

Brown's interest in atomic abundances and his postwar research program were both strongly shaped by his experiences in the Manhattan Project. Born in Sheridan, Wyoming in 1917, Brown had earned a B.S. at Berkeley in 1938 and a Ph.D. from Johns Hopkins three years later, where he had worked under the physicist Robert Fowler. Because his dissertation research had involved the chemical separation of uranium isotopes, in mid-1942 Brown had been recruited to the Manhattan Project's chemical separation group at Chicago. There he had worked on the problem of isolating plutonium, assisting the physical chemist (and later Atomic Energy commissioner) Glenn Seaborg in this effort. In 1946, just after the Institute for Nuclear Studies was founded, Brown, then 29, received an invitation to join this body, becoming one of its fourteen junior charter members. Its senior researchers, including Urey, Enrico Fermi, and Maria Mayer, were all at the forefront of U.S. research in nuclear chemistry.[6]

The first project that Brown took up after arriving at Chicago was an investigation of the natural abundances of meteorites. His decision to work on this problem was shaped by the institute's mandate and by the interests of several senior staff members. The Institute for Nuclear Studies, one of three permanent centers set up by University of Chicago President Robert Maynard Hutchins to retain leading scientists brought to Chicago for the Manhattan Project's "Metallurgical Laboratory," had been designed to address fundamental problems of atoms and isotopes. Institute director Samuel Allison, following Fermi's vision, announced that staff members would focus on "fundamental studies of properties of nuclei," including chemical

[4] Quoted from Wildt (1947, 101 [see also 84, 96–7]). Similar arguments from a European perspective appear in Paneth (1940, 9).

[5] Craig, Miller, and Wasserburg (1964, xvi) and Pettijohn (1984, 201).

[6] Smith, Fesharaki, and Holdren (1986, 3, 77); see also "Research in Atomic Structure," n.d. [ca. 1947–8], Box 107, HCU, and "Appendix to 'Proposal for a grant to study problems of Lunar and Planetary Exploration,'" Oct. 5, 1959, Box 96, LAD.

problems "connected with the separation of the natural isotopes."[7] Several senior researchers, however, soon turned to the application of isotope chemistry to astronomical and chemical problems, including Willard Libby's efforts to establish carbon-14 as a tool for geologic dating. What made these studies attractive, aside from their intrinsic interest and competitive advantages, was the resources on hand to support them. In addition to $53 million the Metallurgical Laboratory had received in government contracts during the war, institute staff members had access to two nuclear reactors developed in the 1940s, housed at what became the Argonne National Laboratory, twenty miles southwest of Chicago. Like the high-precision mass spectrometers developed during World War II, nuclear reactors provided researchers a means to determine chemical concentrations several orders of magnitude finer than by chemical techniques. By the late 1940s reported accuracies moved into the parts-per-ten-million range, roughly a thousandfold increase over prewar accuracies.[8]

Two factors apparently led Brown to consider analyzing meteorite abundances. The first was his confidence that postwar methods of nuclear analysis could pinpoint the age of Earth, a matter of growing interest to astronomers and geochemists. After the 1931 National Research Council study led by Arthur Holmes, Adolph Knopf, and Charles Schuchert was released, arguing that Earth's age was 2–3 billion years, other American geochemists turned to this question. However, these investigations, conducted by such noted chemists as Alfred O. C. Nier, Alfred Lane, Robley Evans, William Urry, and T. W. Richards, did little to alter the 1931 estimate. As we first saw in Chapter 1, astronomers such as Russell initially welcomed the close coincidence between the age of the universe and that of Earth; but by the mid-1940s, this time scale began to trouble leaders of both disciplines. Friedrich Paneth, a widely respected European geochemist, wrote in 1946 that "[i]t has been hailed by some as being satisfactory that the two figures are so close, although it is a rather surprising conclusion that the rocks which we find now in the surface of Earth should have solidified at the same time as that at which the Universe came into existence."[9] In an influential review of the cosmic time-scale issue, also published in 1946, Harvard's Bart Bok noted that astrophysical evidence about the universe's rate of expansion made it "unlikely that we shall soon be permitted a return to the comfortable and spacious long time-scale of cosmic evolution," one in which Earth formed long after the universe originated. The field of stellar evolution, Bok declared, was "in a state of flux" from these conflicting ages. Since the age of Earth was considered one of the strongest arguments that the universe was at least 2 billion years old (and since most

7 "Research in Atomic Structure," n.d. [ca. 1947–8], Box 107, HCU, p. 10. An Institute of Metals and an Institute of Radiobiology and Biophysics were also created; see Fisher (1963, 66) and Craig et al. (1964, vi).

8 Brown and Goldberg (1949, 347), Burke (1986, 267), Hewlett and Anderson (1962), Knopf (1957, 228–30, 232–3), and Pettijohn (1984, 200–1).

9 Quoted from Paneth (1946); see also Bondi (1952, 50, 165).

American astronomers then conceded that meteors originated within the solar system), Bok urged new inquiries into "the probable ages of meteorites" to reassess the age of the solar system.[10]

Apart from the age issue, several institute researchers became interested in the origin and abundance of the elements as a problem of both stellar astrophysics and cosmology. Shortly after the war, new astronomical estimates of the cosmic abundances of hydrogen and helium were published by Albrecht Unsöld in Germany and Menzel and Lawrence Aller in the United States. Their figures indicated that the proportions of these gases to the heavier elements were much greater than those Russell had proposed in the 1930s, thereby requiring further revisions to theories of stellar evolution and nucleogenesis. This finding interested several Chicago astronomers, including Struve, who by late 1947 considered accepting a joint appointment with the Institute for Nuclear Studies in order to study atomic abundances from an astronomical perspective and to encourage such work at Yerkes–McDonald.[11] Institute physicists were also interested in how the heavy elements had originated. What strengthened interest in this problem was that by 1946 two competing cosmological theories for the origin of the heavy elements were being debated among astrophysicists and nuclear chemists. One idea, proposed by Harvard astrophysicist Theodore Sterne and German physicist Karl Friedrich von Weizsäcker, held that the contemporary distribution of abundances had been established when the universe was smaller and at intense temperatures, an argument later expanded by Chandrasekhar. The second, developed by physicists George Gamow and Edward Teller, argued that only light elements had emerged in the initial moments of what became known as the Big Bang. For younger institute staff like Brown, who learned about the Gamow–Teller model of the expanding universe directly from Teller, then also at the institute, the opportunity to use meteorite abundance work to contribute to cosmological theory had considerable intellectual appeal.[12]

Convinced that abundance studies could provide important information on both issues, Brown first sought to measure the age of Earth. In 1946, Brown assigned a graduate student, Claire Patterson, to determine the lead isotope abundances in several meteorites. Since U^{238} and U^{235} decay into Pb^{206} and Pb^{207}, respectively, and U^{235} generates lead isotopes six times faster than U^{238}, geochemists regarded measurements of the absolute abundances of decay-produced lead isotopes in meteorites and terrestrial rock samples as an extremely promising technique for age determina-

[10] Quoted from Bok (1946, 75, 62–3). Harlow Shapley also grew concerned about the time-scale problem by the mid-1930s; see Shapley, "Outline of Project [on Time Scale]," Feb. 17, 1936, Box 36, HS, and DeVorkin (1984, 51). American billions (10^9) are used throughout this discussion.

[11] Struve to J. L. Greenstein, Jan. 30, 1948, Box 6, JLG; see also Brown (1949, 625), Gamow and Hynek (1945, 249–50), and Struve (1947b, 146).

[12] Bok (1946, 72), Brown (1949, 634), Fowler (1956, 89–90), Pettijohn (1984, 200–1), and Stuewer (1972). Teller was familiar with astrophysical research via his participation in a 1938 Yerkes symposium on interstellar lines and his studies with Gamow of the stellar mass–luminosity relationship; see G. Herzberg to H. N. Russell, Aug. 29, 1938, Box 44, HNR, and Brush (1992, 63).

tions. Despite the development of the mass spectrometer during the late 1930s, principally by Nier at Minnesota, sufficiently accurate determinations of these lead isotope ratios seemed beyond reach.[13] However, because Brown and Patterson had access to improved Nier-type mass spectrometers used to monitor the chemical separation of U^{235} from U^{238} during the Manhattan Project, Brown felt confident that a new age estimate for Earth would be "duck soup."[14] Nevertheless, despite considerable effort, Patterson failed to achieve a precise lead measure. The problem, as Brown learned from Nier – whose mass spectroscopy laboratory at Minnesota also grew rapidly after 1945 – was that primordial lead isotopes in his meteorite samples were contaminated by traces of modern lead, including lead produced in chemical plants and by automobile emissions. Instead of a revised age, Patterson found that past lead abundances reported for meteorites were suspiciously high, casting doubt on previous age estimates. This discovery, which kindled Brown's interest in environmental issues, also left him wondering whether other abundance estimates were similarly distorted.[15]

Blocked by this unexpected obstacle, Brown turned his attention to the larger question of the natural abundances of all elements. As we have seen, astrophysicists were deeply interested in absolute atomic abundances by the early twentieth century; following Russell, they accepted that meteorites (as well as natural abundances in Earth's crust) were important sources of geochemical data needed to fill gaps in stellar abundances. Initially Brown himself seemed to see meteoritic studies primarily as a way of contributing to stellar astrophysics and nuclear chemistry. In his first postwar paper with Patterson on meteorite abundances, Brown stressed that "[t]he science of meteoritics has come to be looked upon in recent years as a science of increasing importance, largely because of its recognized bearing upon astrophysical problems."[16]

Soon after becoming involved in nuclear abundance studies, however, Brown became increasingly aware of the implications that this work had for geochemical and geological research. Geochemists considered atomic abundance studies fundamental to their discipline; as Victor Goldschmidt wrote in 1938, "one of the primary aims of geochemistry is the knowledge not only of the chemical composition of our own planet . . . but also the chemical composition of other parts of the universe. Only through comparison with the universe as a whole can we gain an understanding of the chemical evolution of earth." Geochemists viewed meteorites as the best means to study the linkage between terrestrial and stellar abundances, since meteorites could be investigated by the same analytical techniques applied to terrestrial rock

[13] Burchfield (1975, 204), Goodman and Evans (1941, 539), Knopf (1957, 230–2), Paneth (1940; 1946), Tilton and David (1959, 192), and Whipple (1959a, 1662).

[14] Patterson (1986, 10).

[15] Brown and Patterson (1947a,b); Kirk R. Smith, "Introduction," in Smith et al. (1986, 4); see also Patterson (1986, 12).

[16] Brown and Patterson (1947c, 456); see also Brown (1949, 625) and Brown and Goldberg (1949, 347).

samples. Moreover, geologists valued meteorites for providing insight into the chemical composition of the planet as a whole. Since geologists accepted that rocks near Earth's surface did not reflect its general composition, meteorites (widely believed to be fragments of a disrupted planet) remedied this deficiency. Goldschmidt declared that meteorites "give us a picture of the chemical evolution of the earth and some clue as to the composition of its inaccessible interior." Finally, geochemists considered the age-of-Earth problem (whose solution required a precise measurement of certain atomic isotopes) essential for addressing fundamental geological questions, including how long radioactive heating had affected Earth's interior, how much crustal melting had occurred, and whether convection currents in Earth's mantle could explain the formation of mountains and the existence of continents. The gravity of these questions had moved Goldschmidt and Russell to begin exchanges over abundance measurements in the 1930s, an early instance of interdisciplinary cooperation on this problem.[17]

Geologists and geochemists by the late 1940s were already in general agreement about the relationship between meteorite studies and atomic abundance research. Virtually all geologists accepted that meteorites were samples of the cosmic material from which the solar system had formed, an idea framed in the accretion cosmogony proposed by American geologist Thomas C. Chamberlin early in the century.[18] Many also accepted that meteorites were remnants of a disrupted planet whose physicochemical properties had resembled those of Earth. By the late nineteenth century geochemists had developed distinct categories of meteorites, mainly irons and stones. Stony meteorites, or simply *stones,* were mostly composed of magnesium and iron silicates; *irons* typically were 90 percent metallic iron and 8 percent nickel, with trace abundances of minor constituents. Because the mean density of irons was significantly higher than that of stones, and because most geologists accepted the late-nineteenth-century arguments of Emil Wiechert and R. D. Oldham that Earth has an iron core, geochemists readily embraced the idea that irons were derived from the interior of a large planetary body, and stones from its mantle.[19] By the 1920s, several geochemists had developed models of Earth on the basis of these meteoritic categories. In 1922, for example, Goldschmidt had argued that Earth possesses a nickel–iron core surrounded by shells of compressed silicate. Several years later, two geochemists at the Carnegie Institution of Washington, Henry S. Washington and Leason H. Adams, had advanced a similar model. Like Goldschmidt, both men drew an explicit parallel between meteorite studies and planetary structure. Earth, they had argued, was not a huge meteorite, but rather "a body similar to that of which

[17] Quoted from Goldschmidt to Russell, Dec. 22, 1937, and Jan. 25, 1938, both Box 42, Folder 22, HNR; see also Burke (1986), Muir (1954, 69), Paneth (1940, 16–17), Tilton and David (1959, 190).

[18] Paneth (1940, 19); see also Greene (1982, 258–94).

[19] Brown (1949, 627). Stones were subclassified chondrites or achondrites, depending on their crystalline structure, whereas *troilites,* with pockets of iron sulfide in stony and iron phases, were recognized as a third, rarer category of meteorite; see Brush (1982a, 1186), Burke (1986, 145–73), Greene (1982, 269), and Urey and Craig (1953, 37).

meteorites are but fragments." Widely accepted through the mid-1940s, these models remained central to geochemical theories of Earth. Moreover, since most geochemists accepted that phase separations had occurred within Earth, causing its core to form, Goldschmidt's arguments that Earth was once molten gained general acceptance among members of the community.[20]

One problem that geochemists faced in applying meteorite analyses to a general study of atomic abundances was uncertainty about the relative proportion of stones to irons. Geochemists regarded this ratio as critical for determining the absolute abundance of elements as well as for interpreting the structure of planets. Neither astronomical nor geological evidence, however, helped pin down this ratio. Stones flared more brilliantly than irons while passing through Earth's atmosphere, leading to high numbers of reported stony meteorite "falls." Once on the ground, however, irons were more easily distinguished from terrestrial rocks than were stony meteorites, and thus were collected in disproportionate numbers. In 1940, for instance, Paneth estimated that whereas reliable reports existed for just 40 iron falls, compared to 550 witnessed stone falls, museums held irons from over 400 locations for which no falls had been reported.[21] Moreover, through the 1930s, geochemists had generally split with astronomers over the origin of meteorites. While geochemists had regarded meteorites as fragments of a planetary-sized body formed within the solar system, and hence chemically linked to the Sun, most American astronomers until the late 1930s had followed Hoffmeister and Öpik in accepting an interstellar origin for meteors and meteorites, providing no firm links to terrestrial geochemistry. Disagreements over this fundamental issue, for which no common methodological or institutional framework existed to force a resolution, appeared to limit contacts between members of these communities during much of the early twentieth century.[22]

When Brown first began studying atomic abundances in meteorites in 1946, he seemed unconcerned whether his samples had a solar or interstellar origin. In an article published the following year, Brown simply declared that researchers "must make certain important and as yet unproved assumptions concerning the nature and origin of meteors and meteorites, before one is able to apply data on their composition to a compilation of relative abundances of nuclear species."[23] Since Brown initially saw meteorite abundances as a way to address problems in stellar astrophysics and cosmology, this was not a surprising position for him to take. By 1949, however, he accepted arguments that meteorites derived from a former planet similar to other terrestrial planets in size and composition. His increasing attachment to this idea

[20] Quoted in Burke (1986, 259–60); see also ibid., p. 261, as well as Daly (1943), Lewis (1934, 898), Paneth (1940, 16–17), and Servos (1983, 147–52).

[21] Brown (1949, 627), Paneth (1940, 5), and Watson (1938, 199).

[22] Paneth's 1940 Halley Lecture to the Royal Astronomical Society, for example, was largely a challenge to the von Niessl–Hoffmeister–Öpik interpretation for the origin of meteors, and supported Whipple's emerging argument that meteors are solar; see Lankford (1987) and Paneth (1940, 8–16).

[23] Brown and Patterson (1947c, 456).

caused him to see the problem of natural abundances more and more in geological and geochemical terms. This represented a significant shift in his outlook, one that inspired him to increase his contacts with astronomers, geologists, and geophysicists and to reorient his research toward the geochemical evolution of planetary bodies.

Brown advanced his studies of meteorite abundances in two ways. First, using meteorites obtained from the Chicago Natural History Museum and other sources, he assigned a graduate student, Edward Goldberg, to assess their compositions at the nearby Argonne National Laboratory. The process involved irradiating a meteorite sample, then a sample of an isolated element under investigation; by comparing the specific activity of the meteorite and the isolated element, Goldberg calculated the abundance of that element within the meteorite. The nuclear irradiation technique enabled Brown and Goldberg to make measurements of previously unattainable precision. Using the Argonne reactor, Brown and Goldberg estimated that their gallium and palladium abundances reached accuracies of a few parts per hundred million. Nevertheless, the technique was time consuming, and although the Argonne facility was nearby, many researchers competed for time on this reactor. By 1948, Brown and Goldberg had completed only a handful of determinations.[24]

The second and less costly approach that Brown employed was to review existing tables of abundances, with an eye toward updating them with contemporary data whenever possible. Through the 1940s the table of atomic abundances most widely employed by astrophysicists and geochemists was the *Geochemische Verteilungsgesetze der Elemente und der Atomarten,* published by Goldschmidt in 1938. Goldschmidt had prepared his table by combining astronomical estimates of volatile elements (derived from stellar studies) with geochemical analyses he and others had made using meteorites and terrestrial samples for the heavier, nonvolatile elements. A product of over twenty years's labor, begun as a study of mineral resources in Goldschmidt's native Norway after a military blockade of foreign trade had begun in World War I, Goldschmidt's table was highly valued by physicists and astrophysicists as well as geochemists and geologists.[25] Brown himself declared it "the only reasonably adequate table of relative abundances of nuclear species available for comparing theory with experiment."[26] Nevertheless, Brown was aware that new meteorite analyses had been completed following its publication, and that Unsöld, Menzel, and Aller had greatly increased astronomical abundance estimates for hydrogen and helium. Between 1946 and 1947, Brown and Patterson used these new data to reanalyze the composition of the silicate and metal phases of stony meteorites, and the composition of iron meteorites. It seems clear that Brown still intended these redeterminations primarily to aid astrophysicists and theoretical physicists,

[24] In 1948 Goldberg received a doctorate for his measurements of palladium and gallium in meteorites; see Urey to D. H. Menzel, Feb. 3, 1949, Box 57, HCU; see also Brown (1949, 625–6), Brown and Goldberg (1949, 347–9, 353), Burke (1986, 267), Greenstein (1986, 26), and Knopf (1957, 229–30).

[25] Bernal (1949, 2113), Brown (1950a, 11), Oftedal (1948), and Suess and Urey (1956, 53).

[26] Quoted from Brown (1949, 625); see also Brown and Patterson (1947a, 456).

including Teller, drawn to stellar nucleogenesis and cosmology. In a short article he prepared in early 1947 for the *Physical Review*, a favorite journal for nuclear physicists, Brown called attention to the relevance of the work he and Patterson had performed for "such questions as the relationships between nuclear stability and relative abundance."[27]

By late 1947, however, Brown came to see the problem of natural abundances in meteorites increasingly in geochemical terms. In March 1948 Brown and Patterson published an article in the *Journal of Geology* on the phase equilibria and genetic relationships among meteorites. Although ostensibly the third in the series of reports by these scientists on the chemical composition of meteorites, this article, more than three times the length of its predecessors, for the first time attempted to synthesize these new abundance estimates into a geochemical framework. The paper was significant in two respects. First, Brown and Patterson no longer entertained the idea that meteorites originated beyond the solar system; the connection between meteorites and the solar family was implicitly assumed. Second, both scientists argued that the chemical properties of meteorites were best explained by assuming their origin within a planet similar in size and composition to Earth, as most geochemists already argued. Brown and Patterson based this claim on several lines of evidence. Peaks occurred in the frequency distribution of metal phase in stony meteorites; this suggested to them that stones derived from different levels in a parent planet's mantle, higher metal content corresponding to deeper levels in its mantle. Granting this assumption, the abundances of various elements could be interpreted as phase variations in regimes of varying temperature and pressure. Moreover, the presence of oxides in both stony and iron meteorite samples was interpreted to mean that they had formed at temperatures of at least 3,000° K, indicating an origin within a body of moderate internal temperature. Drawing on chemical thermodynamics, Brown and Patterson declared that "the conclusion appears irrefutable that meteorites at one time were an integral part of a planet" whose size, they calculated, was larger than Mars. While basing their arguments on their revised abundance measurements and their studies of phase equilibria in meteorites, it was clear that Brown and Patterson had come to accept Whipple's evidence for the solar origin of meteorites and geophysical models for the internal structure of Earth. Writing about possible parallels between metal phase variations and "the well-known seismic discontinuities of second order existing within the Earth," for example, Brown was appropriating concepts developed far from his field of training.[28]

Brown's reorientation toward planetary geochemistry can be pinpointed, with precision, to the late summer of 1947. It arose through his participation in the Yerkes Symposium on the Atmospheres of Earth and Other Planets, another remin-

[27] Quoted from Brown and Patterson (1947c, 456); see also Brown to F. L. Whipple, May 2, 1949, Box 1, FLW–HU.

[28] Quoted from Brown and Patterson (1948, 108); see also Brown (1949, 626; 1950b, 641–2) and Brown and Patterson (1947b, 510).

der of how strongly Kuiper's conference influenced solar system research in the immediate postwar years. Among American astronomers that Brown met at this conference, he became particularly close to Whipple; through him, Brown learned that U.S. astronomers accepted a solar origin for meteors, and generally embraced Hirayama's evidence for asteroid families. He also learned that leading American astronomers then regarded asteroid research as a high priority. Aware that the German Rechen-Institut had been seriously undermined during World War II, Struve, Shapley, and Brouwer initially endorsed a proposal to shift responsibility for asteroid orbital calculations to the Leningrad Institute for Theoretical Astronomy; but as the cold war intensified in 1947, and Leningrad astronomers moved slowly to fulfill their obligations, Struve and Brouwer instead endorsed a rival plan to coordinate international asteroid research under the direction of Paul Herget at Cincinnati, and Struve endorsed more ambitious plans to create an institute for asteroid research in the United States. Through Kuiper's symposium, Brown found a small but active group of American astronomers interested in the physical and geochemical properties of small bodies. He also realized that geochemical evidence could be used to test competing astrophysical theories of the origin and evolution of asteroids and meteorites.[29]

Brown's increased contact with American astronomers also made him reluctant to deal with leaders of the (U.S.) Meteoritical Society, revealing an important fault line between astronomers and traditional geochemists in postwar America. Founded in the 1930s, this society was led by Lincoln LaPaz, a mathematical astronomer and physicist who chaired the Department of Mathematics and Astronomy at the University of New Mexico. LaPaz, trained at Chicago, had moved vigorously to capitalize on wartime developments in New Mexico, creating an Institute of Meteoritics at New Mexico in 1945. The purpose of the institute, LaPaz announced, was to strengthen university relations with the nearby White Sands Proving Ground. Emphasizing the problem of ballistic flight, LaPaz declared his willingness to provide meteorite samples to "nuclear physicists, ballisticians, aerodynamisticians, and other investigators engaged in research of value to the development of meteoritics."[30] LaPaz's associates, including Frederick C. Leonard, a UCLA astronomer and former Leuschner student, produced catalogues of meteor craters and meteorite falls. Although Brown did obtain several meteorites from the institute, he found little common ground with LaPaz's Meteoritical Society, whose members continued to employ petrographic abundance analyses. He also wearied of LaPaz's fierce, per-

[29] Brown to Whipple, Sept. 19, 1947, and Whipple to Brown, Oct. 7, 1949, both Box 1, FLW–HU. On international asteroid studies after 1945, see Kuiper to Major Fisher, June 30, 1945, and idem to J. Stebbins, Nov. 5, 1946, both Box 18, GPK; Harlow Shapley, "Memorandum on the Support of Astronomical Research by the Office of Naval Research," n.d. [ca. 1946], Box 7, FLW–SI; Brouwer to G. M. Clemence, Oct. 16, 1946, Box 58, Folder 22, DB; Whipple to Brouwer, July 15, 1947, Box 1, FLW–HU; Struve to Brouwer, July 25, 1947, Box 86, Folder 326, OS; see also Brouwer (1956, 943, 947), Delaporte (1950), and Herget (1950, 167).

[30] Quoted in "Meteoritical Activities" (1949).

sonal feud with noted meteorite collector Harvey N. Nininger, an independent researcher who had formerly taught biology and geology at McPherson College. Returning from "meteoriticists's feuding country" after a western trip in 1949, Brown denounced the "unadulterated amateurism" of LaPaz's group. Whipple was similarly annoyed at LaPaz's dismissal of evidence favoring the solar origin for meteors. One result of LaPaz's cool relationship with Whipple and Brown was that traditional meteorite geochemists gained little support from American astronomers. This made it difficult for meteoriticists, who shared few of the resources available to nuclear chemists, to compete for extramural funds. Finding little merit in proposals submitted by traditional meteoriticists that the National Science Foundation had asked him to review, Kuiper lamented that "this extraordinarily interesting field of science is at present very thinly populated by a rather strange assortment of people." Despite his colorful claims, neither LaPaz nor the Meteoritical Society significantly influenced solar system astronomy in America in the postwar era.[31]

By contrast, the geochemical evidence that Brown and Patterson presented in 1948, supporting their claim that meteorites are fragments of a disrupted planet, provoked lively discussions among American geologists, astronomers, and geochemists. Geologists tended to view their evidence favorably, in part because Brown and Patterson's conclusions, which supported a direct link between meteorite abundances and those presumed to exist within Earth, promised to place geochemistry on a firm methodological foundation. Francis Pettijohn, a geologist at Chicago and editor of the prestigious *Journal of Geology,* praised their work in glowing terms, writing that their implications regarding the origin of meteorites and Earth's interior "will be of interest to every geologist and astronomer."[32] At Princeton, Harry H. Hess, a rising presence among its geology faculty, quickly incorporated their conclusions into his studies of Earth's crust, which employed meteorites as a primary source of data. Interest in the Brown–Patterson exploded planet hypothesis grew when both men won a $1,000 prize for "the outstanding paper" of the 1947 meeting of the American Association for the Advancement of Science. Although the detonation of the first atomic bombs two years earlier may have influenced reception of their paper, the increased accuracy of postwar nuclear abundance studies attracted widespread attention. In 1948, the same year in which the Brown–Patterson paper was published, an "International Committee on Meteoritics" was convened at the London meeting of the International Geological Congress. Soon thereafter Pergamon Press launched *Geochimica et Cosmochimica Acta,* the first international journal solely devoted to geochemistry and "cosmic" chemistry.[33]

[31] Quoted from Brown to Whipple, May 2, 1949, Box 1, FLW–HU, and Kuiper to P. van de Kamp, Feb. 2, 1955, Box 28, GPK; see also Leonard to S. Einarsson, Sept. 6, 1947, Box 26, AOL; Brown to Whipple, July 1, 1949, Box 1, FLW–HU; Brown to H. C. Urey, May 2, 1956, Box 51, HCU; as well as Leonard (1946), Mark (1987, 78), and Marvin (1993, 277).

[32] Pettijohn (1948).

[33] Hess to Kuiper, Feb. 25, 1954, Box 11, GPK; see also Davis and Hess (1949) and "International Commission on Meteorites" (1950).

American astronomers interested in solar system research also welcomed the geochemical support that Brown and Patterson gave to the exploded planet hypothesis, seeing it a confirmation of Hirayama's theory of asteroid families. Their results inspired several astronomers to study further the physical characteristics of asteroids. Whipple, already favoring by June 1947 (three months before meeting with Brown) the idea that asteroids and meteorites resulted from "a collision or explosion" of a planet, became particularly interested in such work. Aware that Carl A. Bauer, a Michigan doctoral candidate in astronomy, was working up a thesis on meteorite geochemistry that similarly placed meteorites originating from a parent asteroid, Whipple advised Brown that "the figures are such that a planet the size of Mars can be disrupted without stretching the theoretical circumstances." Kuiper, too, favored increased astrophysical study of asteroids. In 1948, as director of Yerkes–McDonald, Kuiper endorsed a plan proposed by Brouwer to make new orbital studies of asteroids on a Schmidt telescope at McDonald, since it would partially test the geochemical conclusions that Brown and Patterson had advanced.[34]

Support for the exploded planet hypothesis was shaken, however, by criticisms from an eminent physical chemist, Irving M. Klotz. A Northwestern University researcher and author of widely read texts on chemical thermodynamics, Klotz took exception to the certainty with which Brown and Patterson framed their conclusions. In a *Science* article in 1949, Klotz focused on entropy changes at low temperatures, a factor Brown and Patterson had ignored as insignificant. Arguing that such changes indeed had a measurable effect on geochemical evolution, Klotz declared that meteorites could have originated at either higher or lower temperatures and pressures than Brown and Patterson had claimed. "While the hypothesis of a single planet of origin for meteorites may be an attractive one, as indeed its popularity for almost a century testifies," he asserted, "there is as yet *no thermodynamic basis* for justifying this assumption." Despite Whipple's continued support, American astronomers gradually accepted the validity of Klotz's criticisms by the 1950s, particularly after Urey raised new doubts about Brown's thermodynamic work. In 1950 Brown himself hedged his bets on the origin of meteorites, arguing that they resulted "either . . . from the breakup of a relatively small planet, or [are] small fragments of condensed matter which escaped assimilation into a body of planetary size." While geologists were slower to back away from the comfortable certainty of the Brown–Patterson hypothesis, Hess joined astronomers in accepting the uncertainty of meteoritic evidence.[35]

[34] Quoted from Whipple to Brouwer, June 17, 1947, and idem to Brown, Jan. 14, 1948, both Box 1, FLW–HU; Brouwer to Kuiper, Feb. 23, 1948, and Kuiper to Brouwer, Mar. 3, 1948, both Box 29, GPK; Whipple to Kuiper, May 18, 1948, "1948 correspondence file," YO; and idem to Brown, June 30, 1950, Box 1, FLW–HU; see also Bauer (1948) and Burke (1986, 268–9). Whipple was almost certainly aware of a major 1943 article by Reginald V. Daly, his Harvard geology colleague, which also supported the breakup hypothesis for asteroids; see Daly (1943, 416–17) and Chapter 5 of the present volume.

[35] Quoted from Klotz (1949, 251) and Brown (1950b, 645); see Whipple to Brown, Apr. 4, 1949, Box 1, FLW–HU; Urey to O. Struve, n.d. (ca. 1950), Box 16, OS; Harmon Craig OHI (Ronald E. Doel, Nov. 27, 1990, at AIP); K. Aa. Strand to Oort, Oct. 18, 1951, Box 34, JHO; Urey to H. Shapley, July 5, 1956,

By 1950 the precise origin of meteorites was becoming less important to Brown than the related but broader problem of atomic abundances and their significance for studying planetary evolution. In 1949 Brown published in the *Physical Review* a revision of Goldschmidt's table of atomic abundances, utilizing new stellar results for volatile gases and the work he, Goldberg, Patterson, and a third graduate student, George Tilton, had done on nonvolatile elements in meteorites. Then, early in 1950, Brown published a study in which for the first time he turned his attention solely to the planets. Using abundances figures derived from his meteorite analyses, and arguments drawn from geophysical studies of Earth's core, Brown concluded that the "observed physical and chemical characteristics of the planets can be explained in a reasonably satisfactorily manner by assuming they condensed from a medium possessing varying temperatures and densities but fairly uniform composition." More than the actual values he obtained, the importance of Brown's work was his insistence that principles derived from physical chemistry could be quantitatively applied to predict the internal structure of the terrestrial planets.[36]

Brown's work was also important in establishing the Institute for Nuclear Studies as a center for planetary geochemistry. Because the problem of atomic abundances had broad appeal to physical chemists and nuclear physicists in addition to geochemists and astrophysicists, Brown had little difficulty recruiting graduate students to take up this work. He also labored to interest federal patrons in this issue, proposing to the Atomic Energy Commission in 1949 an ambitious plan to measure precise abundances for sixty elements found in meteorites.[37] Finally, Brown's interest in applying geochemistry to planetary geophysics illuminated a broad trend within American geology, for by the early 1950s increasing numbers of geochemists, geophysicists, and geologists showed intense interest in the question of Earth's internal structure. A key role in these developments was played by Urey, Brown's senior institute colleague. His efforts to understand the geochemical evolution of planets and asteroids came to have enormous influence on U.S. solar system astronomy.

HAROLD C. UREY AND THE GEOCHEMICAL EVOLUTION OF PLANETS

Following 1945, American geochemists and geophysicists learned of new challenges to the standard model of Earth. Most earth scientists until this time believed that

Footnote 35 *(cont.)*
Box 84, and idem to L. Aller, Feb. 21, 1957, Box 3, Folder 1, both HCU. Kuiper cited Klotz in his rejection of the Brown–Patterson view; see Kuiper to H. H. Hess, Mar. 9, 1954, Box 11, GPK.

[36] Quoted from Brown (1950b, 641); see Brown memorandum to H. C. Urey, G. P. Kuiper, and C. F. von Weizsäcker, Dec. 21, 1949, Box 15, HCU; see also Brown (1949), Brown, Kullerud, and Nichoporuk (1952), Goldberg, Uchiyama, and Brown (1951), and Greenstein (1986, 11–12).

[37] Harrison S. Brown, "A Proposed Program for the Accumulation of Quantitative Data Concerning: The Chemical Composition of Meteorites and the Earth's Crust; the Relative Abundances of Elements in the Solar System; The Ages of the Elements and Planets," n.d. [1949], Box 15, HCU; and Brown to Whipple, May 2, 1949, Box 1, FLW–HU. This proposal apparently was not funded.

Earth possessed an iron core; yet the 1941 theory of a hydrogen-rich core (proposed by Swiss scientists Werner Kuhn and Arnold Rittmann) and the 1948 theory of a compressed silicate core (advocated by British geophysicist William Ramsey) raised difficult questions regarding what was actually known about the chemical and thermal history of Earth and other planets. The problem had broad relevance to earth scientists, since such concepts as the convective theory of mountain building and continent formation depended on the rate of heat flow and the likelihood of convection within Earth's deep interior. Because Kuhn–Rittmann and Ramsey both made use of recent astronomical and cosmogonal data in their arguments, astronomers had a stake in this work as well. The plurality of models of Earth's structure by the late 1940s owed much to the incorporation of planetary astronomical data into a field formerly dominated by geochemists and geophysicists studying Earth in isolation.[38]

One scientist who deeply influenced these debates in the late 1940s was Urey. A founding member of Chicago's Institute for Nuclear Studies, Urey did much to encourage solar system research within his institute-based laboratory, in addition to training such leaders of planetary geochemistry as Gerard Wasserburg and Harmon Craig, and the biochemist Stanley L. Miller. More important, Urey established interdisciplinary contacts with scientists in neighboring fields, including Kuiper and Whipple, and came to have a decisive influence on their research (in Kuiper's case, on his career and personal life as well). As Hess wrote Urey from Princeton in 1953, "You have stimulated a lot of us to think about problems on the planet-scale which otherwise we would have gone on looking at in small bits and pieces. . . . It would have been quite impossible for any of us to break new ground in the direction in which you are going."[39]

Urey's inclination to interdisciplinary research, a consistent thread throughout his career, reflected his creative powers, his abundant self-confidence, and his intellectual restlessness. Born in 1893 in Walkerton, Indiana, a small town near South Bend, Urey had earned his B.A. in zoology at Montana State in 1917 and a Ph.D. in chemistry at Berkeley in 1923. After a fellowship year at Niels Bohr's Institute for Theoretical Physics in Copenhagen, he had taught at Johns Hopkins and then Columbia. While at Columbia in late 1931, working with Ferdinand G. Brickwedde and George M. Murphy, Urey had found experimental evidence for a heavy isotope of hydrogen, later named deuterium. Three years later, as leader of this team, he had won the Nobel Prize for chemistry. His reputation established, Urey had continued isotope research through the 1930s. In 1940, he had developed the gaseous diffusion method for separating U^{235} from U^{238}, then directed efforts to produce this critical isotope for the Manhattan Project at Columbia, and later at Chicago. By late 1945,

38 Kuhn and Rittmann (1941) and Ramsey (1948; 1949); see also Brush (1982a) and Howell (1990).
39 Hess to Urey, Nov. 19, 1953, Box 42, HCU.

then an outspoken opponent of the use of atomic weapons against Japan, Urey had remained at Chicago at the Institute for Nuclear Studies.[40]

In 1952, Urey declared that he had become interested in the solar system just three years earlier, while preparing for a summer course at Chicago on "Chemistry in Nature" with Brown. In fact, Urey had been involved in astrophysics and solar system geochemistry during much of his career. His 1923 dissertation, on the thermal conductivity of cesium vapor, was a problem of considerable interest to astrophysicists of the time; he later decided against pursuing this problem principally because Indian physicist Megh Ned Saha produced a more elegant and generalized version of the theory first. While at Bohr's institute, Urey had addressed another problem of interest to stellar astrophysicists, the relative numbers of hydrogen atoms occurring in various possible quantum states, publishing this work in the *Astrophysical Journal.* Urey's interest in astrophysics had apparently been kindled at Berkeley by Gilbert N. Lewis, his dissertation advisor and mentor. Throughout much of his career, Lewis, one of America's most influential physical chemists, had been intrigued by the idea of considering stars as chemical "laboratories," principally as a means of studying chemical reactions impossible to duplicate on Earth. In 1922, while Urey was at Berkeley, Lewis had lectured on the possibility of applying astrophysical techniques to launch a new conceptual attack on the problem of radioactivity. In one lecture before the Astronomical Society of the Pacific, Lewis had argued the "laboratory" analogy in quantitative terms: Although the highest pressures then obtainable in research laboratories approached just 30,000 atmospheres (atm), he estimated the pressure at the center of Jupiter as 10×10^6 atm, and of the Sun as 1×10^9 atm. Urey seemed to find Lewis's concept of stars as chemical laboratories a useful concept. His initial models of the Moon and Mars three decades later, which eschewed geological complexities, remained faithful to this conceptual framework.[41]

Urey had retained an interest in solar system studies in the 1930s, even when the focus of his research turned to isotope geochemistry. In 1931, Urey had used the facilities of Columbia's Department of Chemistry to measure the ratio of O^{16} to O^{18} in meteorites in order to determine whether it matched Earth's (indicating a solar system origin for meteorites) or differed from it considerably (suggesting an interstellar origin). Learning of the Harvard–Arizona Meteor Expedition the following year, Urey had asked Shapley for his views of meteorite origins, apparently discovering that Harvard astronomers remained divided on this key question. While his discovery of deuterium momentarily deterred him from this research, in 1934 one of Urey's graduate students, Samuel H. Manian, had completed a doctoral thesis on

[40] Urey, untitled draft autobiography, n.d. [ca. July 1969], and "Personal History/Family History [of Harold C. Urey]", both Box 1, HCU; Aaserud (1990, 189–91), Craig et al. (1964, v–xvii), Rhodes (1986, 380–1), Servos (1990, 315–21), and Tatarewicz (1990b).

[41] Urey (1952b, ix); see also Urey to L. Pauling, Jan. 8, 1951, copy in Box 9, LAD, and idem, untitled draft autobiography, p. 7, n.d. [ca. July 1969], Box 1, HCU. On Saha's theory and stellar astrophysics, see DeVorkin and Kenat (1983a, 126–7) and Urey (1924, 2); on Lewis, see Lewis (1922, 309–311; 1934, 897–8).

this topic. Later that year, Manian, Urey, and Walker Bleakney, a chemical physicist at Princeton, had published a comprehensive review of this problem, claiming that the oxygen isotope ratios in meteorites and Earth matched within the experimental limits of their procedure. Urey's interest in pursuing such problems partly arose from his eagerness to expand the fields covered within Columbia's Department of Chemistry; in 1936, for example, he had alerted Linus Pauling that his empirically oriented department now needed someone "principally interested in a ton of paper and a gross of pencils as his research facilities."[42] It also derived from his role as editor of the *Journal of Chemical Physics*, founded in 1933 to accommodate research on the border between chemistry and physics.

Urey after 1945 continued to interest himself in applications of isotope chemistry to problems of Earth. Although his postwar efforts with Pettijohn, Norman L. Bowen, and Carl-Gustav Rossby to establish a new "Institute for Geophysics" at Chicago ultimately failed, Urey saw the earth sciences as a promising way to distance himself from his wartime research in nuclear chemistry.[43] Between 1946 and 1949, with funding from the Geological Society of America, Urey studied the $O^{16}:O^{18}$ ratio frozen in the calcite skeletons of ammonites and belemnites, extinct cephalopods (squid-type organisms) abundant in shallow oceans roughly 125 million years ago; these measurements provided a means to estimate temperatures within Earth's outer skein, during the Cretaceous period. The value of this work for Urey was that it permitted precise measurements in a field where few quantitative studies had previously been made. Writing the Harvard astronomer Menzel early in 1949, he declared that applying the methods of the exact physical sciences to the "more qualitative sciences" of geology and zoology would lead to "interesting and startling results." Seen from this perspective, Urey's commitment to the earth sciences and astrophysics followed a consistent pattern through much of his career. His discovery of deuterium was simply one of the most fruitful and far-reaching detours from this road.[44]

Urey's interest in the thermal and chemical history of Earth – the subject to which he turned in 1949 after his Paleozoic temperatures program was well underway – arose through discussions with Brown about how meteorite abundances were related to the structure of Earth, as well as of his own work on the history of Earth

[42] Quoted from Urey to Pauling, July 28, 1936, Box 35, LP; see also Urey to Shapley, July 8, 1932, Box 52, DHCOs; idem, untitled draft autobiography, pp. 11–12, n.d. [ca. July 1969], Box 1, HCU; as well as Brickwedde (1982, 35–6), McBain (1931), Manian, Urey, and Bleakney (1934, 2609), Paneth (1940, 10), and Servos (1990, 319).

[43] E. C. Colwell to W. Bartky, Oct. 28, 1946, and L. Kimpton to Colwell, Apr. 25, 1947, Box 32a, both UChi.

[44] Quoted from Urey to D. H. Menzel, Feb. 3, 1949, Box 57; see also Urey to H. Aldrich, Sept. 29, 1947, Box 2, idem to S. Allison, June 14, 1949, Box 2, and idem, untitled draft autobiography, p. 22, n.d. [ca. July 1969], Box 1, all HCU; as well as Urey et al. (1950). Urey was far from the first atomic scientist to turn to the biological and geological sciences after World War II; for another example, see Schrödinger (1945).

temperatures. In contrast to Brown, however, Urey was intrigued by the problem of Earth's geochemical history for what it told about the chemical evolution of planets, rather than its implications for theories of stellar nucleogenesis and cosmology. What seemed to stimulate his thinking about the problem was a 1941 review article on the thermal history of Earth by the geophysicist Louis Slichter, which Urey read while preparing his course lectures in 1949. Based upon then-current estimates of radioactive element abundances, Slichter concluded that "[i]t is unknown whether the earth is heating or cooling at depth." Intrigued that so basic a question remained unanswered, and aware through Teller and Brown that new abundance determinations were being made at the institute, Urey began examining this issue. Using new radioactive abundances derived from iron and stony meteorites, likely provided by Brown, Urey calculated in mid-1949 that Earth is warming, rather than cooling. This central idea led him to what soon became known as the cold accretion theory for the origin of Earth. Challenging then-accepted ideas, Urey proposed that the core of Earth had formed gradually through geologic history, not at its origin; furthermore, he argued that Earth had never become molten, the view shared by Goldschmidt and other leading geochemists. Urey publicly aired these ideas later that year, first at the autumn 1949 meeting of the National Academy of Sciences in Rochester, New York, then at the Rancho Santa Fe Conference Concerning the Evolution of the Earth, hastily arranged in January 1950 in part to provide a forum for discussing Urey's ideas.[45]

Urey's cold accretion model, the basis of his extensive planetary studies in the early 1950s, sought to explain Earth's structure and geochemical history, as well as its unique characteristics among the terrestrial planets. His 1949 model assumed that chemical fractionation had occurred in the initial solar nebula, causing lighter elements, including silicate, to condense before denser materials such as iron. Urey found this process necessary to account for the different properties of Earth and the Moon, granting the widely held assumption that both bodies had formed as part of a common system. In geochemical terms, Urey found the Moon difficult to comprehend. One puzzle was its low density in contrast to that of Earth. Another was the Earth-facing lunar bulge deduced by Cambridge mathematical astronomer Harold Jeffreys, which Urey took as evidence that the Moon had remained at a low temperature (thus retaining its internal rigidity) throughout its history. Moreover, Urey was struck by the lack of evidence for folded mountain chains on the Moon. The presence of a body like the Moon in close proximity to Earth seemed to Urey especially puzzling. Although a tidal origin for the Moon might explain the distribution of densities within a common Earth–Moon framework, Urey accepted, as did most geophysicists and astronomers, that Jeffreys's work ruled out this approach. As he later wrote Kuiper, "it seems rather surprising to me that a moon could form

[45] Urey to L. Pauling, Jan. 8, 1951, Box 9, LAD; see also Brush (1981, 93–4), Slichter (1941, 598), and Urey (1952c, ix). As late as 1947, for example, Urey was not regarded as interested in atomic abundances among his Chicago colleagues; see Struve to J. Greenstein, Jan. 30, 1948, Box 6, JLG.

within about eight diameters of the earth and yet have such a markedly different density."[46]

Urey resolved these difficulties by assuming that the Moon was a primitive body (composed of lighter elements present in the early solar nebula), whereas Earth was an object of later origin, formed after chemical fractionation had permitted iron and other heavy elements to condense. Urey saw at least two advantages to this idea. First, while it allowed both Earth and the Moon to develop silicate-rich centers, only the larger, more massive Earth would accumulate heavier materials as the solar nebula's chemical makeup evolved, accounting for the Moon's lesser density. Moreover, Urey believed that the slow migration of iron from Earth's surface to its interior would create a heat engine within the planet, supplementing internal heat from radioactivity. Aware that geophysicists such as David T. Griggs and Felix Vening Meinesz believed that mountains resulted from convection currents in Earth's mantle, Urey argued that the sinking of massive amounts of iron would fuel such convection. As he declared at Rochester, "[t]he forces are adequate to account for the formation of mountains and roughly the time of convection estimated is approximately that required by geological evidence." Urey felt confident about his theory because it predicted not only the general chemical makeup of Earth and the Moon, but the appearance of mountains on Earth and their absence on the Moon.[47]

What attracted many geologists, geochemists, and geophysicists to Urey's ideas was that they promised to shed new light on what had suddenly become a controversial and unsettled subject, the nature of Earth's core. Until the early 1940s, as we have seen, most earth scientists had accepted that an iron core existed, in part because of its utility in explaining seismological studies of earthquake waves, the high overall density of Earth, and the existence of Earth's magnetic field. In the years immediately following the war, however, American geophysicists had become aware of new interpretations that challenged this orthodoxy, based on wartime increases in understanding of the behavior of solids and fluids under extremely high temperatures and pressures. The first had come from the Swiss geochemists Kuhn and Rittmann at the University of Basel. In a *Geologische Rundschau* article in 1941 (but not generally seen by Allied scientists until 1945), Kuhn and Rittmann had calculated that the viscosity of inner Earth was too high to permit the gravitational separation of light from dense materials in a time as short as several billion years, thus preventing the formation of an iron core. Instead, they had argued that underlying a thin solid crust and shallow magmatic zone of molten silicates was a large core consisting of undifferentiated solar matter, rich in hydrogen and helium. Seismic reflections revealed not the core boundary, but rather a phase transition between gaseous

[46] Quoted from Urey to Kuiper, Feb. 2, 1950, Box 51, HCU; see also idem to E. C. Bullard, July 7, 1950, Box 16, HCU, as well as Brown and Patterson (1948, 110–11), Slichter (1950a, 511), and Urey (1949; 1953, 281).

[47] Urey (1949, 446); see also Urey to G. P. Kuiper, Feb. 2, 1950, Box 51, HCU, and Brush (1982b, 891–92).

and liquid states. While Wildt condemned this "radically different hypothesis" for failing to predict the sharpness of seismic wave reflections or the extremely low densities of the giant planets, Kuhn and Rittmann were considered reputable scientists, and many scientists, aware that Kuhn had investigated viscosity since the 1930s, regarded the challenge seriously.[48]

A second, more formidable challenge came from William H. Ramsey, a theoretical physicist at Manchester, in 1948. Unlike Kuhn and Rittmann, Ramsey did not deny the existence of a core. He did, however, argue that the internal composition of the terrestrial planets was essentially homogeneous, and that compression effects, rather than a concentration of iron or nickel at Earth's center, were responsible for the density distribution within Earth. For Ramsey, the advantage of this theory was that it could account for the densities of Mercury, Venus, Mars, and the Moon then standard in astronomical texts, all of which seemed lower than Earth's. No less than the Kuhn–Rittmann hypothesis, Ramsey's theory received serious review; as Ramsey himself noted, rejecting the iron core model "has far-reaching implications, not only in geophysics but in planetary astronomy." Although Brown and Walter Elsasser, then developing his dynamo theory for the origin of Earth's magnetic field, rejected the idea, it gained wary acceptance from Caltech's Beno Gutenberg, whose seismic wave studies were often cited as the best available evidence for an iron core. Even Urey was at first attracted to Ramsey's hypothesis before Teller and Fermi persuaded him that Ramsey's high-pressure physical transformations were improbable.[49]

The clearest evidence of the importance American geophysicists and geochemists attached to these ideas emerged at the Rancho Santa Fe Conference Concerning the Evolution of the Earth, held at Rancho Santa Fe, California, on January 23–5, 1950. Supported by the National Academy of Sciences and the Institute of Geophysics of the University of California, the meeting was one of the first interdisciplinary conferences organized around the theme of planetary structure and evolution. Slichter, who as director of UCLA's new geophysics institute arranged the conference, declared that its specific inspiration was Urey's "stimulating paper" on the formation of Earth's core at the October 1949 Rochester meeting of the academy. It was evident, both from the eminence of individuals invited to the conference and the alacrity with which it was arranged, that uncertainties about Earth's interior were an overriding concern for many of its participants. Among the twenty-four individuals who took part were Brown, Urey, and Pauling (physical chemistry), Gutenberg, Griggs, and Patrick M. Hurley (geophysics), Francis Birch and W. F. Latimer

[48] Kuhn and Rittmann (1941, 252–4), Brush (1982a, 1187), and Wildt (1947, 99–101); another critical review was Brown (1950b, 641, 649). For American reactions to Kuhn–Rittmann, see "Werner Kuhn" file, Box 13, LP, Kuhn to Urey, June 10, 1947, Box 51, B. Gutenberg to Kuhn, Feb. 27, 1947, Box 2, Gutenberg to A. Eucken, Apr. 13, 1948, and W. Elsasser to Gutenberg, Feb. 16, 1950, both Box 1, all BG.

[49] Elsasser to Gutenberg, Mar. 28, 1949, and Feb. 16, 1950, both Box 1, and Gutenberg to Urey, June 12, 1950, Box 4, all BG; Urey to Kuiper, July 19, 1950, Box 51, HCU; see also "Meeting of the Royal Astronomical Society . . ." (1949), Ramsey (1951, 427), and Urey (1951a, 250), as well as Brush (1982a, 1187).

(geochemistry), James Gilluly, Adolph Knopf, and William W. Rubey (geology), and Teller and Howard P. Robertson (physics). Astronomy was represented by Whipple.[50]

Despite its brevity, the Rancho Santa Fe conference came to have enormous influence on planetary research and solar system astronomy in the United States during the 1950s. A striking feature of the conference, which included no prepared papers but rather informal discussions among its invited participants, was the wide range of subjects considered: Topics treated ranged from the origin of Earth and the solar system to the Earth's chemical history, the creation of continents and mountain building, and the evolution of Earth's atmosphere and oceans. Despite their galvanizing effect on American geochemists and geophysicists at large, neither the Kuhn–Rittmann nor Ramsey hypothesis gained adherents at this meeting; instead, all present accepted that an iron core existed at Earth's center. At the same time, conference participants came to favor a central tenant of Urey's model of Earth's formation: that the gravitational separation of heavier and lighter elements through geologic time helped furnish the energy necessary to set up convection in the outer mantle. One factor behind this emerging consensus was the strong commitment of several conference participants, particularly Urey and Griggs, to the idea that convection played an important role in both continent building and mountain formation. Although Griggs's ideas on orogeny ran counter to traditional assumptions that mountains were caused by the shrinking of a cooling Earth, conference members saw several lines of evidence favoring this approach. One was the considerable research by Norman L. Bowen, the venerated American geochemist then at Chicago, that indicated the basaltic composition of the upper mantle was consistent with convective processes. Another was evidence that orogenic activity on the ancient Canadian Shield, containing some of the oldest rocks then known, pointed to "an orderly process of growth of the continent," a pattern expected of convective activity. One result of their focus on mantle convection was that conflicts between participants arose, not over the constitution or existence of a core, but over such issues as the degree to which Earth became molten during its formation. Geologists and geochemists at the conference disagreed over the possible range of temperatures involved, but generally came to believe a cool origin for Earth was preferable. Most accepted, more hesitantly, Bowen's argument that if the crust had melted on a worldwide scale, a more complete layering of the crust and mantle would have resulted than what was observed. Nevertheless, the overall harmony of participants on these issues convinced them that convective models were on the right track. In his summary report Slichter quoted a participant that "[o]ne of the striking results of the conference seemed to be

[50] Slichter (1950a, 511–12, 514). Over half of the conference participants in geochemistry and geophysics held university affiliations, indicating the degree to which these fields, still centered in private or governmental research institutions circa 1900, had gained the resources, training facilities, and prestige associated with traditional academic disciplines; see Doel (in press-b).

the emergence of general agreement concerning the ideas presented, and the way these all point to a reasonable hypothesis for the growth of the continents."[51]

Several researchers, inspired by the consensus that took shape at the meeting, began research programs designed to find evidence for these views. Rubey, then a senior scientist at the U.S. Geological Survey, expanded his investigations of the geochemical evolution of Earth's atmosphere and oceans. Hugo Benioff, a Caltech geophysicist intrigued by Urey's theory, began studying fault movements along the Pacific Ocean margin to find possible evidence of Earth's continued warming. At Urey's urging, Pauling and Teller produced new theoretical calculations of the compressibility of silicates at high pressures that challenged Ramsey's compressed silicate model. Elsasser, in close communication with Urey, worked with renewed vigor on his dynamo theory of Earth's magnetic field. Even Harvard's Francis Birch, least enthusiastic among conference attendees about the existence of mantle convection currents, undertook new efforts to assess the composition of Earth's mantle; by the end of the decade he would become the leading American opponent of this concept.[52] Writing Urey, Gilluly declared that the Rancho Santa Fe meeting had been "one of the most noteworthy events of my experience in scientific work," their new consensus "truly surprising to us all."[53]

Most stimulated by the meeting was Urey himself. At first he seemed discouraged by the reception of his ideas. His proposal that an intact silicate core had been displaced by pooling concentrations of iron, he wrote in February 1950, had been rejected by geologists as not "having anything to do with past geological history." Moreover, he worried that his chemical explanations were insufficiently supported by astrophysical evidence; he admitted to Kuiper that he had first considered such methods as turbulent eddies and the effects of winds on small particles, and had discarded them "only because I thought the chemical fractionation looked so much better. I do not know any reason why that might not be the mechanism. Perhaps I forgot my role of the humble student and confused myself with the Creator instead."[54] But Urey soon became convinced that his chemical fractionation model was correct. He was encouraged by learning through Whipple that most astronomers believed Mars had neither high mountains nor volcanoes, regarding this as further evidence that both Mars and the Moon were chemically homogeneous "fossil" planets, bereft of the core-forming activity that powered continent growth and

[51] Quoted from Slichter (1950a, 514); on Griggs, see LeGrand (1988, 116–17).

[52] R. Wildt to F. A. Paneth, Sept. 20, 1951, Box 5, RW; Benioff to Urey, Apr. 21, 1950, Box 13, Urey to Teller, July 19, 1950, and Teller to Urey, July 21, 1950, Box 90, all HCU; Kuiper to Slichter, Dec. 15, 1950, Box 31, GPK; Elsasser to B. Gutenberg, Feb. 7, 1951, Box 1, BG; and John Verhoogen OHI (Ronald E. Doel, Nov. 6, 1990, at AIP, pp. 6–7). See also Birch (1951; 1954), Brown (1950b, 647), Elsasser (1963, 6), Rubey (1951; 1955), Slichter (1950a, 513).

[53] Quoted from J. Gilluly to Urey, Feb. 14, 1950, Box 38, HCU; on the conference and its perceived significance, see Gutenberg to L. Kober, Feb. 27, 1950, Box 2, BG; and Leon Knopoff OHI (Ronald E. Doel, Apr. 27, 1990, at AIP), as well as Slichter (1950a, 511–14; 1950b) and Urey (1951a, 249).

[54] Quoted from Urey to Kuiper, Feb. 2, 1950, Box 31, GPK, and idem to Kuiper, Oct. 23, 1951, Box 51, HCU.

mountain building on Earth. These two ideas – that silicates had condensed before iron in a low-temperature solar nebula, and that the sinking of iron in larger terrestrial planets caused mantle convection and hence orogeny and continent formation – became the basis of his numerous articles in the early 1950s as well as his 1952 monograph *The Planets,* the first study of cosmogony and planetary evolution written from a chemical rather than an astronomical perspective.[55]

To gain additional evidence for his model of planetary evolution, Urey established contacts with a number of leading geologists, geophysicists, and astronomers, thereby opening important new lines of communication among these disciplines. For instance, in 1952 Urey began an exchange with Hess over the implications of an unusual metallic rock discovered on Disko Island in Greenland. Since the early 1930s, Hess had been interested in the chemistry of the mantle as a means of understanding geotectonics, and in 1941 had helped set up a cooperative program between the Carnegie Institution of Washington and the Department of Geology at Princeton to study the radioactive content of certain mantle rocks. Like Urey, Hess was interested in the role of convection in fueling geological and geochemical processes, although his primary focus remained radioactive heating and its influence on mantle geochemistry; it was for this reason that Hess was interested in the radioactivity of the Disko Island basalts. Urey, by contrast, saw the existence of this nickel–iron body high in the mantle as evidence for convective currents, and found additional support for this idea from Vening-Meinesz in 1953. Although Urey did not launch cooperative projects with either researcher, he became the center of an informal network of scientists interested in planetary evolution. He urged Hess, for instance, to spend time with Kuiper, a connection that reinforced Hess's interest in ocean floor topography and global geotectonics, leading to his proposed theory of seafloor spreading in 1960.[56]

Urey also initiated new contacts with American astronomers. He sought from them evidence of two kinds: that the surface of the Moon indeed showed little sign of geologic evolution (as predicted by his convective theory), and that the diameters and densities of other terrestrial planets conformed to his model of planetary evolution. In 1950, Urey began using the 40-inch refracting telescope at Chicago's Yerkes Observatory for visual inspections of the lunar surface. These studies, together with Ralph B. Baldwin's investigation of the Moon (see Chapter 5), convinced him that

55 Urey to F. Birch, Nov. 8, 1950, Box 13, HCU; idem to Kuiper, Feb. 2, 1950, and idem to Slichter, Jan. 26 and Nov. 10, 1950, all Box 23, HCU; idem to Gutenberg, June 7, 1950, Box 4, BG; Urey (1949, 446; 1950; 1951a, 209, 212–13, 225, 247, 275; 1951b; 1952a; 1952c, 49–56, 142–7). For Urey on Rancho Santa Fe, see Urey (1951a, 275).

56 Urey to Hess, Dec. 31, 1952, Box 42; Urey, "Disko Island Basalts," in "Conference on the Abundance of the Elements," Nov. 6–8, 1952, MS report, n.d. [1952], Box 107; Urey to Vening Meinesz, May 4, 1953, and Vening Meinesz to Urey, June 5, 1953, both Box 57, and Urey to Chandrasekhar, Aug. 25, 1953, Box 18, all HCU; see also Davis (1950, 107–9) and Davis and Hess (1949). On Hess's early career, see Holland (1974) and Shagan (1972).

the lunar landforms were caused by meteorite impact rather than volcanic explosions. He also solicited from Kuiper, Brouwer, and other astronomers new estimates of the diameters of Mercury and Mars in order to calculate their likely chemical constitution; through Brouwer, Urey learned that Cincinnati astronomer Eugene Rabe's revised mass for Mercury no longer supported Ramsey's core model. At the same time, Urey sought advice from astronomers about physical conditions in the early solar nebula, a topic (as we shall see in Chapter 4) to which a number of astronomers and astrophysicists turned in the early postwar period. By mid 1950, Urey came to accept Kuiper's cosmogonal model as a "most convincing analysis" of the system's origin, and used it as the starting point for his own theory of chemical fractionation and planetary formation. By 1953, following new research into the problem of thermal convection by Chandrasekhar, Urey became even more convinced that convection in the mantle caused mountains to develop on Earth. He also welcomed the opinion of Lick astronomer Robert J. Trumpler that visual and photographic studies had ruled out tall mountains on Mars.[57]

For Urey, the most persuasive evidence in favor of his theory of planetary evolution came not through geological or astronomical investigations, but rather via his own geochemical studies of meteorite abundances. Beginning in 1951, when Brown left Chicago for Caltech, Urey launched a program for determining natural abundances using meteorite samples. His project differed from Brown's in two key respects. First, Urey was interested in atomic abundances principally to confirm a low-temperature origin for the planets and the solar nebulae; Brown, by contrast, had initially sought such values to test theories of stellar nucleogenesis. Second, instead of using nuclear reactors at Argonne to make these measurements, Urey, for several reasons, chose mass spectrometers. He had worked with these throughout much of his professional career, including on the Manhattan Project; he had grown even more familiar with them after initiating his postwar Paleozoic program, for which he had helped develop new models more sensitive and durable than their wartime predecessors. As Urey advised Chicago administrator Walter Bartky, he regarded his new mass spectrometers as ten times more accurate than those previously available. In addition, with support from the Geological Society of America and the Office of Naval Research, Urey had built several mass spectrometers for use by his graduate students at Chicago. His regular access to these instruments, as well as the new levels of sensitivity they obtained, made it possible for him to study a wider

[57] Urey to Brouwer, July 7, 1950, and Brouwer to Urey, July 23, 1951, both Box 15, HCU; Urey to Trumpler, Aug. 14, 1950, and Apr. 18, 1952, and Trumpler to Urey, Sept. 25, 1950, all Box 90, HCU; Urey to Whipple, May 2, 1952, Box 102, and idem to Hess, May 4, 1953, Box 42, both HCU; Kuiper to Urey, Sept. 15, 1950, Box 18, and idem to H. P. Berlage, Feb. 8, 1952, Box 30, both GPK; Whipple to Urey, Dec. 3, 1951, Box 3, FLW–HU; see also Urey (1951a, 211, 215–16, 271; 1951b, 3; 1952b, 228–9). In 1971 the U.S. *Mariner 9* spacecraft discovered immense volcanic mountains on Mars, overturning this long-held view. Despite overlapping interests, Urey had limited contact with Rupert Wildt, partly because of personal frictions; see Wildt to F. A. Paneth, Sept. 20, 1951, Box 5, idem to Ramsey, May 21, 1951, and idem to Urey, Apr. 21, 1953, all Box 1, RW.

range of natural abundances than Brown had been able to do just several years before.[58]

Urey's research soon convinced him that meteorites had originated in a low-temperature environment, and experienced little internal heating either as parts of larger bodies or as independent planetesimals, as Brown had argued. His discovery of iron sulfide in meteorites, for example, persuaded him that the exploded planet hypothesis for asteroids was incorrect. Since iron sulfide is unstable in the presence of cosmic proportions of hydrogen and sulfur at temperatures over 600° K, the presence of this compound within iron meteorites seemed significant. Urey was also gratified to find that the ratio of iron to silicate differed among chondritic meteorites. His graduate student, Harmon Craig, inferred in 1952 the existence of two distinct groups of chondrites. This finding contradicted earlier reports of greater chemical uniformity, and suggested to them that at least two "parent" asteroidal bodies once existed. The discovery of the Urey–Craig groups, later regarded as the starting point for modern meteorite classifications, reinforced Urey's faith in the view of Rancho Santa Fe participants that Earth showed no evidence of global melting. He used these two findings to argue against a competing high-temperature theory for the origin of the solar system championed by German geochemist Arnold Eucken. Urey's arguments became even more widely known when Whipple wrote a laudatory review of *The Planets* in *Sky and Telescope,* announcing his own support of Urey's geochemical orientation.[59]

At the same time, however, Urey became more and more uneasy over the accuracy of atomic abundance estimates in general, including modern estimates published by Goldschmidt and Brown. Since the absolute abundances of radioactive elements such as uranium and thorium were essential for calculating the degree of radioactive melting possible within planetary bodies, Urey found that uncertainties in these data threatened his program of research. Moreover, since scientists regarded absolute abundances of any one element interrelated with the abundances of all others, the matter interested physicists, astrophysicists, and others who used the abundances of particular elements as evidence for or against theories of stellar nucleogenesis, terrestrial and stellar age determinations, and the cosmological origin of matter.[60]

Urey's doubts about the accuracy of existing meteorite abundances was heightened by a number of factors. In October 1951 Urey wrote Caltech's Jesse L. Green-

[58] Urey to D. White, Mar. 23, 1948, Box 2, and idem to Bartky, May 5, 1949, Box 12, HCU; H. Aldrich to Urey, July 22, 1953, and Urey to G. Randers, Oct. 29, 1954, and idem to S. K. Allison, June 14, 1949, all Box 2, HCU; Urey, [Budget proposal, re. Development of High Precision Mass Spectrometers], accompanying W. B. Harrell to Atomic Energy Commission, Apr. 14, 1949, Box 1, HCU; see also Craig et al. (1964, xvi).

[59] Harmon Craig OHI (Ronald E. Doel, Nov. 27, 1990, AIP); see also Craig (1953), Patat (1944), Urey (1951a, 274; 1952c, 110–12), Urey and Craig (1953), and Whipple (1952c), as well as Burke (1986, 286).

[60] Brown to Urey, June 10, 1951, Box 15, and Urey to Ramsey, Jan. 16, 1953, both Box 77, HCU.

stein, a noted authority on atomic abundances found in various stellar classifications, to inquire about a recent boast by Kuiper that many astronomical abundances were known to an accuracy of about a tenth of a logarithm. Urey felt "exceedingly skeptical" about such claims of precision, but wanted the views of another astronomer on the matter. In his reply, Greenstein expressed surprise at Kuiper's claim, noting that a more reasonable figure for him was a log of ±0.3, or a factor of 2. He went on to express doubt about the prospects for obtaining absolute abundances for any stars other than the Sun, adding that, for certain important elements, "disagreements by about 0.6 in the log are quite common." Urey had already heard similar doubts regarding the accuracy of heavier-element abundances determined from meteorites from Brown, who advised him that many older measurements for rare elements were accurate to an "order of magnitude only in most cases and frequently not even that." Doubts about terrestrial abundance estimates inclined E. C. Bullard, Keith Bullen, and other eminent geophysicists to accept Ramsey's compressed silicate core model as possibly correct, and Urey found that the uranium, potassium, and thorium abundances he had employed in his cold accretion theory now seemed too low to sustain levels of convection required by Elsasser's dynamo model of the magnetic field. Finally, Urey heard from Russell, the veteran researcher of astronomical abundances, who cast doubt upon the entire postwar program of abundance determinations. "I am more than ever unwilling to accept the meteorites – an extreme instance of segregation – as cosmically typical," he wrote Urey, "especially as marked chemical segregation is found in the stony and iron phases." Like Brown, Russell joined other researchers in agreeing that a review of existing abundance data was needed to achieve a new consensus on atomic abundances.[61]

Seeking to address these difficulties, and to create a new forum for chemists, physicists, and astrophysicists to discuss the chemical evolution of planets and stars, Urey organized what became known as the Williams Bay Conference on the Abundance of the Elements. Held in Williams Bay, Wisconsin, on November 6–8, 1952, and funded by the National Science Foundation, the conference brought together fifty-four astronomers, physicists, geologists, geochemists, and geophysicists from over a dozen institutions in North America. Coorganizers with Urey were Rubey, Nier, and Struve, who had unsuccessfully sought AEC support for such a meeting three years earlier. The conference was organized around three main themes: astronomical problems and techniques of measurement, chemical and geological problems related to the abundance of elements, and the origin of the elements. As at Rancho Santa Fe two years before, these issues attracted distinguished contributors and guests from the upper echelon of American science, among them Maria Mayer, Teller, Hess,

[61] Quoted from Russell to Urey, June 23, 1952, Box 28, HNR; see Urey to Greenstein, Oct. 11, 1951, Greenstein to Urey, Oct. 16, 1951, and L. Goldberg to Urey, Jan. 11, 1952, all Box 39, HCU; Menzel to Urey, Feb. 26, 1951, Box 57, HCU; and Urey to Bullard, July 7, 1950, idem to Bullen, Mar. 14 and July 9, 1952, both Box 16, HCU.

Birch, Brown, and senior astrophysicists Chandrasekhar, Wildt, Kuiper, Leo Goldberg, and Martin Schwarzschild.[62]

Urey intended the Williams Bay conference to achieve several aims. One was to restore greater consensus and order to the critical field of abundances. Increasingly convinced that Brown's postwar abundance determinations were flawed, and that Brown's 1949 attempt to revise Goldschmidt's classic *Geochemische Verteilungsgesetze der Elemente* had been misguided, Urey used the meeting to announce new abundance determinations he had obtained jointly with Hans Suess. An Austrian-born physicist in the German uranium project during World War II who had become a research associate at Chicago in 1950, Suess computed new abundance tables based on theoretical constraints and on new meteorite abundances measured in Urey's laboratory. Another aim was for Williams Bay participants to continue the dialogue about planetary evolution begun at Rancho Santa Fe. Much discussion at this 1952 meeting indeed focused on implications of abundance research for cosmogony and planetary evolution, including Rubey's study of continental structure and the physical significance of the Urey–Craig groups.[63]

At the same time, Urey intended the conference to accomplish a more practical aim: to remind museum curators and administrators that advanced analytical techniques to study meteorite abundances now rested in the hands of atomic scientists, not academic geochemists. Certain that new meteorite values were inaccurate in part because museum directors offered only "a little nondescript chunk" of their vast collections to academic geochemists, Urey voiced anger at the British Museum's M. N. Hey and others. If inaccuracies persisted, he irritably declared, physicists and astronomers might conclude "that all attempts to derive abundances from meteorites are so much eyewash that they can ignore them completely." While Hey ignored Urey's carping, the Williams Bay conference illuminated the degree to which meteorite abundance studies had shifted from museums to academic departments after World War II, where Manhattan Project veterans enjoyed access to costly analytical techniques. Researchers at the Smithsonian Institution's National Museum (later Natural History), by contrast, faced limited staff, obsolete equipment, and tight budgets after 1945. Despite possessing the largest collection of meteorites in North America, this facility had lost its pre–World War II status as a leading pro-

[62] Belgian astronomer Pol Swings, then visiting the United States, was also present; see "Attendance at Williams Bay Conference, November 6–8, 1952," and "Conference on the Abundance of the Elements," n.d. [ca. Nov. 1952], pp. 2, 3–4, 6, both Box 24, HCU. For Struve's views, see, e.g., Struve to I. S. Bowen, Jan. 29, 1949, Box 14, ISB; idem to Urey, Aug. 1, 1952, Box 16, HCU. Geophysicists making similar statements include Brown and Goldberg (1949, 347) and Elsasser (1950, 1). Struve, offered telescope time at Mount Wilson during the conference, did not attend; see Struve to Urey, Oct. 6, 1952, Box 16, HCU.

[63] "Conference on the Abundance of the Elements" file, ca. Nov. 1952, and Urey to Seeger, July 2, 1952, both Box 107, HCU; Urey to Whipple, Oct. 17, 1950, idem to Kuiper, Nov. 26, 1951, June 9 and June 13, 1952, all Box 51, HCU; idem to B. Gutenberg, Jan. 17, 1951, Box 4, BG; see also Brush (1981, 93), Suess and Urey (1956), and Urey (1952b, 268–9). Suess's wartime work is traced in Walker (1989, 141–3, 146).

ducer of meteorite abundance estimates. Significantly, no museum-based scientist spoke at Williams Bay.[64]

In several respects the Williams Bay conference fulfilled Urey's ambitions. It stimulated additional interdisciplinary research between astronomers and geochemists as the Rancho Santa Fe meeting had two years before – and indeed like the 1947 Planetary Atmospheres symposium at Yerkes – underscoring the contributions of Chicago researchers to this field, and the importance of these transient institutions. Urey worked even more closely with Hess to study geological evidence of convection after this gathering, while Kuiper began extended exchanges with Suess as well as with Hess. Intrigued by the astronomical and chemical implications of Urey and Craig's evidence for meteorite families, Whipple and Hess helped coorganize a symposium on the origin of meteorites in 1953. Held at the American Association for the Advancement of Science meeting, it featuring twenty papers from an international field of speakers.[65]

It is also clear, however, that Urey's ultimate aim for the Williams Bay meeting, to reach consensus on natural abundance curves, was not met. In contrast to the strong consensus that emerged at the Rancho Santa Fe meeting, results from Williams Bay were rarely mentioned in subsequent letters from Kuiper and Urey, or in papers published by attending scientists. Indeed, the problem of atomic abundances seemed ill resolved to many participants. An observer at the 1953 AAAS meteorite conference revealingly noted that this meeting "brought out the uncertainties and even reversals of opinion in the minds of many authorities on the origin and age of meteorites." Continued disagreements among geochemists over the absolute abundances of radioactive elements (and hence the resultant melting of planetary interiors) hindered efforts to build an "astro-chemical" approach to planetary evolution. For Urey and Kuiper, as we shall see in Chapter 4, disagreements over this issue proved fatal to their efforts to create an interdisciplinary home for studies bridging their disciplines.[66]

This failure should not overshadow the importance of the geochemical programs that Brown and Urey launched at Chicago in the years following World War II. The Institute for Nuclear Studies, because of its concentration of researchers, resources, and instruments, as well as its success in securing patrons, was a singularly influential center for planetary and meteorite geochemistry throughout the early 1950s. As former members and their students diffused to other institutions across the United States, the influence of the Chicago school of nuclear geochemistry similarly

[64] Urey to Seeger, July 2, 1952, Box 107, idem to H. S. Ladd, Feb. 27, 1952, Box 51, idem to Brown, May 6, 1958, Box 15, and idem to M. N. Hey, Jan. 24, 1958, Box 42, all HCU; Hey to Whipple, July 20, 1959, Box 8, FLW–SI. On meteorite research at the Smithsonian Institution, see Yochelson (1985, 350).

[65] Bauer (1954).

[66] Urey to S. Chandrasekhar, Nov. 25, 1952, Box 18, and idem to Hess, Dec. 12, 1952, and May 13, 1953, Box 42, all HCU; see also Greenstein to NSF Astronomy Panel members, Dec. 13, 1953, Box 28, GPK. No recollections of Williams Bay discussions appear in letters between Kuiper and Urey, or among Urey, Struve, and Rubey; an exception is Urey to W. H. Ramsey, Jan. 16, 1953, Box 77, HCU.

expanded. In 1951, for instance, Brown and Patterson left Chicago to build a program of planetary geochemistry at Caltech. The swift development of this new Caltech venture illustrates how planetary geochemistry continued to influence solar system studies and American astronomy in general. It also illustrates one way in which nuclear geochemistry took root within academic departments of geology in the United States, a development no less important for the earth sciences than for the growth of geophysics.

WESTWARD SHIFT: DETERMINING THE AGE OF EARTH

Between 1952 and 1955, Brown and Patterson both returned to an old and difficult problem: determining the age of the solar system. This project was one of several they initiated in the new geochemical laboratory of Caltech, which Brown had fostered within its Division of Geological Sciences after his appointment to its faculty in 1951. Their work was significant for several reasons. Astronomers and geochemists both saw Earth's age as a critically important datum. For astronomers, determining the age of the solar system promised to put the evolution of the universe itself in clearer perspective. For geochemists, it offered a means to calibrate the rate of geochemical processes, including the effect of radioactive heating on chemical evolution in planetary interiors. Brown and Patterson's work, however, also illuminates important developments in American academic geology. While their effort to expand geochemistry was uniquely influenced by Caltech's connection with Mount Wilson–Palomar, Brown's new geochemical laboratory illustrated a trend toward quantitative techniques within departments of geology after World War II. While Caltech was among the first centers of academic geology to emphasize planetary geochemistry in the early 1950s, others, including Yale and Wisconsin, soon followed Caltech's example in introducing mass spectrometers and nuclear chemistry as research tools. The development of Brown's research program thus casts light on an important transition in the earth sciences during the 1950s, paralleling the tremendous expansion of academic geophysics stimulated by cold-war anxieties and the International Geophysical Year.[67]

By the early 1950s, even more than in 1946, many astronomers regarded the age of Earth as a significant problem. Their concern rose as geochemists declared with increased confidence that Earth was indeed at least 2–3 billion years old, a figure that cast doubt on the age of the universe indicated by the Hubble constant (the rate of expansion of the universe as a whole). Refined Hubble constant estimates yielded an

[67] Doel (in press-b). By "geochemistry" I mean its post-1945 usage as the practice of mass spectrometry by individuals skilled in nuclear chemistry, not its older embrace of crystallography, geomorphology, and mineralogy; see D. Jerome Fisher (1963, 52–63), Pettijohn (1984, 194–201), and Skinner and Narenda (1985, 371); for a pre-1945 definition, see Servos (1983, 150).

age for the universe of just 1.8 billion years, ending the rough overlap of estimated ages for Earth and universe that geochemists and astronomers had accepted since the early 1930s. Because the Hubble constant was critical to studies of stellar luminosity, as well as of stellar and galactic evolution, geologic evidence suggesting that this constant was in error disturbed astronomers. At Cambridge, the mathematical physicist Hermann Bondi seemed ready to blame the geochemists, arguing that statistical evaluations of isotope ratios used to find Earth's age "are a matter of some controversy." He nevertheless admitted that the lower bound for this value exceeded that of the Hubble constant "possibly by a substantial margin." Bondi called for a redetermination of Hubble's constant and new efforts to determine the age of terrestrial rocks and meteorites.[68]

The age of the universe particularly concerned astronomers at Mount Wilson–Palomar. One reason was that these jointly managed facilities were identified with the discovery of the Hubble constant. Edwin P. Hubble had first determined its value using Mount Wilson's 100-inch Hooker telescope in 1929. A fundamental intellectual achievement that provided the framework for modern cosmology, the Hubble constant was also indelibly linked to the unparalleled Mount Wilson–Palomar telescopes, for Hubble, his student Allan Sandage, and other staff subsequently used the Palomar 200-inch to refine it. Astronomers at Mount Wilson–Palomar were also interested in atomic abundances for another reason, however. In 1948, seeking to build an astronomy department at the California Institute of Technology, Caltech leaders had hired Greenstein from Yerkes to develop a graduate program. Long fascinated with atomic abundances and their implications for stellar astrophysics, Greenstein sought to expand abundance studies, a plan fellow Caltech researchers supported. Already in 1946, the physicist William Fowler joined Ira S. Bowen – about to become Mount Wilson–Palomar director – in leading informal seminars on "nuclear problems in astrophysics and astronomy" at Bowen's Pasadena home. Moreover, these researchers claimed that natural abundance work was critical for cosmology. Expanding cosmologies (later called the Big Bang) predicted slightly different curves of natural abundances than the "steady state" model developed by Bondi, Fred Hoyle, and Thomas Gold in 1948, spotlighting abundance studies.[69]

The possibility of appointing Brown to Caltech, first mentioned in 1951, thus appealed to Greenstein, who had become aware of Brown's "astro-geochemical" work while at Chicago. By February of that year, Greenstein began lobbying members of Caltech's geological community, including Gutenberg, on Brown's behalf. The opportunity also appealed to Caltech's new president, Lee A. DuBridge. A former Rochester physicist who had led MIT's Radiation Laboratory during World War II, DuBridge belonged to the generation of American researchers who had partici-

[68] Bondi (1952, 40, 50–2, 140, 165).
[69] Struve to Greenstein, Jan. 30, 1949, Box 6, JLG; see also Bondi (1952) and Fowler (1956).

pated in interdisciplinary research programs in the 1930s and had recognized the interest of the Rockefeller Foundation in pursuing such work. In 1946, after becoming Caltech's third president, DuBridge moved to strengthen its science departments, in part by encouraging new cooperative research initiatives. DuBridge – like Greenstein, Fowler, and Bowen – believed that an interdisciplinary program in atomic abundances promised to facilitate interactions between Caltech physicists and astrophysicists. They also hoped this research could resolve the discordant ages for Earth and the universe, for Mount Wilson–Palomar astronomers a difficult and increasingly embarrassing issue.[70]

Yet the decision to expand geochemistry at Caltech was not imposed from above; it came, rather, from Caltech geologists themselves. The rapidity with which members of Caltech's Division of the Geological Sciences came to favor such a program reflected both the changing focus of geology in the United States in the postwar period, as well as the advantages that academic geologists believed geochemistry would bring. In 1946, for example, John P. Buwalda, the eminent paleontologist and division chair, had proclaimed as his department's most pressing need additional people in mineralogy, petrology, and economic geology. By early 1951, however, following the 65-year-old Buwalda's retirement and the death of Chester Stock, a senior departmental faculty member and internationally recognized paleontologist, attitudes shifted. Ian Campbell, a Harvard-trained economic geologist who became the division's acting chair in November 1950, argued that "both now and in the future the greatest advancement in the geological sciences is to be expected from emphasis on origins, on processes and interpretations." Campbell identified geochemistry as an especially promising field, since new concepts and techniques in physics and chemistry were applicable to fundamental geological problems. In May 1951, he supported Brown as the leading candidate to develop such a program, and welcomed Brown's plan to use mass spectrometry "for geochronology and the study of the distribution of trace elements in matter."[71]

Campbell's decision to make geochemistry a major field at Caltech was based on two distinct but interrelated factors. One was his awareness that isotopic analysis – particularly the age and temperature measurements developed by Libby, Mayer, Urey, Brown, Patterson, and others at Chicago – were judged far superior to those made by fossil identifications. Better still, they could be used for older Precambrian rocks in which no fossils appeared. Already by the 1930s, most American geologists had accepted that the Precambrian era comprised about 95 percent of all geologic time, leaving them keenly interested in quantitative techniques that promised to

[70] Urey to Greenstein, Oct. 11, 1951, and Greenstein to Urey, Oct. 16, 1951, both Box 39, HCU; Greenstein to Gutenberg, Feb. 13, 1951, Box 5, BG; see also Brown (1949, 632) and Greenstein (1986, 18). On DuBridge, see Goodstein (1991, 265–78) and Kohler (1991, 366, 384–5).

[71] Quoted from Campbell, memo, staff of the Division of the Geological Sciences to President DuBridge, Jan. 27, 1951, and idem to J. Gilluly, May 5, 1951, both Box 12, LAD; see John P. Buwalda, "Aims and Objectives of the Division of the Geological Sciences," Dec. 1946; H. C. Urey to L. Pauling, Jan. 8, 1951; and Brown to Campbell, June 1, 1951, all Box 12, LAD.

yield dates for these ancient rock sequences.[72] Moreover, Campbell was aware that Buwalda's retirement and Stock's death eliminated Caltech expertise in paleontology. Although paleontology had been a chief means of determining geological ages for Cambrian and younger rocks, Campbell soon learned from Rubey and Gilluly that senior vertebrate paleontologists of Stock's caliber were few in number and difficult to recruit. Gilluly frankly advised Campbell "to write off the investment in vertebrate paleontology" in favor of securing a foothold in geochemistry. Establishing geochemistry at Caltech therefore involved a conscious decision to adopt a new analytical tool for determining absolute ages of rocks, as well as stratigraphic relationships among them. Since Brown was known for his geochemical studies at Chicago, and personally known to such faculty members as Pauling and Gutenberg through the Rancho Santa Fe conference, it is not surprising that Brown became the principal candidate to lead this work at Caltech.[73]

A second factor that encouraged Campbell and his colleagues to embrace geochemistry was that mass spectroscopy promised support from numerous private patrons, potentially allowing department members to strengthen contacts with institute colleagues in physics, astronomy, chemistry, and biology. Because academic geology had benefited little from the growth of foundation support for physical and biological sciences in the interwar years or from the dramatic expansion of federal science funds following World War II, geologists at many universities, including Caltech, remained closely tied to mining and petroleum firms for financial support. Compared to the physical and biological sciences, where nearly 20 percent of American graduate students received federal funds, only 5 percent of geology graduate students enjoyed government support in the early 1950s. Caltech geologists thus looked favorably at new private patrons. In 1952, aware of the increasing importance of geochemistry to petroleum firms (and their unwavering demand for Caltech geology graduates), division staff initiated an "Industrial Liaison Program in the Earth Sciences," emphasizing to prospective oil company sponsors the value of nuclear chemistry to their industry. Backed by Brown, the plan was designed to secure continued and stable funding for geochemical research at Caltech.[74]

At the same time, however, Brown and Greenstein recognized that the Atomic Energy Commission had emerged as a major patron of natural abundance research. Through Struve – who had sought such funds for astronomical abundance work at Chicago – Greenstein knew that AEC Commissioner David Lilienthal was interest-

[72] Goodman and Evans (1941, 493) and Knopf (1957, 227–8).

[73] Quoted from Gilluly to Campbell, May 14, 1951, Box 12, LAD; see DuBridge to Brown, June 11, 1951, Box 9, and R. Sharp to DuBridge, memo, "Industrial Liaison Program in Earth Sciences," June 16, 1952, Box 12, both LAD. The Yale paleontologist G. Evelyn Hutchinson was nevertheless considered to replace Stock; see Sharp to L. Pauling, Dec. 16, 1952, Box 101, LP. DuBridge later declared that Caltech paleontology "has been abandoned entirely"; see Dubridge to A. Wasem, Jan. 15, 1960, Box 9, LAD.

[74] Brown to Messrs. Pettijohn et al., "Formal Establishment of Training in Geochemistry," May 17, 1950, Box 7, HCU; Sharp to DuBridge, memo, "Industrial Liaison Program in Earth Sciences," June 16, 1952, Box 12, Folder 1, LAD, p. 1; see also Angel (1958, 16), Hubbert, Hendricks, and Thiel (1949), Nace (1958), and Pettijohn (1984, 202).

ed in "astro-geochemical" studies. Soon after arriving at Caltech, Brown secured AEC support to study the geochemistry of uranium isotopes. These funds in part built the lead-free mass spectrometer laboratory at Caltech that he and Patterson employed during 1953–4 to determine the age of Earth. For Campbell and Robert P. Sharp, another Harvard-trained geologist at Caltech, geochemistry not only added rigor to geology but increased their access to the prestige and patronage of physics and chemistry, then among the fastest-growing disciplines at Caltech. "Just as geophysics advantageously links us with physics," Campbell wrote in 1951, "so geochemistry offers the opportunity of closer association with our sister science of chemistry." Backed by DuBridge, eager to increase federal patronage to Caltech geologists, Brown's plans to expand geochemistry thus enjoyed support on several levels. Since this was the kind of interdisciplinary effort that he had failed to initiate at Chicago in 1950, the opportunity to build a major geochemistry program at Caltech likely appealed to Brown as much as it did to his new colleagues in geology.[75]

Although Brown arrived in Pasadena in 1951, work on geochronology and the age of Earth did not begin for more than a year. This delay largely resulted from Brown's increased social activism over nuclear disarmament and population control, concerns he first voiced in 1945. Although he remained attentive to meteorite geochemistry and continued studying uranium geochemistry for the AEC, by the mid-1950s Brown took up new studies of arms control, peaceful applications of atomic energy, and world politics. Precise uranium and lead measurements were instead made by Patterson, who transferred from Chicago to California in 1952 after AEC support for Brown's research was expanded.[76]

Patterson felt convinced that uranium–lead isotope ratios offered the most accurate means for age determination. Moreover, the lead contamination problem that had thwarted attempts to employ this method at Chicago were eliminated by Brown's success in creating the Caltech lead-free laboratory. As a result, he assembled a team of scientists to investigate lead isotope concentrations within meteorites and in Earth's upper crust. In a 1953 paper coauthored with his former graduate student Goldberg and Mark G. Inghram, a young physics professor at Chicago, Patterson attempted to determine the average amount of lead in seawater and ocean floor clays, sources considered stable through geologic time. After finding mean averages of isotopic abundances within these environments, Patterson compared them with new lead abundances derived from meteorites. In a second 1953 article, Patterson concluded that the age of Earth appeared to be at least 4 billion years, nearly twice the 2–3-billion-year estimates then accepted. He further noted that nearly all of Earth's uranium seemed concentrated in the outer crust, arguing that it had been established there when Earth formed. This was important not only for understand-

[75] Brown to Campbell, June 1, 1951, Box 12, LAD; Sharp to Messrs. Allen, Brown, et al., "Excerpts from Division Annual Reports," Aug. 17, 1966, CITG; see also Greenstein (1986, 22).

[76] Brown to DuBridge, Jan. 14, 1952, Box 9, LAD; see also Patterson, Tilton, and Inghram (1955, 75) and Smith et al. (1986, 73).

ing the degree of radioactive heating within Earth but also, as Caltech geologists noted, for the mining of uranium.[77]

Patterson's team announced a more refined estimate of Earth's age at a conference on the Application of Nuclear Processes to Geological Problems, held in Lake Geneva, Wisconsin, in September 1953. At this meeting, Patterson reported that his group's lead isotope ratio measurements yielded an approximate age of 4.5 billion years, if lead studies involving the oceans, Earth's crust, and meteorites were all taken into account. Patterson used the occasion to criticize past, lower estimates of Earth's age derived from other techniques, including isotope studies employing lead ores. He nevertheless showed restraint in announcing this important finding. Emphasizing the limited number of lead measurements made on Earth, Patterson declared "[i]t should be recognized than an approximate age value is sufficient and should be viewed with considerable skepticism until the basic assumptions that are involved in the method of calculation are verified."[78]

Two factors may account for Patterson's surprising caution, given his experience in age measurements. His prudence apparently had less to do with worries about methodology or data than the pressure of working at the interstices of distinct disciplines, heightened by the unusually interdisciplinary environment of Caltech. One was that his upward revision of Earth's age, nearly doubling past estimates, placed the age of the solar system squarely at odds with the age of the universe derived from the Hubble constant. Despite an upward revision in this time scale from 1.8 to 3.6 billion years made by Walter Baade at Mount Wilson–Palomar in 1952, Patterson's new age for Earth still exceeded that for the universe by a billion years. This discrepancy thus posed an institutional dilemma for Caltech scientists, one that resembled the famous age-of-Earth controversy fought between physicists and geologists a half-century earlier. As in that conflict – which pitted Lord Kelvin's confident assertion that Earth's internal heat stores indicated an age of less than a million years against the deduction of geologists and biologists that evolution and erosion were processes requiring many millions of years – the question was which set of evidence and assumptions seemed the more reliable and trustworthy to members of both communities.[79]

A second, perhaps more immediate worry for Patterson was that his new age for Earth conflicted with unannounced age determinations of the solar system by Urey and his Chicago graduate student, Wasserburg. In June 1954, Urey advised Greenstein that studies he and Wasserburg had made of another radioactive decay pair gave an age of 5.8 billion years for the Beardsley meteorite, a newly analyzed stone. The problem was not simply that Urey's age diverged from Patterson's – Patterson

[77] Elsasser (1950, 17) and Patterson, Tilton, and Inghram (1953a,b).

[78] Patterson et al. (1955, 75).

[79] On the Hubble constant, see Harlow Shapley, "The Distance of the Magellanic Clouds," MS, n.d. [1952]; "Complete Official Minutes of the Meeting of Commission 28," IAU Rome meeting, Sept. 4–13, 1952; and Baade to D. H. Menzel, Jan. 29, 1953, all Box 28, GPK; on the earlier age-of-Earth controversy, see Burchfield (1975).

had concentrated on iron phase in meteorites, and Urey on stone, so one could resolve this discrepancy by imagining that irons and stones had formed at different times – but that incongruent ages for these objects contradicted the Urey–Craig theory of asteroid formation.[80] Urey favored this idea because it predicted the common creation of stones and irons. As he advised Greenstein:

> I draw a long breath whenever I see the figure [of 5.8×10^9 years]. We are not too sure of this age, but we are sure that chondritic meteorites are very old . . . this means, together with recent data secured from Patterson and Inghram on the lead abundances, that it is impossible to put together a story for the origin of these objects that leaves the lead ages intact. I do not believe that the 4.5 billion year figure for the age of the solar system has any meaning at the present time.[81]

In 1954, therefore, Patterson's figures opposed not only cosmic time scale favored by his astronomical colleagues at Caltech, but new evidence from his former senior Chicago colleague as well.

By early 1956, however, a resolution was reached. It came about through revisions announced by Caltech astronomers and by Urey himself. In 1956, Sandage joined other Mount Wilson–Palomar colleagues in a comprehensive review of redshift-apparent magnitude determinations made between 1935 and 1955. Although Sandage made no firm recommendation to alter the Hubble constant, the fact that it was discussed at all, within three years of Hubble's 1953 death, signaled to American astronomers that the once firm constant was in doubt. Soon afterward, Sandage announced a new, greatly revised Hubble constant that yielded an age of the universe of 13 billion years, eliminating the conflict with radioisotope determinations of the solar system's age. By then Urey and Wasserburg had already revised their age of the Beardsley meteorite downward to about 4.6 billion years, blaming their former value on uncertainties in the decay constants of the potassium–argon system. Writing Whipple just five months after questioning Patterson's work, Urey now allowed that "I rather think this age [4.5 billion years] is good." After additional analyses of stony meteorites, Patterson himself offered new evidence to justify his lead isotope ages. By then debate over the age of the solar system was over. In 1959, geophysicist John A. Jacobs asserted that the figure of 4,500 million years for Earth's age "is generally accepted and is satisfactory to all branches of science," revealing how uncontroversial this datum had become.[82]

This finding provided the first modern figure for Earth's age, and forged a consensus among astronomers and geochemists that all bodies within the solar system had formed essentially at the same time – both significant results. However, it also

[80] F. Paneth to Urey, Jan. 16, 1954, Box 73, HCU.

[81] Quoted from Urey to Greenstein, June 16, 1954, Box 39, HCU.

[82] Quoted from Urey to Whipple, Nov. 17, 1954, Box 102, HCU, and Jacobs, Russell, and Williams (1959, 7); see Urey, "The Age of the Solar System," [unpub. report], Aug. 1955, Box 102, and Patterson, [untitled discussion of lead measurements], Apr. 24, 1956, Box 21, both HCU; see also Ahrens (1956, 60), Bondi (1952 [2d ed., 1961, 39]), Brush (1989), Dalrymple (1991, 318–24), Greenstein (1986, 19), Humason, Mayall, and Sandage (1956, 97), Knopf (1957, 225), and Tilton and Davis (1959, 191–2).

demonstrated the extent to which American astronomy became influenced by the work of nuclear chemists, particularly Urey and Brown, by the mid-1950s; the impact of their studies on the discipline was considerably greater, for example, than that of planetary meteorology in this same period. Moreover, the research facilities that enabled Brown and particularly Patterson to pursue this work were far from an inevitable outgrowth of disciplinary research programs. The willingness of geologists and administrators to embrace nuclear geochemistry at Caltech, a decision motivated by economic as well as methodological factors, proved important in providing this inherently interdisciplinary activity a secure institutional home.

In retrospect, one of the most important achievements of the "geochemical revolution" led by Brown, Urey, and their students at Chicago was the degree to which it fostered cooperation among a wide range of scientific fields. As mass spectrometers made increasingly exacting measurements of atomic abundances, astronomers, geophysicists, geochemists, and geologists sought to study planetary evolution in much the same way astrophysicists sought to explore stellar evolution. Hess put it clearly in 1954: "So long as astronomers were only concerned with the very light elements from H to O and lumped the rest, there was no overlap with the interests of geologists. When you begin to consider Si, Al, Mg and Fe then we have common ground and can contribute to the picture."[83]

These techniques, aided by the scientific infrastructure of the Manhattan Project, enabled geochemists to influence developments far beyond their field. The fact that Brown, Urey, and Patterson all shared close ties with the American physics community, and employed quantitative techniques, limits the validity of comparing interdisciplinary skirmishes over the age of the universe in the 1950s with the famous turn-of-the-century controversy over Earth's age. Yet geochemistry's influence on its disciplinary neighbors reveals how complex the earth sciences had become in the late 1940s, more than a decade before continental drift and plate tectonics were to create still more revolutionary changes. Despite their reputation as an intellectual backwater with limited intellectual authority, the geosciences by the early 1950s became a strong presence in the political economy of twentieth-century U.S. science.[84]

The traditional flexibility of academic science departments and patrons in the American context also aided the growth of this field. This is not to diminish the unique importance of instruments like mass spectrometers, or texts like Urey's singularly influential *The Planets,* in creating the seeds for interdisciplinary cooperation; but it is also important to note how Chicago's decision to create the Institute for Nuclear Studies – and Caltech's to embrace geochemistry – stimulated a renewed burst of interdisciplinary experimentation. The rapid blooming of new tran-

[83] Hess to Kuiper, Feb. 25, 1954, Box 11, GPK.

[84] On the inferior status commonly accorded geology and chemistry in contrast to physics and astronomy through the first half of the twentieth century, see Brush (1978b), Burchfield (1975, 212–15), Greene (1982, 276–9), Greenstein (1986, 18–19), and Menard (1971).

sient institutions, including the Rancho Santa Fe conference, testifies to the enduring faith American scientists had in interdisciplinary approaches, undiminished from the interwar era of foundation funding. Moreover, as during the 1920s, university administrators embraced modifications to existing disciplinary structures when this promised to increase significantly their access to extramural patronage and to retain prominent faculty.

Until the mid-1950s, astronomers and geochemists continued to cooperate in studies of the origin and evolution of the solar system. New attention to cosmogony and to the origin of comets encouraged American researchers to pursue planetary evolution as an "astro-chemical" problem. By the early 1950s, Kuiper and Urey had emerged as American leaders of these efforts. Their attempts to maintain borderland research between astronomy and chemistry, and their shipwreck in the Hot Moon–Cold Moon controversy, are the subjects of the following chapter.

Raymond Lyttleton, ca. 1963. Lyttleton completed a dissertation on cosmogony under Russell's direction in 1937, then joined Fred Hoyle, Hermann Bondi, and Thomas Gold at Cambridge University in developing the steady-state theory of cosmology. Though he believed his model of comet formation supported the steady-state theory, American astronomers soon came to support a rival theory by Whipple, leading to bitter controversy. (Photograph courtesy Raymond Lyttleton and the Niels Bohr Library, American Institute of Physics.)

Jan Oort *(center right)* at the Harvard College Observatory with Bart J. Bok *(far left)* and Harlow Shapley *(right)*. In the late 1940s Oort became interested in the problem of cometary origin, and with his students Maarten Schmidt and A. J. J. van Woerkom developed ideas about a distant reservoir of primordial comets, later known as the Oort cloud. (Photograph by Dorothy Davis Locanthi, Dorothy Davis Collection, courtesy Niels Bohr Library, American Institute of Physics.)

(Above) Gerard P. Kuiper at the 40-inch refractor of Yerkes Observatory, ca. 1955; *(facing)* Harold C. Urey in his laboratory at the Institute for Nuclear Studies, University of Chicago, ca. 1955. Leaders of astronomical versus geochemical approaches to solar system research, Kuiper and Urey tried, but failed, to sustain interdisciplinary collaboration on planetary evolution. Their bitter conflict strongly influenced the subsequent development of solar system research, intellectually and institutionally. (Kuiper photograph by Stephen Lewellyn, courtesy University of Arizona Library; Urey photograph courtesy University of Chicago Archives.)

Frederick E. Wright *(second from left)* at Mount Wilson. From 1925 to 1940 Wright chaired the Moon Committee of the Carnegie Institution of Washington, spearheaded by Carnegie President John C. Merriam. Analysis of data from the Moon Committee lagged, and Wright's "Moon House" at Mount Wilson was partially destroyed in a windstorm in 1937. Others *(left to right):* Paul W. Merrill, Roscoe F. Sanford, and Ferdinand Ellerman. (Courtesy the Huntington Library.)

Ralph B. Baldwin, ca. 1960. After obtaining critical data on bomb explosions from the U.S. Army and Navy while pursuing military research during World War II, Baldwin argued that the depths and diameters of lunar and meteor craters shared an exponential relationship with chemical explosion craters. His 1949 book *Face of the Moon* caused many American astronomers to accept the meteoritic hypothesis. Remarkably, Baldwin was then a full-time manager in his family's firm, and prepared his lunar work in free time. (Courtesy Ralph B. Baldwin and the Niels Bohr Library, American Institute of Physics.)

Carlyle Beals *(right)* at the New Quebec Crater, northern Canada, in 1963. Assisted by Peter M. Millman, Beals began searching the Canadian Shield for evidence of ancient meteorite scars in the mid-1950s, stimulated by Baldwin's work. By 1960 Beals's group announced the discovery of a half-dozen craters of probable impact origin. (Courtesy Richard A. F. Grieve and the Niels Bohr Library, American Institute of Physics.)

Eugene Shoemaker, ca. 1965. A geologist trained at Caltech and Princeton, Shoemaker demonstrated that Meteor Crater and nuclear explosion craters had equivalent structures. His codiscovery of coesite in Meteor Crater and in the 60-mile-wide Ries Basin in Germany enabled him to create the Branch of Astrogeology within the U.S. Geological Survey in 1960. (Photograph courtesy Jody Swann / USGS.

The Jangle U *(center background)* and Teapot Ess *(right background)* nuclear explosion craters at the Nevada Test Site, dwarfed by the 1,200-ft-wide Sedan Crater created by a 100-kiloton bomb in 1962. Shoemaker's access to the Jangle U and Teapot Ess craters through the then-secret MICE program of the Atomic Energy Commission aided his attempts to compare their structure to that of Meteor Crater in Arizona. (Lawrence Livermore National Laboratory photograph, courtesy National Archives, Washington, D.C.)

The 101st meeting of the American Astronomical Society, held at the University of Florida campus in Gainesville, December 27–30, 1958. With just 3 percent of the number of physicists, astronomy was one of the smallest science disciplines in the United States in the 1950s; yet it experienced a spurt in its growth rate from 7 to 20 percent per year soon after the launch of *Sputnik* in October 1957. As new funds for expanding facilities and graduate programs came from government patrons, tensions over allocating instrument time and resources rose. (Courtesy *Sky and Telescope*.)

Fred L. Whipple, Cecilia Payne-Gaposchkin, and Donald H. Menzel at the Harvard College Observatory. In contrast to most American astronomers, Menzel and Whipple by the mid-1950s advocated using rocket- and satellite-based instruments for astronomical research. Their bold embrace of "astro-geophysics" raised conflicts at Harvard and nationally over the future direction of American astronomy. (Courtesy Fred L. Whipple and the Niels Bohr Library, American Institute of Physics.)

Yerkes Observatory staff, autumn 1958. Kuiper built a large program of solar system research at Yerkes–McDonald with funding from the Air Force, the Office of Naval Research, and the National Science Foundation. After escalating institutional tensions forced him from the directorship in 1960, Kuiper transferred his staff and contracts to the University of Arizona. *Front row (left to right):* Kevin Prendergast, D. Nelson Limber, Soviet visitor, Margaret Burbidge, Joseph W. Chamberlain, Kuiper, Geoffrey Burbidge, and unidentified visitor. (Courtesy Dale P. Cruikshank, Yerkes Observatory and the Niels Bohr Library, American Institute of Physics.)

4

Consensus, then Controversy: Interdisciplinary Turmoil, 1950–1955

After 1949, American astronomers increasingly drew on geochemical research to study the origin of the solar system. Although astronomers still considered cosmogony a part of their own discipline, a number of them believed that the postwar geochemical work by Brown and Urey offered a promising key to this problem. "I am sure," Kuiper wrote Urey in late 1949, "that with a pooling of information and methods, the solution can be made to tie in with the segregation of the planets."[1] Since planetary formation involved the chemistry of cold matter – the borderland between chemistry and astronomy – astronomers interested in the solar system and interstellar matter saw geochemistry as a natural extension of their discipline, as necessary for developing quantitative models of cosmogony and the chemistry of interstellar solids as physics had been for interpreting the expanding universe.

Cooperative research between astronomers and geochemists burgeoned in the early 1950s. In less than five years, American astronomers achieved consensus on the nature of the solar system's birth, the structure of comets, the existence of distant comet reservoirs, and the formation of comets, all outstanding problems of twentieth-century astronomy. The fruits of such postwar programs as Whipple's Harvard-based meteor study and Kuiper's asteroid project at Yerkes–McDonald, as well as such interdisciplinary agencies as the Rancho Santa Fe conference, these results were welcomed by astronomers for providing new insights into star formation and cosmogony. Researchers also saw these results relevant to the emerging debate over steady-state versus expansionist cosmogonies.

Cooperative research between astronomers and geochemists involving cosmogony, however, was short-lived. By 1955 the respective leaders of astronomical and geochemical approaches to solar system research, Kuiper and Urey, became involved in a bitter and protracted debate over the thermal history of planetary bodies. The Hot Moon–Cold Moon controversy, as it became known, was among the most vitrolic, costly debates in solar system astronomy to that time. Its effects upon this emerging field were immediate and significant. The conflict not only illustrated that astronomers and geochemists were far from agreement over the thermal evolution of plan-

[1] Kuiper to Urey, Oct. 19, 1949, Box 31, GPK.

ets (which members of both disciplines agreed was a key test of cosmogonic theories) but also that astronomers, geochemists, and geologists lacked a common methodological and conceptual foundation for evaluating solar system research. Their intense controversy showed the influence of personality on the reception of scientific ideas, as well as the problem of authority in relations between members of different disciplines. It also revealed that, good intentions aside, scientists working at the professional margins of their discipline remained vulnerable to methodological and personal disruptions.[2]

FINDING COMMON GROUND: THE ORIGIN OF THE SOLAR SYSTEM

Agreement among American astronomers by 1940 that the encounter theory failed to explain the birth of the solar system encouraged them to reconsider alternative cosmogonal explanations. Because cosmogony was central to understanding the evolution of the solar system and Earth, many astronomers found the subject an irresistible challenge; but because cosmogonic theories were required to account for the dynamic properties of the solar family as well as the chemical and physical properties of such diverse objects as planets, asteroids, and meteors, developing an acceptable cosmogony seemed a daunting task. One major problem was that no single theory seemed adequate. Criticisms of encounter theories advanced by Russell and Spitzer in the 1930s convinced scientists that this approach was no longer valid. Whipple himself accepted in 1941 that no hypotheses satisfactorily explained the origin of the planets, although he argued that "the evidence of their physical composition points to an origin in the Sun itself." Nonetheless, nebular theories, the only widely accepted alternative to encounter theories, still seemed flawed by the same angular momentum problem that had forced astronomers to accept the necessity of encounter scenarios in the first place. Russell's 1935 comment, that "[n]o one has ever suggested a way in which almost the whole of the angular momentum could have gotten into such an insignificant fraction of the mass of an isolated system," seemed no less valid to many astronomers in the early 1940s.[3]

Despite Russell's criticisms, a number of American and European astronomers had attempted to rescue the encounter theory during the late 1930s. One of the most highly regarded of these efforts was a model proposed by Raymond Lyttleton, a reader of mathematics at Cambridge University. Lyttleton had become interested in the self-styled "wildish" idea that Russell had offered in his 1934 lectures at the University of Virginia that the Sun originally possessed a binary companion, whose disrup-

[2] Burchfield (1975, x, 215–16).

[3] Quoted from Whipple (1941, 243) and Russell (1935b, 96); see also Struve (1950, 1) and Struve and Zebergs (1962, 171).

tion during an encounter with a third star provided the material from which the planets coalesced. In 1936, while still a graduate student, Lyttleton had studied with Russell at Princeton, and later that year had published a quantitative version of the theory. However, few astronomers in the United States were willing by then to accept a collision-based cosmogony. By 1940 Russell himself had discounted Lyttleton's theory. Spitzer's argument on the thermal diffusion of gases, he advised Struve, "appears to me to be so conclusive that it is not really worthwhile to spend much space on any form of the encounter theory of the origin of planets unless some fallacy in his physical argument and conclusion can be shown." Moreover, at a time when most astronomers believed binary stars were relatively infrequent and grazing collisions between stars more infrequent still, cosmogonic theories that required both seemed ad hoc. The reaction of American astronomers to Lyttleton's theory therefore signaled an important shift in attitude away from dualistic to monistic theories of the solar system's origin.[4]

Few American scientists contributed to the problem of cosmogony during the early 1940s, as their involvement with wartime military research increased; but shortly after the end of hostilities in Europe, Allied astronomers learned of a new attempt to revive the Laplacian nebular hypothesis. In 1943, Karl Friedrich von Weizsäcker, a young German physicist then holding a chair in German-occupied Strasbourg, proposed that the planetary system had developed from an extended cloud of dust and gas surrounding the newly formed Sun. Possibly drawing on experience gained while in the wartime German nuclear project, von Weizsäcker relied on turbulence theory to explain both the formation of planets and the previously perplexing distribution of angular momentum within the system. In von Weizsäcker's theory, large vortices with dimensions equaling the Titius–Bode distances between the planets formed and began rotating. Dust and gas forced into small whirlpools between the vortices, acting like ball bearings, initiated the accretion of this matter into planetary bodies. Von Weizsäcker raised two arguments favoring this arrangement. First, rotation of the primary vortices would transfer sufficient angular momentum from the Sun to the planets to explain observed distributions. Second, he argued that angular momentum would also be lost to the Sun through the dissipation of large volumes of hydrogen and helium not accreted into the planets, assuming the solar nebula had a composition similar to the Sun's. It was the first time that the discovery of hydrogen's predominance in the universe influenced theories of cosmogony, stimulating new research in this area much as it did geochemical studies of planetary evolution.[5]

[4] Quoted from Russell to Struve, June 10, 1940, Box 27, Folder 70, HNR; see Spitzer to Lyttleton, Apr. 20, 1941, Box 63, Folder 47, and idem to Russell, Apr. 21, 1941, Box 63, Folder 47, both HNR; see also Brush (1978a, 86–97) and Lyttleton (1936).

[5] C. F. von Weizsäcker OHI (Mark Walker, May 1985, courtesy of Walker); see also Brush (1981, 91), and Cassidy (1992, 464–5, 538).

Von Weizsäcker's theory came to be greeted warmly by astronomers and physicists in the United States, who generally viewed it as vindicating nebular approaches to the solar system. One reason that it fared well was that it addressed two neighboring fields then of considerable interest to many astrophysicists: the natural abundance of the elements and turbulence theory. This was evident in a comprehensive review that George Gamow and J. Allen Hynek published in the *Astrophysical Journal* early in 1945. During the early 1940s, Gamow, then at George Washington University in Washington, D.C., began collaborating with Hynek, on wartime assignment at the Applied Physics Laboratory in nearby Silver Spring, on the problem of stellar evolution. This effort, which married Gamow's experience in theoretical physics with Hynek's familiarity with observational stellar data, made them particularly interested in von Weizsäcker's new interpretation. Their review – which also introduced von Weizsäcker's cosmogony to Allied astronomers, since his theory had appeared in the *Zeitschrift für Astrophysik*, then unavailable in the United States – focused on the significance of cosmogony for understanding stellar evolution. Gamow and Hynek called attention to the high abundances of hydrogen and helium in the Sun that von Weizsäcker's theory required. Recently published results by Menzel, Goldberg, and Aller in the United States, they noted, agreed with those of Unsöld in Germany (see Chap. 3), indicating that hydrogen and helium make up 99 percent of the Sun's total mass. For Gamow, also working with Edward Teller on nuclear fission and fusion in stars and the origin of the elements, von Weizsäcker's sharp distinction between solar and terrestrial abundances was especially welcome. As a line perhaps written by Gamow stated, "this change in the constitution of stellar matter will help to improve the agreement between the calculated rate of energy production in the carbon cycle and the observed stellar luminosities." Other reviews (all positive) emphasized similar points. Writing from Yerkes, Chandrasekhar praised von Weizsäcker's application of turbulence theory to the evolution of the solar nebula. Chandrasekhar had made major contributions to the study of turbulence during World War II, a fact known to American astronomers and physicists. His endorsement thus helped convince fellow researchers that von Weizsäcker's arguments had merit.[6]

Two additional factors caused rapid acceptance of von Weizsäcker's nebular theory in the United States. By 1945, many astronomers had grown comfortable thinking in terms of numerous planetary systems, in sharp contrast to opinions held before Russell made his sweeping criticisms of encounter theories ten years earlier. Several distinct developments were behind this shift, but one was certainly a series of announcements in the early 1940s that other planetary systems had been discovered indirectly, through precise measurements of stellar motions. In 1943 the astronomer Kaj Aa. Strand, then away from Swarthmore College on war-related work, reported

[6] Quoted from Gamow and Hynek (1945, 250); see Russell to Chandrasekhar, Jan. 26, 1945, Box 5, Folder 81, HNR; Kuiper to Oort, Jan. 14, 1950, Box 30, GPK; see also Chandrasekhar (1946), DeVorkin and Kenat (in press), and Struve and Zebergs (1962, 175).

what he termed "planet-like companions" orbiting the double star 61 Cygni. A similar discovery of an object with one-hundredth the mass of the Sun circling the star 70 Ophiuchi was announced that same year by Dirk Reuyl and Erik Holmberg. Although these discoveries were not confirmed (and Russell cautioned that objects greater than one-three hundredth of the Sun's mass would contain degenerate matter and therefore have stellar as well as planetary characteristics), most astronomers accepted that these results required an upward revision in the total number of planetary systems. Even before these discoveries were announced, however, the rejection of encounter theories alone seemed to encourage astronomers to reconsider the assumption commonly held during the early twentieth century that planetary systems were rare. Already in 1942, for example, eminent British astronomer James Jeans revised his previous estimate of only a few planetary systems in the galaxy to one for every six stars. In 1943 Russell also adjusted his own expectations sharply higher. These new predictions, he noted, represented "a radical change – indeed practically a reversal – of the view which was generally held a decade or two ago." In 1945, Russell expressed enthusiasm for the general contours of von Weizsäcker's theory. This renewed acceptance of planets as a common celestial phenomenon was an important development in itself, and encouraged astronomers to develop theories for their origin that, like von Weizsäcker's, accounted for their plenitude rather than exceptionality.[7]

Another reason that von Weizsäcker's theory intrigued scientists was the rapid increase of knowledge about the physical characteristics of stars, strengthening the conviction of astronomers that planetary systems are common and tied to processes that guide stellar evolution.[8] By the 1940s, astronomers had at hand such statistical data as the rotational velocities, average proper motions, surface temperature, and abundance ratios for each of the major classifications of stars. The absence of feasible alternatives to the encounter theory thus became a catalyst for applying these data to the problem of planetary formation as astronomers searched for new sources of information relevant to its solution. Leading Soviet astronomer Viktor Ambartsumian wrote, for instance, that since observational evidence on other planetary systems in different stages of evolution was not available, "[i]t is far more productive to base one's conclusions [about cosmogony] on the large body of facts which have been accumulated concerning the stars," rather than on the observed properties of the solar system alone. American astronomers also accepted the idea that planetary systems were the routine product of stellar evolution rather than the chance result of improbable collisions.[9]

In the United States, the chief advocate of this approach to cosmogony initially was Struve, the Russian-born director of Yerkes–McDonald. Struve's interest in

[7] Quoted in Wildt (1947, 92); see also Russell to Chandrasekhar, Jan. 26, 1945, Box 5, Folder 81, HNR, and Brush (1981, 91).

[8] Bobrovnikoff (1951, 304) and Keenan and Morgan (1951, 12–29).

[9] Quoted from Ambartsumian (1948); see also Graham (1987, 380–427), Levin and Brush (1994), Struve (1945).

cosmogony was apparently sparked by the 1944 discovery by Kuiper, his Yerkes colleague, of Titan's atmosphere. It was reinforced by his reading in original Russian a book on the solar system by Soviet astronomer Vaseilii Grigor'evich Fesenkov late in 1944. Early the following year, he published a detailed three-page review of Fesenkov's cosmogony in the *Astrophysical Journal.* Struve did not much care for Fesenkov's theory, which held that the Sun, while undergoing episodes of extremely rapid rotation, had shed a ring of nebular material that collapsed into planets. He found more useful Fesenkov's insistence on "considering the evolution of the Sun as a member of the galaxy" in evaluating theories for the origin of the solar system, and used the occasion to review his own extensive studies of stellar characteristics and binary stars. What Struve found particularly interesting was the fact that virtually all G-type stars like the Sun had slow rotational velocities, generally about 2 km per second (kps). The majority of stars classified as B, A, and early F (hotter and more luminous than the Sun), however, had rapid rotational velocities approaching 120 kps, close to the limit of instability. This transition between F- and G-type stars, Struve wrote, "is very sudden and does not resemble the gradual changes observed in most (but not all) physical characteristics." Struve also calculated that the angular momentum possessed by the planets, if added to that of the Sun, nearly equaled the quantity typically observed in early-type stars. Although the angular momentum problem had hindered all previous attempts to develop nebular cosmogonies, this correlation convinced Struve that nebular theories were essentially correct, even if the causal mechanism remained unknown.[10]

Struve soon presented these arguments about planetary origins more broadly to American astronomers. In his Vanuxem lectures, delivered at Princeton in 1949 and published the following year as *Stellar Evolution,* Struve declared that all the observed properties of the Sun were also found in a large group of stars of solar class. It was therefore more reasonable to assume that planets normally accompany such stars than to argue that planetary systems are very rare. Adopting this new hypothesis, Struve observed, would lead to "a small revolution in our process of thinking about the problem." As Struve's work was held in wide respect among stellar astronomers in the United States, his argument that planetary systems were a common phenomenon carried considerable weight.[11]

Other researchers also turned their attention to the problem of nebular cosmogony in the mid and late 1940s, including Swedish astrophysicist Hannes Alfvén. One of the most influential was Kuiper, whose 1949 nebular theory soon surpassed von Weizsäcker's as the most highly regarded among American astronomers. What focused Kuiper's attention on cosmogony was not von Weiszäcker's theory but rather

[10] Quoted from Struve (1945, 265); see Struve to Russell, July 27, 1945, Box 64, Folder 43, HNR; see also Graham (1987, 394–6), Mikaylov et al. (1964, 563–93), Struve (1949, 303), and Struve and Zebergs (1962, 167).

[11] Struve (1950, 49).

subsequent criticisms made of it by his Yerkes colleague, Chandrasekhar.[12] In 1948, Chandrasekhar concluded from new studies of turbulent flow that the large vortices on which von Weizsäcker depended to initiate planetary formation would not be of regular size, but instead vary widely in width. Astronomers soon regarded this as a serious limitation to von Weizsäcker's theory. Kuiper's close contacts with Chandrasekhar and Struve, his research in stellar spectroscopy, and his rapidly expanding programs in solar system astronomy all seemed to convince him by 1948 of the desirability of addressing this problem. Moreover, as director of Yerkes–McDonald, Kuiper had access to institutional resources to initiate new research efforts. Acting quickly, Kuiper secured von Weizsäcker's appointment as Alexander White Visiting Professor in the Committee on Social Thought at Chicago in the fall and winter of 1949–50. As a result, virtually all the principal contributors to cosmogonic research outside the Soviet Union were briefly assembled in Chicago. Moreover, although Kuiper did not begin his famous collaboration with Chicago colleague Urey until the end of 1949, at which time his own cosmogonic theory was drafted, Urey also influenced its later development.[13]

Kuiper's theory, presented in a series of articles from 1950 through 1954, came to influence U.S. solar system research deeply during the 1950s. It differed fundamentally from von Weizsäcker's in a single but key respect: Whereas von Weizsäcker had proposed that planets originated between adjoining vortices, Kuiper, heeding Chandrasekhar's criticisms, proposed instead that the planets had formed in regions of gravitational instability. In subsequent articles, Kuiper extended this theory to the formation of satellite systems, asteroids, and comets. He declared that his theory predicted not only the general distribution of the asteroids, but also zones within the primordial solar nebula, including beyond the orbit of Pluto, in which comets were likely to form.[14] His argument that regions of gravitational instability could explain the observed size and spatial distribution of the planets was adapted from studies of density distributions and motions of stars about the nuclei of galaxies. Kuiper used this approach because it allowed him to sidestep the inherent difficulty of applying turbulence theory to the development of the solar nebula, but it also illustrated his commitment to employ concepts firmly embedded in astronomy and astrophysics to explain the origin of the solar system. Importantly, Kuiper continued to emphasize this approach even when Urey argued that geochemical explanations were more appropriate to theories of planetary origin.[15]

Kuiper also followed Struve's lead in arguing that a high incidence of planetary systems was likely, judging by stellar characteristics alone. In particular, he asserted

[12] Kuiper to J. Millis, Nov. 24, 1947, File 232:7, YO; see also Brush (1990, 87–8) and Ter Haar (1948, 407–10).

[13] Kuiper to Zachariasen, Oct. 7, 1949, Box 29, GPK, and University of Chicago Round Table, [radio] program no. 598, Sept. 4, 1949, UChi; see also Brush (1981, 92).

[14] Kuiper (1950a; 1951a,c,d; 1953a,b; 1954). A summary appears in Kuiper (1956a); see also Brush (1981).

[15] See, for instance, Kuiper (1956a, 57–68, 105–21).

that planetary formation was simply a "special case of the very general process of binary star formation," occurring whenever conditions did not permit the growth of binary companions. His arguments revealed the degree to which Struve and Kuiper had stimulated and influenced one another between 1937 and 1941, when both researchers had worked jointly at Yerkes on the problem of binary stars. Resurrecting his spectroscopic binary data, Kuiper challenged the prevailing idea, based on a 1935 statement by Russell, that binary pairs are uncommon; he argued instead that at least 50 percent of all stars of classes A–K (including solar-class stars) are binary. Extending this theme, Kuiper employed his data to show that spectroscopic binary periods ranged from several months to several years, like those of the planets, suggesting to him that planetary and binary star formation were closely related. The strong respect accorded to Kuiper's stellar studies helped give his cosmogonal theory added visibility and support. It is important to observe that his reasoning was analogous to that which Struve had employed in arguing a strong correlation between solar-type stars and slow rotation periods. Acceptance of the binary star analogy thus increased acceptance of the more fundamental idea that planetary systems were a common byproduct of stellar formation among certain class stars.[16]

By the early 1950s, Kuiper's cosmogonic theory emerged as the leading model among American astronomers. Several European astronomers favored it as well. Jan Oort, director of the Leiden Observatory and one of Western Europe's most eminent astronomers, declared, for instance, in 1950 that this theory "offers for the first time a plausible basis for the origin of the solar system." Yet Kuiper's theory had not been without its competitors as of the late 1940s. In addition to a model proposed by Dutch astronomer Dirk ter Haar, which had retained much of von Weizsäcker's approach, Whipple had introduced his "dust" theory in 1948, which held that the sun and solar family had formed within a gradually condensing cloud of dust and gas. What paved the way for consensus behind Kuiper's nebular cosmogony was, above all, a short but intense debate waged over the origin of comets between 1949 and 1951. This debate not only addressed such fundamental questions as the relation of minor bodies in the solar system with each other, but also their relationship with the interstellar medium. It also established Kuiper as one of the leading solar system astronomers in the United States, and set the stage for the confrontation between Kuiper and Urey over the evolution of the solar system.[17]

THE ORIGIN OF COMETS

The nature and origin of comets remained among the most perplexing problems of astronomy in the late 1940s. Although a number of astronomers who had worked on

[16] Kuiper (1951b); see also Cruikshank (1993).

[17] Oort to Kuiper, Jan. 6, 1950, Box 30, GPK; see also Jeffreys (1952, 283–4), ter Haar (1948, 407; 1950), ter Haar (1950), and Whipple (1948a,b).

comets before 1940 returned to cometary studies after the war, a continuing puzzle was whether comets are original members of the solar system, or had been captured in some way from interstellar space. If actual members, this meant they had formed at the same time as the planets, requiring cosmogonal theories to explain their existence. If interstellar bodies, this raised the possibility that comets had formed either before or after the planets. Moreover, it remained uncertain whether comets had originated from a cloud of solar composition or from one with radically different elemental abundances. Astronomers from various fields thus saw advantage in pursuing the problem of comet origins.

Evidence accumulated since the turn of the century had shed little light on this fundamental issue. In 1914 (as noted in Chapter 1), Danish astronomer Elis Strömgren had presented what many astronomers regarded as convincing evidence that no comets had hyperbolic orbits, thus ruling out a direct interstellar origin. Later, however, astronomers began to wonder whether this precluded such an origin entirely. In 1949 Bobrovnikoff, the Ohio Wesleyan astronomer actively engaged in cometary studies since his work with Leuschner at Berkeley two decades before, pointed out that Strömgren's result merely meant no comets *currently* observed were hyperbolic. It did not rule out initially hyperbolic orbits that had been subsequently changed, perhaps through gravitational perturbations caused by Jupiter or other large planets in the solar system.[18] What made Bobrovnikoff, Whipple, and others question Strömgren's conclusion was the observed fragility of comets. The destruction of a number of short-period comets like Biela's after several passages by the Sun suggested that they were not as old as the planets, since the calculated lifetimes of these comets appeared far shorter than the age of the solar system. Another argument against including them as genuine members of the solar family was the chaotic distribution of comet orbits through space, in sharp contrast to the regularity of planetary orbits along the plane of the ecliptic. Finding a way of explaining how both kinds of objects could have originated at the same time and in similar environments puzzled astronomers. After wrestling with this issue unsuccessfully in the early 1930s, Russell had termed it "the greatest unsolved problem in astronomy." American astronomers seemed disposed toward an interstellar origin. At Harvard, both Watson and Whipple favored this view. Although Bobrovnikoff preferred to "explain the presence of all small bodies in the solar system by one hypothesis" – already his guiding principle in 1930 – by 1951 he too argued that only asteroids and meteorites belonged to the solar system, with comets captured later from interstellar space long after the sun and planets had formed.[19]

In 1948, Lyttleton had proposed just such an explanation for the origin of comets. Lyttleton's idea, based on earlier suggestions by Bobrovnikoff and German astrono-

[18] Bobrovnikoff (1951, 351–2), Clemence (1951, 229), Kuiper (1951b, 359), and Russell (1935b, 35).

[19] Quoted from Russell (1935b, 9); see also Bobrovnikoff, "The Origin of Minor Planets," draft, Oct. 25, 1928, Box 1, AOL; Russell to Kuiper, Jan. 25, 1944, Box 28 GPK; Bobrovnikoff (1951, 353), Lyttleton (1953b, 159–60), Watson (1938, 295, 302), and Whipple (1941, 232).

mer Friedrich Nölke a decade earlier but presented by him in quantitative terms, was that comets formed in great numbers whenever the Sun passed through clouds of interstellar matter. Particles in these clouds, perturbed by the Sun's gravity, would converge along the axial line of the Sun's motion. Streams of particles accumulating behind the Sun, he argued, would become more and more concentrated and come to share the Sun's relative motion through space. Comets created by this process would thus have elliptical orbits, satisfying Strömgren's 1914 finding. Noting that Russell in 1935 had argued that the nuclei of comets likely consisted of loosely compressed dust grains mixed with gases, similar to what his theory predicted, Lyttleton further claimed that Russell's solar system research supported his interpretation.[20]

Lyttleton's growing interest in comets had less to do with his earlier advocacy of dualistic cosmogonies than with his wartime studies of interstellar dust grains, as well as his emerging ideas about cosmology. In 1939, Lyttleton had begun collaborating with Fred Hoyle, an applied mathematician also at Cambridge, to study the infall of interstellar particles into the Sun's atmosphere. Their work convinced Hoyle – although few American astrophysicists – that the accretion of meteoritic particles by stars was sufficient to maintain their energy output over many millions of years. After the war, Hoyle had incorporated this idea, together with other concepts developed by fellow Cambridge associates Hermann Bondi and Thomas Gold, into what became known as the steady-state model of cosmology, one of the major cosmological theories of the twentieth century. This model held that the observed expansion of the universe was compensated by the continuous creation of new atoms in the interstellar medium. Strongly attracted to the steady-state idea, Lyttleton argued that the origin of comets in interstellar space validated the Cambridge approach to stellar energy production and the expansion of the universe. "One of the principle successes of the New Cosmology," Lyttleton declared, was that "without having any idea of an attack on the cometary problem in view," its required high levels of interstellar dust had led "quite naturally to a straightforward, and indeed necessary, explanation of the presence of comets." A product of the heavily mathematical environment at Cambridge, Lyttleton, like Hoyle, Bondi, and Gold, saw much merit in applying broad deductive theories to astronomy. They found observational astronomers – including most senior American astronomers – too willing to advance theories covering only limited kinds of astrophysical phenomena.[21]

American astronomers initially gave Lyttleton's comet theory a mixed review – an important point, for it indicated no consensus then existed within this community regarding the nature of comets. In 1949, for instance, Bobrovnikoff noted warmly that "[p]erhaps the passage of the solar system through such clouds might at the same time explain other events in the history of the earth, such as the recurrence of glacial

[20] Russell to Whipple, Jan. 20, 1950, Box 29, Folder 95, HNR; and Lyttleton (1948; 1952; 1953b, 146–8). Nölke's theory was developed in Nölke (1926).

[21] Quoted from Lyttleton (1953b, vii); see Bondi (1952, 140–56), Graham (1987, 381–4), and Hoyle (1949; 1951, 35–6, 83–4).

periods." Whipple was less enthusiastic about the idea than Bobrovnikoff, but unwilling to dismiss it entirely. In early 1949, apparently before learning of Lyttleon's ideas, Whipple had sought to begin new physical chemistry studies with Harrison Brown, convinced that comets were composed of interstellar material condensed when the solar system had formed; the Lyttleton alternative gave him pause. Lyttleton's theory, he advised Urey in January 1950, was "the only sensible theory for comets evolving later than the solar system itself." Although he did not believe Lyttleton's particular idea would stand up numerically, he added, "It is quite conceivable that the comets have evolved at a later period." To Oort he declared that Lyttleton's argument for the capture of material from interstellar clouds "is surprisingly good." If Lyttleton had not ignored intrinsic motions in these clouds, he continued, the theory "would appear to be satisfactory and complete."[22]

Within ten months, however, most American astronomers involved in solar system research came to reject the possibility that comets were formed from interstellar dust clouds, favoring instead their origin in the early solar system. In the autumn of 1950, for example, Whipple wrote that "it seems almost an incontestable conclusion that the comets must have originated with the planets or at an earlier period."[23] Several specific factors led to this shift. One was new research by Whipple himself indicating that the composition and structure of comet nuclei contradicted the model proposed by Lyttleton. Another was a theory of cometary origin proposed by Jan Oort in January 1950, which offered a means for understanding how comets could have originated with the planets and yet survived in great numbers until modern times. A third factor was new theoretical research on the origin of comets and asteroids announced by Kuiper in mid-1950. With the exception of Kuiper's studies, none of these projects had been intended to address the origin of comets, and each was initiated independently of the others. Nevertheless, as these researchers struggled to make their ideas mutually compatible, they came unanimously to accept that all comets were formed at the origin of the solar system. By early 1951, American astronomers came to reject the arguments about interstellar matter advanced by Lyttleton and other members of the informal Cambridge school, embracing instead competing ideas advanced by Oort and his colleagues in Leiden. To see how this happened, and to appreciate the significance of this new consensus, we must examine the work of Whipple, Oort, and Kuiper in turn.

Whipple's interest in comets, latent in the early 1940s, had grown stronger after 1945. His interest rose within the context of his Navy-funded study of the upper atmosphere, which employed meteors as atmospheric probes. By the late 1940s, after reviewing new meteor project reductions, Whipple had become convinced that individual meteor streams contained irregular distributions of meteoric particles

[22] Quoted from Bobrovnikoff (1951, 354), Whipple to Urey, Jan. 12, 1950, Box 102, HCU, and idem to Oort, Nov. 2, 1949, Box 32, JHO; see also Whipple to H. Lansberg, Mar. 17, 1949, and idem to Brown, Apr. 4, 1949, both Box 1, FLW–HU.

[23] Whipple to Urey, Sept. 28, 1950, Box 102, HCU.

within their orbits. Since stream meteors were believed to derive from disintegrating comets, Whipple looked to the structure of cometary nuclei to account for these differences. In 1949, he proposed that the size and rotation of the nucleus were responsible: Large and rapidly rotating nuclei would spray meteoric particles along meteor streams at a higher velocity than smaller nuclei, causing them to become uniform over short intervals. He believed this explained why meteor streams associated with Halley's comet, such as the Perseids, had nearly uniform stream distributions, whereas those tied to smaller comets like Biela's were tightly focused. To account for the production of these particles, and to understand better their density and interactions with the upper atmosphere, Whipple advanced the idea that comet nuclei were compacted amalgamations of gas and dust. This soon became known as the "snowball" or "icy" model of comets, which Whipple developed in a series of widely read papers in 1949 and 1950.[24]

Whipple found support for his comet model from a number of sources. One was a study of comet orbits by A. J. J. van Woerkom of Leiden, which he had begun shortly after World War II as a means of studying interstellar matter, on the assumption that comets had an interstellar origin. In his doctoral dissertation, published late in 1948, van Woerkom investigated how planetary perturbations affected comet orbits. He calculated that comets had formed either within an interstellar cloud sharing the same motion as the Sun, or within the solar system. The former struck van Woerkom as "highly improbable"; as a result, he rejected Lyttleton's comet model. He proposed instead that comets had originated together with the planets, and that their seemingly chaotic distribution above and below the plane of the ecliptic resulted from gravitational perturbations introduced upon them by the major planets.[25]

Whipple also gained support for his model from Russell, who cared for Lyttleton's comet theory no more than his cosmogony. He received an even stronger endorsement from Kuiper, just then working on comet formation as a cosmogonal problem. Kuiper found that his own theory predicted the formation of cometary bodies at the periphery of the solar nebula, which he placed roughly at the present distance of Pluto. Stimulated by Whipple's new theory, Kuiper quickly endorsed it, arguing that this independently derived concept, backed by the abundant new data of the Harvard meteor project, favored his new cosmogony. In addition, Kuiper voiced support for a 1947 study by Hedrick van de Hulst, then at Yerkes, that indicated the zodiacal light was caused by fine meteoric matter. Kuiper thus concluded that Whipple's cometary model accounted for two kinds of small bodies within the solar system: comets and diffuse meteoritic debris. He quickly communicated this model to other astronomers. Writing Oort on a wintry day early in 1950, he declared that Whipple's comets looked "like Chicago snow, with dirt in it." Although Whipple

[24] Whipple to Oort, Feb. 28, 1950, Box 33, JHO; idem to Urey, Dec. 3, 1951, Box 102, FLW–HU; see also Whipple (1949a; 1950a; 1951).

[25] Oort (1950, 91, 96–7) and van Woerkom (1948, 445).

felt disturbed by the alacrity with which Kuiper subsumed this work into his cosmogonal framework, complaining to Urey that "[e]verything seems to be accelerated these days," Whipple was soon encouraged by the compatibility between his comet model and Kuiper's cosmogony, which gained increasing acceptance by American astronomers.[26]

Until this point, however, Whipple still had not excluded the possibility that comets formed according to Lyttleton's hypothesis. This was largely due to Whipple's continued attachment to the "dust cloud" theory of cosmogony that he had put forth in 1948. Whipple recognized that his theory had difficulty accounting for the planets themselves, admitting to Kuiper that he had developed it "solely to avoid the difficulty of the sun's rotation." Yet he also believed that his theory explained the origin of comets and meteors rather well, advising Urey that "[t]he cosmic cloud hypothesis for the origin of the sun and planets is completely adequate for the comet evolution." Basic similarities existed between his theory and Lyttleton's in this respect, for Whipple believed that the origin of comets occurred at least two orders of magnitude further from the Sun than Kuiper's own predicted "zone" of cometary formation, estimating its distance as "many thousands of astronomical units" and hence far beyond the orbit of Pluto. In addition, Whipple found useful new results by ter Haar, another young Leiden-trained astronomer, who had investigated the properties of the interstellar medium and developed a nebular cosmogony of his own. In late 1949 Whipple declared that observed chemical abundances in comet nuclei "are roughly consistent with certain of ter Haar's calculations for molecules formed from interstellar atoms." This interpretation further encouraged him to consider an interstellar origin for comets.[27]

Two new results, both published in the first half of 1950, helped to change Whipple's mind on the birthplace of comets. The most important was a long, quantitative paper by Oort in January 1950 entitled "The Structure of the Cloud of Comets Surrounding the Solar System, and a Hypothesis Concerning Its Origin." In this paper, Oort extended the comet studies by his student van Woerkom, and developed a theory to explain the formation and distribution of comets throughout the solar system's history. The central part of Oort's theory was a detailed analysis of the orbits of twenty-two comets initially studied by Elis Strömgren. From this analysis, Oort concluded that each of these twenty-two comets had initially approached the solar vicinity from a distance of 20,000–150,000 astronomical units (AU). Next, Oort considered what caused comets at such enormous distances to return to the neighborhood of the Sun. Since 200,000 AU was roughly half the distance from the Sun to the next nearest star, Oort concluded that gravitational perturbations caused by

[26] Quoted from Kuiper to Oort, Jan. 14, 1950, Box 30, GPK, and Whipple to Urey, Sept. 28, 1950, Box 102, HCU; see Whipple to Russell, Mar. 2, 1950, Box 8, FLW–HU; idem to Oort, Sept. 28, 1950, Box 33, JHO.

[27] Quoted from Whipple to Kuiper, Feb. 7, 1950, Box 3, FLW–HU; idem to Urey, Jan. 12, 1950, Box 102, HCU (where see also idem to Urey, Sept. 28, 1950); and Whipple (1950b, 83).

passing stars were sufficient to nudge individual comets from a diffuse sphere or cloud of comets (50,000–200,000 AU in diameter) toward the solar system. This accounted for the continued existence of comets in the present. Oort then offered a description of the life cycle of comets: After their formation within the evolving solar system, comets were removed to great distances through perturbations introduced by the massive outer planets (as van Woerkom had demonstrated). Once in the comet cloud, comets remained at this remove, still attached to the gravitational field of the Sun, until occasional members experienced perturbations from other stars and were redirected toward the Sun.[28]

In addition to his analysis of these well-studied comets, Oort soon added further observational evidence to support his claim. Working with graduate student Maarten Schmidt, Oort concluded that "new" comets (those approaching the Sun for the first time after leaving the cloud) had different photoelectric and spectral characteristics than "old" comets (those in short-period orbits about the Sun). "New" comets, Oort and Schmidt argued, deteriorated rapidly on approaching the Sun, expelling vast amounts of dust and gas in the form of extended comas and tails; comets already subjected to close approaches past the Sun showed less intense brightening, which they interpreted to mean that gases near the comet's core were depleted after successive perihelion passages. That such differences existed strengthened Oort's conviction that a comet cloud (later called the *Oort cloud*) existed. Oort used these arguments to challenge Lyttleton's hypothesis, which called for comets to form between 20 and 2,000 AU from the Sun, far closer than the computed aphelion distances for "new" comets. Moreover, Oort argued that what was known about frequency and motions of interstellar clouds indicated there was only a "slight chance" that the Sun would have passed through such a cloud at sufficiently low velocity to form comets. Oort therefore declared that a solar origin for comets was required.[29]

As one of Western Europe's most eminent astronomers, Oort's arguments about comet origins carried considerable authority; yet astronomers came to accept his conclusions not only because of his recognized skills in celestial mechanics, but because of his work on the interstellar medium, a field he had pursued alongside his better-known investigations of galactic structure.[30] Under Oort's influence, Leiden had emerged as a center for the study of the interstellar particles in the 1940s. Astronomers had first become attracted to the problem of interstellar matter when Struve, Trumpler, and other astronomers demonstrated in the 1930s that a significant gaseous substratum existed, perceiving its relevance to theories of stellar formation, galactic structure, and cosmology. Oort's own interest in the interstellar medi-

[28] Although Oort's influence on his research is evident, van Woerkom credited the dissertation to Hertzsprung, who presumably wished to find comets of undisputed interstellar origin; see Oort (1950) and van Woerkom (1948, 472).

[29] "Meeting of the Royal Astronomical Society . . ." (1949), Oort (1950, 109), and Oort and Schmidt (1951).

[30] J. M. Burgers to Oort, Jan. 27, 1949, and Burgers to C.-G. Rossby, May 18, 1949, both Box 32, JHO; see also Bok (1946, 72–3) and Oort (1946).

um had initially been to trace the distribution of hydrogen in galaxies, including the Milky Way. By the mid-1940s, however, he had became convinced that interstellar particles were not composed of "meteor dust" and required further investigation. Other members of the Leiden community had become interested in interstellar particles for similar reasons. Van de Hulst had tackled the problem of the zodiacal light in a successful bid to win a prize offered by Leiden University in 1941 on the composition and structure of interstellar solids. Similarly, van Woerkom had launched his study of cometary orbits in the hope that an interstellar origin for comets was possible, thus making them neighborhood samples of interstellar particles. Astronomers therefore understood that Oort's criticisms of Lyttleton's comet theory were not limited to cometary formation alone, but instead to broad assumptions about the interstellar medium critical to the arguments that Hoyle, Bondi, and Gold employed in promoting their steady-state model. It was also another demonstration that divisions between fields characteristic of post-1960 astronomy were not yet evident in 1950.[31]

Whipple and Kuiper, already aware of Leiden research on comets and interstellar particles through informal discussions with Oort at the Zurich meeting of the International Astronomical Union in 1948, read Oort's arguments for a distant comet cloud in proof copies sent by Oort the following summer. Both scientists were favorably inclined to his argument that such a cloud existed and to the mechanism he proposed for sending its comets back to the inner solar system. Whipple wrote Oort that he was "very excited" by his paper, adding, "Frankly I feel that you are on solid ground and that the arguments are fundamentally sound." Their acceptance of his general theory was aided by other factors. Oort had been a mentor to Kuiper during his graduate years at Leiden, and Whipple recognized similarities between Oort's comet cloud ideas and previous but less precise arguments developed in 1932 by his own Harvard mentor, Öpik. Moreover, favorable reviews of van Woerkom's and Oort's dynamical arguments were soon published by such American experts in celestial mechanics as Clemence, and word of these reached them by 1950.[32] Furthermore, close associations existed between members of the Leiden school of interstellar solids and major centers of American astronomy. Oort was a frequent visitor in America, van Woerkom had accepted a position under Brouwer at Yale by 1950, and van de Hulst became a visiting research associate at Yerkes, teaching at the Harvard summer school in 1948. Finally, virtually no astronomers in the United States inclined toward the Cambridge school's view that interstellar matter could produce stellar radiant energy, a fundamental plank of the steady-state model. In 1950, for example, Struve declared his opposition to this idea. Numerous interrelated factors,

[31] P. Swings to Oort, Dec. 21, 1949, Box 32, JHO; Oort OHI (David DeVorkin, 1977, SHMA at AIP, pp. 41–2); see also Greenstein (1951, 563), Oort (1946, 165–6, 173), van Woerkom (1948, 445).

[32] Quoted from Whipple to Oort, Aug. 16, 1949, Box 32, JHO; see Oort to Whipple, June 8 and 23, 1949, both Box 32, JHO; Whipple to Urey, Sept. 28, 1950, Box 102, HCU. Whipple accepted that van Woerkom's dissertation was "excellent indeed"; see Whipple to Oort, June 15, 1949, Box 32, JHO. See also Clemence (1951, 229), Oort (1950, 93).

from celestial mechanics to accepted theories of stellar energy production, thus caused American astronomers to be predisposed to a solar origin for comets.[33]

What kept Whipple and Kuiper from initially embracing Oort's theory in full, however, was that Oort proposed an origin for comets that both astronomers rejected. Oort argued that since comets and asteroids had formed side by side in the same region of the solar nebula, both were essentially the same kind of object. He had been led to this argument by several considerations, foremost among them its theoretical simplicity and elegance. (Comets, asteroids, meteors, meteorites, and zodiacal light were then accounted for by one common process.) Using Russell's estimate that the total mass of all asteroids was roughly one-thousandth that of Earth's mass, and observations of Comet 1843 I (which passed through the solar corona), Oort concluded that comet nuclei, like the minor planets, "seem to consist of solid blocks of considerable dimension." While emphasizing that these speculative discussions on the origin of comets "in no way" affected his dynamical arguments, Oort nevertheless saw them as a strong point in his favor. Lyttleton's theory, he wrote, requires "that we must accept an entirely different origin for meteorites and comets," which he revealingly termed its "general difficulty." Oort also favored the disruption of asteroids as the starting point for his comet cloud. Writing Whipple in June 1949, Oort disclosed, "I am getting more and more attracted by the idea that the comets have had the same origin as the minor planets and that only those minor planets which moved in 'neat' orbits between Mars and Jupiter have remained there, while the majority of the remnants were gradually thrown out by the action of Jupiter."[34]

Yet this was a point of view to which almost no astronomers in the United States subscribed in 1950, and one that caused Whipple and Kuiper considerable distress. Although the exploded planet hypothesis had enjoyed support among American astronomers in the late 1940s, most researchers engaged in asteroid, meteor, or meteorite studies came to reject this interpretation on several grounds. "I believe with you, from evidence as presented by Whipple, that most if not all ordinary *meteors* are derived from comets," Kuiper advised Oort. "But I was not aware that *meteorites* had any proven relation to comets. Whipple . . . puts the meteorites with the asteroids, and that is what I supposed to be correct."[35] Convinced by his photographic meteor studies that meteors were far more fragile bodies than meteorites and thus structurally dissimilar, Whipple wrote Oort that his argument for the common origin of comets and asteroids had left the Harvard scientist in an unhappy frame of mind. "I am convinced from all of the evidence from physical studies of meteorites and from the orbital characteristics of the photographic meteors," he declared, "that we are really dealing with two different types of objects." The more he thought

33 Whipple to Oort, Feb. 28, 1950, Box 33, JHO; Bart J. Bok, "Confidential memorandum to the Harvard Observatory Council on the summer's activity at Cambridge and at Oak Ridge," Sept. 22, 1948, Box 6, FLW–HU; see also Brouwer (1950).

34 Quoted from Oort (1950, 107, 110, 109) and idem to Whipple, June 23, 1949, Box 32, JHO.

35 Kuiper to Oort, Oct. 30, 1949, Box 30, GPK; emphasis in original.

about his icy model of comet nuclei, the more displeased with Oort's ideas he became. In early 1950 he wrote Oort, "I must admit that my confidence in doubting the 1 : 1 relationship between comets and meteorites and asteroidal bodies rests largely on the difficulty of understanding the physical behavior of comets were they constructed of nothing other than the heavier elements, non volatile at room temperature."[36] These discussions illustrate the degree to which astronomical research in this field in America had been influenced by meteorite abundance studies by geochemists such as Brown and Urey. They also show how the long-running Harvard photographic meteor program and the Yale asteroid project had focused American attention on small bodies of the solar system.

Kuiper, more than Whipple or Oort, played the pivotal role in resolving this conflict over the origin of comets. As we saw, Kuiper's theory of cosmogony, already in preparation by late 1949, had predicted that billions of cometlike bodies would form along the periphery of the solar nebula at distances of 35–60 AU (later termed the "Kuiper belt"). Kuiper believed that securing additional proof for the physicochemical condensation processes he called on to produce asteroids and comets would be time consuming and (as Urey had convinced him) difficult. He therefore turned, as he had in the past, to astronomical means of considering the problem. Kuiper argued that Pluto (which at 40 AU moved through what he calculated to be the outer fringe of the original solar nebula) had gravitationally diverted comets toward the massive outer planets, thereby providing an alternative mechanism to Oort's for supplying comets to a distant comet cloud. This argument depended, however, on whether Pluto was sufficiently massive to play this role, and no accurate measurements of its mass were available. As a result, Kuiper secured time on the 200-inch telescope at Palomar Mountain in March 1950 to perform this study. His measured diameter (0.4 that of Earth) led to a calculated mass of one-tenth of Earth's, smaller than he had expected but still sufficient for Pluto to perturb comets from its vicinity. Considered an impressive accomplishment by contemporary astronomers and widely reported, Kuiper's measurement not only reinforced the general outlines of Oort's comet theory among American astronomers, but indirectly his own views on accretion and condensation within the solar system.[37]

Within months, Kuiper buttressed his cosmogonal arguments with further interpretations. In June 1950, at a symposium devoted to asteroidal research, Kuiper announced that collisions between asteroids were likely responsible for producing the "families" of asteroids initially discovered by Hirayama, one focus of Brouwer's re-

[36] Quoted from Whipple to Oort, Aug. 16, 1949, Box 32, and idem to Oort, Feb. 28, 1950, Box 33, JHO.

[37] Kuiper had wanted to measure Pluto's diameter since discovering Titan's atmosphere in 1944, to determine if Pluto possessed an atmosphere as well; see Kuiper to Struve, Aug. 13, 1944, and Russell to Kuiper, Mar. 8, 1948, both Box 28, GPK; Kuiper to W. Baade, Nov. 10, 1949, Box 29, and idem to Science Service, May 31, 1950, Box 31, GPK; see also Clemence (1951, 228), Hoyt (1980a, 243), and Kuiper (1950b). While contemporary measurements indicate Pluto's mass is but two-thousandths of Earth's, the Hubble Space Telescope confirmed the existence of the Kuiper belt in mid-1995, a major development of twentieth-century cosmogony; see Weissman (1993) and Wilford (1995).

search at Yale. Kuiper's collisional hypothesis did not explicitly contradict the unity of asteroids and comets that Oort proposed. Nevertheless, it supported the link between asteroids and meteorites that geochemists had accepted since the beginning of the twentieth century, and soon gained Urey's backing as a likely means for producing chemically distinct groups of asteroidal meteorites. It also gained support from the mathematical astronomers Brouwer and Clemence. Moreover, Whipple approved this general framework, since it supported the distinctions he wished to draw between cometary and asteroidal meteors. Kuiper's argument that comets and asteroids had distinct origins thus reinforced Whipple's extensive meteor studies as well as Urey's geochemical research.[38]

This growing body of evidence convinced Whipple to reject Lyttleton's arguments for the interstellar origin of comets. By late 1950 he accepted that Oort's and Kuiper's theories provided an adequate explanation for the formation and dispersion of comets; furthermore, he acknowledged that Urey's arguments on condensation within the early solar nebula supported his idea that comet nuclei were icy conglomerates of dust and gas. The second factor may well have been more important to him, for the chemical composition of comets was relevant to studies of solar system evolution, assuming they had a solar origin. Whipple had previously nurtured this hope but not publicly declared it, owing to his doubts on the origin of comets within the solar nebula. In any case, Whipple now chose to emphasize his comet studies over his cosmogonal theory. In September 1950, Whipple voiced doubts about his own dust cloud hypothesis for the solar system, declaring to Urey he was "not at all sanguine with regard to the formation of the planets themselves by the method I have suggested." And in a late-1950 article on cometary origins published the following year, Whipple made clear the underlying cause of his shift in emphasis: "If the comets were acquired by the system since its origin, possibly in the manner suggested by R. A. Lyttleton, no implications concerning the origin of the system would be involved." In that case the material of the comets would typify interstellar solids, possibly modified. "But if the comets were formed with the planets *and* with the Sun, limits on the physical surroundings of the protosun become established."[39] The arguments Whipple presented concerning the icy comet nucleus, made with greater quantitative rigor by 1951, convinced many American astronomers. Struve, whose own research included the interstellar medium and the natural abundance of the elements, declared that "[t]he principal reason for preferring [Whipple's] new hypothesis is that it accords better with the chemical construction of comets." No less important, Struve and other astronomers read Whipple's new results as providing additional support for Kuiper's ideas on the formation of comets and asteroids.[40]

[38] Urey to Kuiper, June 13, 1952, and Kuiper to Urey, June 21, 1952, Box 51, HCU; see also Brouwer (1950, 162), Clemence (1951, 229), Kuiper (1950a; 1951b, 363), and Urey and Craig (1953, 74).

[39] Quoted from Whipple to Urey, Sept. 28, 1950, Box 102, HCU (emphasis in original), and idem (1951, 471); see also Whipple to Urey, Dec. 3, 1951, Box 102, HCU.

[40] Struve (1952, 273) and Watson (1953).

Consensus, then Controversy: Interdisciplinary Turmoil, 1950–1955

By the early 1950s, this debate over the origin of comets was over. Astronomers in North America and Europe published nine articles on cometary origins in 1951 and eight in 1953, but only two in 1955 and none in 1956. No Americans contributed to these totals after 1952.[41] Although interest in comets remained high – Pol Swings organized the fourth Colloque international d'astrophysique at Liège in 1952 on the physics of comets, for instance – American research came increasingly to focus on comets as representative samples of the solar system, and Oort largely returned to stellar and galactic studies, satisfied that his ideas of cometary origins were correct. Yet consensus about cometary formation did not extend significantly beyond the American astronomical community. Few Soviet astronomers embraced the arguments by Oort, Whipple, and Kuiper, and Lyttleton mounted an increasingly shrill defense of his comet model during the 1950s, describing Oort's comet cloud model as "absurd" and lacking even "a vestige of tenability." Nonetheless, his comet formation ideas no longer received a sympathetic hearing from American astronomers.[42]

The importance of this controversy was threefold. First, it settled (for American astronomers, at least) an old and significant debate about the origin of comets. Second, although centered on comets, it illuminated a profound faith among American astronomers in the accuracy and significance of their immense store of observational data, leading them to distrust the sweeping deductive theories that Cambridge school researchers sought to impose on astrophysics. The depth of these feelings were revealed in a far-ranging, late-night argument between Kuiper and Bondi in 1954. Under the dome of the 82-inch McDonald telescope, during a scheduled observing run, an agitated Kuiper told Bondi that American astronomers would not trade data for theoretical elegance, further declaring that the steady-state theory's embrace of the continuous creation of matter to achieve strict uniformity in space and time undermined "the most fruitful guiding principles both in physics and astronomy for the last decades." While many American astronomers echoed Kuiper's distrust of the steady-state theory – and geochemists complained that Hoyle failed to understand how planetary geochemistry constrained physical theory – Lyttleton, Bondi, and Hoyle nevertheless continued to claim that Lyttleton's model of cometary formation demonstrated the steady-state cosmology. This divide limited productive contact between these communities.[43]

[41] Article counts are derived from the *Astronomischer Jahresbericht* for 1950–3, 1955, and 1956. Articles on the structure of comet nuclei, although related to the question of cometary origin, are not included in this count.

[42] Swings to Whipple, Nov. 20, 1951, Box 8, FLW–HU; "Report for Commission No. 16 of the International Astronomical Union," n.d. [1955], Box 11, GPK; see also Beyer (1956), Jones (1953), Lyttleton (1953a), and Paneth (1954). Lyttleton's sharpest attacks appeared in Lyttleton (1953b, vii, 157–8).

[43] Quoted from Kuiper to S. Chandrasekhar, Feb. 1, 1954, Box 10, GPK; see also Kuiper to B. Strömgren, Feb. 2, 1954, Box 14, and idem to G. Gamow, Feb. 12, 1954, Box 30, both GPK; and Craig to Urey, Sept. 10, 1956, Box 23, HCU. This rift widened further in the 1960s; see Kuiper to Oort, June 19, 1964, Box 9, GPK, and Lyttleton (1964).

Finally, the debate over comet origins convinced most American astronomers that Kuiper's nebular cosmogony was essentially correct. Between 1951 and 1955, Kuiper worked to refine this theory by interweaving astronomical evidence on the solar system's origin with studies of planetary evolution by Brown and Urey into a common tapestry. Although Kuiper and Urey sought further interdisciplinary cooperation between their fields, this cooperation soon dissolved into the largest conflict to affect mid-twentieth-century American astronomy.

THE KUIPER–UREY CONTROVERSY: ENDING COOPERATIVE RESEARCH

Already by 1950, as we have seen, geochemists like Urey had come to regard the Moon as a promising subject from which to gain information on the early environment of the solar system. A number of astronomers after 1945 had also become interested in the Moon. In part they wished to understand the processes that had sculpted the lunar surface, but an even stronger aim was to determine whether the Moon showed evidence of global melting early in its history. The latter issue was important for Kuiper, who wanted to assess the significance of radioactive melting for this body to understand how planets and asteroids had evolved through melting. Lunar research became for Kuiper a way to assess the prevalence of radioactive elements in the original solar nebula and to calibrate its role in his cosmogonal theory. Kuiper believed this problem could be best addressed through collaboration with Urey, providing a foundation for broad investigations in planetary geochemistry.

Most scientists in the early 1950s regarded the evidence for global lunar melting as mixed. In his 1949 *The Face of the Moon* (see Chapter 5), the Michigan-trained astronomer Ralph Baldwin argued that the Moon had melted at least partially in its early history. Baldwin pointed to such lunar features as the lava flows of Mare Tranquillitatis and isolated lava-filled craters as evidence that molten rock had once existed just below the lunar surface. This argument gained support from Dinsmore Alter, the former Leuschner student then at the Griffith Observatory and Planetarium in Los Angeles. Urey nonetheless continued to argue that the Moon was a geochemically primitive body accreted at low temperatures. While agreeing with Baldwin that the lunar maria appeared covered with lava flows, he ascribed them to the impacts that had caused the maria themselves. As evidence, Urey pointed to radial grooves in the mountains ringing Mare Imbrium, declaring that these could only have been caused by dense iron–nickel objects from the original impactor slicing through solid, cold material.[44]

Among the more significant astronomical investigations of the Moon at this time – particularly in terms of their subsequent influence on lunar research – were those

[44] Baldwin (1949, 178–99); see also Urey to Alter, June 9, 1954, Box 3, HCU; and Alter (1948; 1955), Wilhelms (1993, 19).

of Kuiper. His arguments about the Moon's thermal history – and the reasons that Urey so vehemently opposed them – can only be understood in terms of Kuiper's commitment to his cosmogonal theory, which convinced him that a molten origin for Earth and the Moon was required. As we have seen, Kuiper approached the birth of the solar system as a special case of binary star formation, and thus sought to determine the initial conditions in the solar nebula in terms of astrophysical processes. In 1951, Kuiper concluded that planetary systems would develop whenever the density distribution within a contracting gas nebula was too low to permit the formation of a companion star, but not so low as to prevent the production of protoplanets from which planets would condense. This theoretical constraint implied that the mass of the solar nebula was only marginally less than that needed to produce two binary stars, and impressed upon Kuiper the idea that the Sun's original nebula had contained large amounts of hydrogen and helium since expelled from the solar system. At the same time, Kuiper welcomed this predicted high abundance of hydrogen, for without this excess mass, his proposed condensation mechanisms for producing protoplanets failed, resulting instead in the production of billions of cometlike bodies.[45]

A central problem that Kuiper and others interested in cosmogony faced was explaining how solid particles had condensed from gases in the original nebula. One particular difficulty was finding a way to account for the ratio between iron and silicates in Earth compared to that measured in the Sun. By the late 1940s, most astronomers and geochemists agreed that Earth's silicates were depleted in contrast to the Sun's. This troubled members of both disciplines, committed to the idea that the Sun and its planets shared a common origin. It was in order to resolve this problem that Kuiper initially approached Urey in late 1949. From their early exchanges, it is clear that Kuiper regarded his collaboration with Urey with greater anticipation and anxiety than that experienced in other professional relationships. Although a senior American astronomer, Kuiper was acutely conscious of Urey's standing as a Nobel laureate and leader among atomic scientists. In 1950, Kuiper began a letter to Urey with "Dear Harold," exclaiming, "I hope that you will let me address you as I have done above! And I hope you will want to call me by my first name." Urey responded with equal enthusiasm and warmth. With few exceptions, relations between these two scientists remained friendly and cordial through the early 1950s.[46]

In part their enthusiasm rested on a foundation of close intellectual cooperation. As we first saw in Chapter 3, Urey accepted Kuiper's nebular cosmogony as the starting point for his own geochemical studies, and worked to develop his ideas of planetary geochemistry within the broad outlines of Kuiper's failed-binary-star model. For his part, Kuiper initially accepted Urey's argument, developed in detail at the

[45] Kuiper to W. Baade, Nov. 10, 1949, Box 29, and idem to B. Strömgren, Nov. 13, 1950, Box 31, both GPK; see also Brush (1981, 92), Kuiper (1951a, 3, 11; 1951b, 383–4).

[46] Quoted from Kuiper to Urey, Jan. 29, 1950, Box 51, HCU; see idem to Urey, Oct. 19, 1949, Box 51, HCU; idem to H. P. Berlage, Feb. 8, 1952, Box 30, GPK.

1950 Rancho Santa Fe conference, that chemical fractionation – particularly a brief high-temperature phase before Earth had fully accumulated – caused silicate particles to be preferentially lost compared to iron. He believed this would account for the observed depletion of silicates to iron within Earth, compared to the abundance ratios of these materials within the Sun. In early 1950 Kuiper advised Urey that their recent discussions left him feeling "that our respective studies concerning the origin of the Earth are in effect mutually consistent."[47]

An error Urey discovered in his own chemical studies soon caused Kuiper difficulty, however. In April 1952, Urey advised Kuiper that he had omitted solution effects in his initial calculations about the condensation of silicates in the solar nebula. When applied, these factors eliminated the method of chemical fractionation that Urey had proposed the previous year to explain the relative abundance of silicates compared to iron. Urey found that this error posed no significant problem for his study of planetary evolution, as the presence or absence of this chemical process did not affect the general model of chemical evolution that he proposed for the planets. Indeed, Urey first believed that his correction brought his chemical ideas closer in line with astronomical evidence that Kuiper held to be critical constraints on cosmogonal theory; but Kuiper now found that he could not use Urey's chemical arguments to justify removal of excess hydrogen and helium that his theory predicted would exist in the solar nebula, assuming planetary formation to be a special case of binary star formation. This troubled Kuiper, who knew that Whipple and other astronomers regarded the excess hydrogen issue as the most troubling aspect of his cosmogony. Kuiper therefore began to seek an alternative method of accounting for silicate concentrations.[48]

By the summer of 1952, to deal with this unexpected setback, Kuiper began sorting again through the leading hypotheses proposed for the origin and structure of Earth. He had already done so to prepare his studies of planetary atmospheres in the late 1940s, and was preparing to review them again before editing *The Earth as a Planet,* the second volume of his *Solar System* series. Now, however, Kuiper found the matter an urgent undertaking. He continued to reject the Kuhn–Rittmann hypothesis, believing that astronomical studies of asteroids ruled it out. Since iron and stony phases were often separated in large meteorites, which he argued were derived from astronomical bodies "of roughly 500 kilometers in diameter," Kuiper believed a similar differentiation had occurred in larger bodies such as Earth, producing an iron core.[49]

Kuiper also showed no willingness to reconsider the 1949 theory by William Ramsey that high densities at the centers of terrestrial planets resulted not from iron but rather a silicate phase change. Modern astronomical evidence, Kuiper believed,

[47] Kuiper to Urey, Jan. 17, 1950, Box 51, HCU.

[48] Urey to Kuiper, Apr. 10, 1952, Box 51, HCU; Kuiper to Wilson, June 19, 1952, Box 32, GPK; Whipple to W. M. Reed, Oct. 14, 1952, Box 8, FLW–HU.

[49] Quoted from Kuiper to J. T. Wilson, July 2, 1952; see idem to J. Jacobs, July 2, 1952, both Box 31, GPK.

fatally undercut Ramsey's argument: Ramsey had developed his theory partly to account for Mercury's low density, then widely reported as 2.86; but new measurements by Rabe at Cincinnati had increased its accepted density to 5.46.[50] Kuiper also pointed to new geophysical evidence to reject Ramsey's arguments, including Elsasser's efforts to link Earth's magnetic field with an iron core, and Bridgman's high-pressure studies at Harvard, which did not yield the phase changes predicted by Ramsey's theory.[51]

In rejecting the Kuhn–Rittmann and Ramsey models for Earth's interior, Kuiper was breaking no new ground. As we saw in Chapter 3, geologists and geophysicists at the 1950 Rancho Santa Fe conference had all reached similar conclusions, and believed the existence of Earth's iron core was certain. Whipple, Wildt, and Urey had also rejected these ideas. By embracing Elsasser's latest conclusions, however, Kuiper found himself reconsidering the question of Earth's thermal history. As late as 1951, Kuiper seemed to have no firm opinion on whether Earth had formed completely solid or completely molten, although his investigations of Titan's atmosphere inclined him toward the latter view. However, perhaps as a result of reading Elsasser, who referred to the initial origin of Earth from "hot and fluid matter," Kuiper came to reconsider a theory of Earth by Arnold Eucken, an eminent German physical chemist. In 1944, responding to the compressed hydrogen core model of Kuhn and Rittmann, Eucken had asserted that an iron core existed; to support this argument, he had suggested that iron had condensed before silicate in the early solar nebula, assuming that the temperature of the initial solar nebula was high. Kuiper had known about Eucken's ideas since at least 1950. What now attracted him to Eucken's work was that it allowed him to resolve the silicate deficiency and simultaneously rid the solar system of excess hydrogen and helium. Kuiper began leaning toward the idea that Earth had become molten during its formation, and simultaneously began expanding his network of contacts with other researchers in the geological sciences.[52]

The longer Kuiper examined the idea, the more pleased with it he became. In addition, he was soon heartened to learn that not all American geologists subscribed to the Rancho Santa Fe consensus that Earth had never experienced a molten phase. One dissenter was J. Tuzo Wilson, the University of Toronto geophysicist whom Kuiper had come to know while preparing *The Earth as a Planet.* A Princeton-educated geologist also trained in physics and exposed to oceanography through the mentoring of Princeton's Richard Field, Wilson was among the more broadly trained geologists of his generation. By the early 1950s, Wilson's research interests included aerial geology and the structure of the Canadian (Laurentian) Shield, whose exposed rocks were among the oldest on Earth. Wilson advised Kuiper that

[50] Kuiper to O. Struve, Aug. 15, 1955, Box 28, GPK; Wildt to Kuiper, Oct. 29, 1952, Box 1, RW; see also Kuiper (1952b, 308, 339) and Wildt (1947, 85).

[51] Kuiper to Urey, Feb. 2, 1952, Box 51, HCU; idem to E. C. Bullard, June 30, 1952, Box 29, GPK; see also Elsasser (1950, 10, 13; 1951).

[52] Quoted from Elsasser (1950, 8); see Kuiper to Urey, Aug. 28, 1950, and Urey to Kuiper, June 9, 1952, Box 51, HCU; see also Eucken (1944, 112, 116, 120), and Kuiper (1951b, 416; 1952a, 12).

he felt "very much relieved" to learn that astronomical evidence favored the view that Earth had formed as a molten globe, a conclusion he had reached through his studies of the shield. Kuiper came to value Wilson's support. Although Wilson was somewhat of a maverick among North American geologists – he also defied the widely held view that mountain building resulted from convection currents in Earth's mantle, favoring instead the Earth contraction theory – his work on the ages and origins of rock units in the Canadian Shield was highly regarded by American geologists. Moreover, Wilson had made Toronto Canada's leading center for geophysical research, and moved freely between its physics and geology departments. His views thus had a stable institutional foundation and considerable authority.[53]

The principal reason Kuiper felt the need to secure an ally for his molten Earth idea was that Eucken's ideas of condensation challenged Urey's model of chemical fractionation. In his major 1951 *Geochimica et Cosmochimica Acta* article, as well as in *The Planets,* Urey had drawn a clear line around his German colleague's chemical ideas. While calling Eucken's theory of planetary formation "most impressive," adding that he had not known of it before completing his *Geochimica* paper, Urey finally dismissed this concept on several grounds, including its incompatibility with the consensus on Earth's cold origin forged at Rancho Santa Fe and Urey's own argument against the presence of an iron core in Mars. Nevertheless, Urey found Eucken's alternative a considerable challenge, advising Kuiper that he was "seriously jarred by Eucken's work." Perhaps in the hope of convincing Urey that a molten origin for Earth was justified, but more likely as a way of countering opposition, Kuiper asked Wilson's opinion of a particular geochemical argument that Urey had used to defend his cool origin for Earth. What had impressed Urey and other participants at the Rancho Santa Fe meeting was Bowen's contention that basalts, granites, and other comparatively light materials would be more abundant at Earth's surface than observed to be had Earth actually become molten. Urey regarded this as firm evidence for his ideas, whereas Kuiper considered the matter far less clear-cut, speculating that Earth's distribution of crustal minerals possibly owed to celestial debris striking its early surface. Although Wilson apparently failed to respond, Kuiper increasingly drew on astrophysical concepts to guide him past what seemed to him errant geological conclusions. With his key disciplinary ideas threatened, Kuiper found himself increasingly unwilling to yield to the authority of interdisciplinary collaborators on problems in cosmogony.[54]

The importance of disciplinary constraints was evident in Kuiper's response to Urey's claims about Earth's thermal history. In letters to members of Wilson's group in Toronto, Kuiper revealed his increased displeasure with Urey's 1949 position that Earth might be warming, noting that this "strikes an astronomer as going

[53] Quoted from Wilson to Kuiper, July 8, 1952, Box 31, GPK; see also Kuiper to Wilson, June 19, 1952, Box 11, GPK. Bullard and Birch similarly opposed the idea of mantle convection; see J. Tuzo Wilson OHI (Ronald E. Doel, Feb. 16, 1993, at AIP), Strangway (1980, vii–viii), and Wood (1985, 160–1).

[54] Eucken to Urey, Apr. 1, 1949, Box 31, and Urey to Kuiper, Aug. 3, 1950, Box 102, both HCU; Kuiper to Wilson, June 19, 1952, Box 31, GPK; and Urey to Wilson, May 21, 1954, Box 21, JTW.

against the trend of nature." John A. Jacobs, an applied mathematician and geophysicist, agreed, declaring, "Like you, I am against the concept of the Earth heating up, which seems to go against the trend of nature." While Kuiper appreciated Urey's creative thinking, and was stimulated by it, Urey's training in physical chemistry allowed him to explore pathways of chemical evolution that astrophysicists frequently found foreign. Although this particular disagreement stemmed from uncertainties in the absolute abundances of elements, it hinted at the difficulties ahead for those who sought institutional, methodological, and professional alliances between astronomy and geochemistry.[55]

Kuiper also gained encouragement for a molten origin of Earth from other leading American geologists, including Hess and Rubey, whose study of the origin of sea water had come to Kuiper's attention. In May 1953, Hess welcomed Kuiper's emphasis on a molten origin for Earth, finding that it promised to resolve problems in mantle geochemistry. Although Kuiper was unable to answer specific questions that Hess posed, Hess's own embrace of a molten Earth encouraged Kuiper, then preparing to publish a final installment of his cosmogonic research.[56]

His December 1953 publications, "Note on the Origin of the Asteroids" and "Satellites, Comets, and Interplanetary Material," more than his better-known "On the Origin of the Lunar Surface Features," marked the beginning Kuiper's intellectual break with Urey over the chemical evolution of the solar system. Because these papers did not explicitly deal with the formation of Earth or the Moon, Kuiper made no direct criticisms of Urey's ideas; but Kuiper, more than ever concerned to demonstrate the evolution of planetary systems as a special case of binary star formation, used these papers to emphasize a preeminent role for astronomical evidence in developing models of planetary origin. Specifically, Kuiper introduced two new constraints into his theory of cosmogony. First, he argued that the planets had taken an appreciable fraction of the system's lifetime (100 million years) to accrete, owing to the time required for newly born solar-type stars to begin emitting strong ultraviolet and corpuscular radiation. Second, he insisted that global melting of the terrestrial planets was required to account for depleted silicate abundances according to Eucken's condensation theory. Kuiper's ideas thus disagreed with Urey's on three points. Urey had not only argued for a cold origin for Earth and his own model of chemical fractionation but believed (from studies of fractionation processes) that Earth had formed in as little as several thousand years.[57]

In late 1953, for the first time, Kuiper directed his attention to the Moon. His principal reason for doing so was to seek signs that the Moon had undergone global melting, perceiving it as another source of evidence for his cosmogonal theory. In-

55 Quoted from Kuiper to Wilson (and Jacobs), June 19, 1952, and Jacobs to Kuiper, June 26, 1952, both Box 31, GPK.

56 Kuiper to Rubey, Feb. 1, 1952, Box 31, and Hess to Kuiper, May 18, 1953, and Kuiper to Hess, Mar. 9, 1954, both Box 30, all GPK; see also Kuiper (1955, 822) and Rubey (1951; 1955).

57 Struve to Urey, Jan. 7, 1953, and Urey to Struve, Jan. 15, 1953, both Box 87, and Urey to Kuiper, June 13, 1951, Box 51, all HCU; see also Kuiper (1953b) and Urey (1952c, 111, 147).

creasingly Kuiper sought to relate geologic features on Earth to those on the Moon in support of this idea. To Hess, Kuiper declared that the Mid-Atlantic Ridge (which Hess had begun to study) reminded him of ridges on the Moon's Mare Imbrium. Kuiper's attempts to synthesize geological phenomena within his theoretical framework seemed naïve and presupposing to many American geologists; but this work convinced him that systematic investigations of the Moon would yield important new information on planetary evolution. It was at this time that Kuiper constructed his binocular eyepiece for visual study of the Moon with the McDonald 82-inch reflector, and laid the groundwork for a photographic lunar atlas to permit immediate reference to lunar surface features. The Moon, for Kuiper, had become a means to study the thermal history of Earth.[58]

Kuiper's arguments for a molten origin of the Moon appeared in a short paper in the *Proceedings of the National Academy of Sciences* in December 1954. Kuiper argued that the Moon's thermal history could be reasoned from studies of asteroids: Since meteorites showed clear signs of melting, the larger and better thermally isolated Moon was likely to have melted to an even greater degree. He presented calculations to show the Moon "was nearly completely melted by its own radioactivity" between 0.5 and 1 billion years after its formation. He then proceeded to criticize two fundamental observations Urey had made about the Moon: (1) that the lunar highlands exhibited only telltale signs of meteorite infall without melting, and (2) that the lava flows covering the maria were local, formed by the intense heat of major impacts. Rebutting Urey, Kuiper asserted that melting had also occurred in the lunar highlands, pointing to the outlines of "ghost"-like craters visible only under high illumination, which he interpreted as the remains of ancient craters partially drowned by the flow of a molten surface. That central peaks were present in younger but not older craters on the highlands Kuiper also argued as evidence for a changing thermal environment. Since he believed most parts of the lunar surface had melted, Kuiper was then led to maintain that the formation of mare regions resulted when large infalling bodies pierced a thin crust, releasing floods of molten lava from just under the surface. Kuiper admitted two weaknesses to his model. He found it hard to account for the mountains ringing Mare Imbrium, which, like Urey, he believed were great blocks thrown out from the force of the collision; he argued that they were partially submerged in frozen lava, preventing them from sinking completely through the partially melted crust. He also noted, in an important concession to geochemistry, that his arguments rested on the actual abundances of radioactive elements, a field where Urey and his colleagues remained more expert than astronomers. Geochemists, he defensively noted, still viewed these values with "considerable uncertainty."[59]

[58] Kuiper to B. Strömgren, Feb. 2, 1954, Box 14, idem to Hess, Mar. 9, 1954, Box 30, and idem to N. Mayall, Mar. 23, 1955, Box 12, all GPK; and Harold Masursky OHI (Ronald E. Doel, June 17, 1987, AIP).

[59] Kuiper (1954, 107, 102); see also Kuiper to Hess, Mar. 9, 1954, and idem to Gamow, Apr. 30, 1955, both Box 30, GPK.

Since most scientists already interested in the Moon's origin, including Baldwin and Alter, had already published estimates of the degree of lunar melting, Kuiper's article stimulated no immediate response. The following July, however, Urey published in the *Proceedings* a sharply critical reply. His seven-month delay, Urey later claimed, was so that he could write a civil response. It seems more likely that Urey decided to write against Kuiper after he received a copy of Kuiper's edited *The Earth as a Planet* in March 1955. Soon afterward Urey dispatched a letter to David T. Griggs, the UCLA geophysicist and former Bridgman student instrumental in developing the convection current theory of mountain building and its principal advocate in North America. In his letter to Griggs, Urey charged that Wilson's article about Earth's mantle, which Kuiper had commissioned for this volume, shortchanged the convection theory. Urey's complaint had merit: Although Wilson briefly discussed the leading ideas developed by Griggs and Vening-Meinesz for mountain building, the article stressed his favored contraction theory. In this blunt letter of three single-spaced pages, Urey outlined several counterarguments to Kuiper's claims. All of these were incorporated into the famous article Urey submitted to the National Academy of Sciences (NAS) less than six weeks later.[60]

In "Some Criticisms of 'On the Origin of the Lunar Surface Features' by Gerard P. Kuiper," Urey raised three arguments against Kuiper's claim that the Moon had experienced a molten phase. First, declaring that meteorite abundances of radioactive elements were much lower than Kuiper claimed, he argued that the Moon was unlikely to have melted from internal heat. Second, Urey rejected Kuiper's argument that the observed lunar bulge was caused by a preferential accumulation of infalling debris, arguing that the tensile strength of a partially melted Moon would be insufficient to sustain the lunar figure so far from isostatic equilibrium. Finally, he took sharp aim at Kuiper's idea that a solid lunar crust could "float" atop a melted layer. As a chemist, this idea struck Urey as chemically absurd, since among common substances solid water alone floats atop its liquid. Complaining that "[i]t would be a thankless task to review adequately this paper in all its details," Urey closed the first round of this exchange.[61]

Previous accounts have interpreted the heated Kuiper–Urey controversy as resulting from their divergent views on radioactive heating and silicate condensation processes; yet Urey's 1955 attack on Kuiper's melted Moon arguments was not extraordinarily harsh. Inclined toward confrontation, Urey had directed similar rebuttals against colleagues in the past, only to resume cordial relations once his intellectual passions cooled. As colleagues quickly discovered, however, this new clash was hardly limited to Urey's testy criticisms in the *Proceedings,* or Kuiper's subsequent, shrill reply. In the seven short months that separated Kuiper's 1954 lunar article and the

[60] Urey to Griggs, Mar. 31, 1955, Box 40, HCU; the article in question was Wilson (1954). By then Urey had written a *Proceedings* paper criticizing Kuiper's idea that tektites came from the Moon, although its tone was reserved compared to subsequent attacks; see Urey (1955a, 31).

[61] Quoted from Urey (1955b, 427).

General Assembly of the International Astronomical Union (IAU) in Dublin the following summer, Urey abrogated his agreement to write on natural abundances for Kuiper's *Solar System* series, petitioned National Science Foundation director Alan Waterman to end all subsidies for Kuiper's editorial projects, and lobbied senior American and European astronomers to remove Kuiper as President of IAU Commission 16 (Planets and Satellites). For a time, Urey seriously considered leveling a charge of plagiarism against Kuiper for sections of his edited volume, *The Atmospheres of the Earth and Planets,* reissued by the University of Chicago Press in 1952. Although the Moon's thermal history was clearly important for these men, the Hot Moon–Cold Moon controversy, as it later became known, was more than an intellectual disagreement. Rather, it was a painful methodological and professional divorce that temporarily derailed interdisciplinary research between astronomers and planetary geochemists. Deeper roots for this conflict must be sought.[62]

Several overlapping factors influenced the course of Urey's relationship with Kuiper after 1954. All were intensified by the transient nature of interdisciplinary research programs and the lack of clear professional and institutional support for the research field they sought to create. The first was Urey's increasing distrust of Kuiper as a committed partner in geochemical studies, initially caused by Kuiper's embrace of Eucken's high-temperature theory of chemical separation. Urey's commitment to a low-temperature chemical fractionation matched Kuiper's devotion to his astronomically based cosmogony, and through the early 1950s Urey had worked to find added geochemical evidence for his interpretations. For instance, Urey had sought evidence that natural abundances of uranium, thorium, and potassium were lower than Kuiper believed, limiting their influence on planetary melting. In 1953, using new meteorite analyses done at his Chicago laboratory, Urey had declared that radioactive elements were preferentially concentrated at Earth's surface, and that meteorite geochemistry was consistent only with a low-temperature origin. Measured concentrations of radioactive elements in meteorites, he had argued, were too high by a factor of 3. Revealingly, Urey had also sought astronomical evidence favoring a brief high-temperature outburst from the Sun that his theory required, rather than the dim Sun that Kuiper's astronomical model predicted. Writing Chandrasekhar as well as Princeton's Martin Schwarzshild, Urey pressed them to investigate possible astrophysical mechanisms that could cause a high-temperature spike, advising Chandrasekhar that he could not fit "the chemical facts as I see them" with a protracted, cool accretion. Kuiper, perceiving this as an affront to his disciplinary authority, complained that astronomers regarded Urey's search for such a mechanism as "*ad hoc* and contrary to our understanding of stellar structure."[63]

62 Brief summaries of this controversy include Brush (1981, 93–4; 1982b, 891–3), Tatarewicz (1990a, 24), and Whitaker (1985). On related disputes see Urey to Rubey, Nov. 23, 1954, Box 80, HCU, and Harmon Craig OHI (Ronald E. Doel, Nov. 27, 1990, at AIP), as well as Urey to Waterman, Feb. 9, 1955, Box 101, and idem to R. Hemens, draft, n.d. [July or Aug. 1955], Box 107, both HCU.

63 Quoted from Urey to Chandrasekhar, Aug. 25, 1953, Box 18, HCU, and Kuiper to W. W. Rubey, Dec. 5, 1954, Box 31, GPK; see Urey to Whipple, Nov. 17, 1954, Box 102, and idem to R. Revelle, Nov.

Urey was frustrated with Kuiper's enthusiasm for Eucken's ideas for another reason: They undermined results obtained through Urey's chemical fractionation research, including discovery of the Urey–Craig meteorite groups. As he angrily advised Brown, Kuiper's embrace of Eucken's chemistry

> throws away Craig's and my data, for we thought we had found in the two groups of chondrites an example of the fractionation that was general in the solar system. Apparently Kuiper would not fractionate them at all, and for some reason God made them that way in the beginning. I do not like to throw away data; there is too little of it around already.[64]

Writing Griggs, Urey grumbled that "Kuiper's chemistry has an Alice in Wonderland quality. Indeed he must pay off the elements on Saturday night just as Humpty Dumpty paid the words." But professional and disciplinary issues irritated Urey above all: Kuiper "would not regard anything that a mere chemist would say as important." In part a reflexive response to the traditional Comtean rank ordering of scientific disciplines, Urey's remark also reflected his pained, accurate recognition that he and Kuiper no longer shared a common methodological foundation for "astro-geochemical" studies.[65]

Compounding tensions over disciplinary standards was Urey's increasing distrust of Kuiper's abilities as an observer. In particular, he grew frustrated with Kuiper's measurements of planetary diameters. Urey had initially accepted Kuiper's revised value for Mars of 0.532 Earth's radius, which generally supported Urey's argument that the interior of Mars was homogeneous and chemically undifferentiated. By the early 1950s, however, Urey had became aware that all of Kuiper's planetary radius determinations were larger than previously accepted values, and sought reassurance that systematic error did not affect them. Dissatisfied by Kuiper's curt, annoyed reply that previous measures had been made with inferior instruments, Urey had turned to Whipple, Greenstein, and other U.S. astronomers for independent appraisals. Although none found merit in his concerns, Urey's residue of doubt came to influence his perception of Kuiper's measurements in other fields, including lunar studies; to Alter, Urey declared, "I think that all his conclusions are colored by his thinking, so I doubt his observations very much." Exacting in his own experimental work, Urey increasingly found his quarrel to be the standards of astronomical measurement.[66]

Footnote 63 *(cont.)*
24, 1954, Box 78, both HCU; and Schwarzshild to Urey, May 29, 1956, Schwarzshild Papers, Princeton, noted courtesy of David DeVorkin.

64 Urey to Brown, Jan. 13, 1954, Box 15, HCU.

65 Quoted from Urey to Griggs, Mar. 31, 1955, Box 40, HCU; see also Urey and Craig (1953, 36).

66 Quoted from Urey to Alter, Feb. 9, 1955, Box 3, HCU; see Urey to Kuiper, Aug. 3, 1950 and June 13, 1951, and Kuiper to Urey, Nov. 14, 1951, all Box 51, HCU; Urey to Whipple, Nov. 26, 1951, Box 102, idem to K. E. Bullen, Mar. 14, 1952, Box 16, idem to Greenstein, June 23, 1953, Box 39, and idem to E. C. Slipher, Aug. 25, 1953, Box 85, all HCU; see also Urey, "The Radius of Mars," MS dated Apr. 6, 1954, Box 135, HCU.

At the heart of the controversy between Kuiper and Urey, however, was a more common but fundamental issue: recognition of scientific priority. Within their respective disciplines, Kuiper and Urey could count on customary reward systems, including discipline-centered prizes and journals of record, to take notice of their achievements and claims of priority. At the interstices of astronomy and geochemistry, however, no such structures existed, making priority disputes a poignant issue. Tensions over priority had built up gradually between both men, even when their work remained mutually stimulating. In 1952, for instance, after reading page proofs of Kuiper's first major contribution to cosmogony, Urey had written Kuiper, "You italicize the statement that the Moon has been fractionized by a factor of 2. I agree that it is important and it was important to me two years ago. I suggest a footnote and reference in regard to this." Kuiper's 1954 paper on the Moon and the release of his edited *The Earth as a Planet* reignited this long-smoldering issue. Writing Waterman, Urey complained that "Kuiper in his writings minimizes the work of other people in the field by referring to minor things they do with pinpointed references, and then includes their important work at other places without references, so that the reader infers it is his own work." By the early 1950s Kuiper, like Clemence, Rubey, Edward Bullard, and others, had become aware of Urey's unusual sensitivity to proper citation practices, perhaps reflecting Urey's own priority conflicts as a junior scientist (his mentor, Gilbert Lewis, had used Berkeley's well-equipped laboratories to edge out Urey's own Columbia team to develop a method for preparing pure heavy water). Whatever its cause, Urey demanded fastidious attention to priority in all published work.[67]

Until the summer of 1955, the matter of citations remained a troubling problem between these men, but not one that threatened their professional or intellectual relationship. Circumstances changed dramatically late that July, shortly before the scheduled Dublin meeting of the IAU. At that time, Urey received a copy of Kuiper's draft "Progress Report" for the IAU's Commission 16. As commission president, Kuiper was obliged to prepare a summary of planetary research accomplished since the last edition of the IAU *Transactions.* The draft that Kuiper prepared was a curious document. While it emphasized contributions such as Herzberg's studies of planetary atmospheres, the work of the Soviet astronomer V. V. Sharonov on Venus and Mars, and Kuiper's own investigations of the asteroids and the Moon, it mentioned only two contributions by Urey. Nor did it cite works by Urey's coworkers, nor those of Brown's team in California. Furthermore, it eschewed references to cosmogony. In it Kuiper declared, "The author first considered analyzing these [cosmogonic studies] here in brief form, but was forced to abandon this plan when it

[67] Quoted from Urey to Kuiper, Jan. 16, 1952, Box 51, and idem to Waterman, Feb. 9, 1955, Box 101, both HCU; see Urey to Jeffreys, July 5, 1951, Box 47, idem to Rubey, Nov. 23, 1954, Box 22, idem to Bullard, Aug. 8, 1955, Box 16, and idem to Clemence, Feb. 14, 1955, Box 23, all HCU; see also idem, untitled biographical Statement, July 1969, Box 7, pp. 5, 6, 12–13; and Harmon Craig OHI (Ronald E. Doel, Nov. 27, 1990, at AIP).

became clear that such an analysis, to be written convincingly, would require too much space." Kuiper later defended the peculiar contours of his summary by claiming that he had mailed a questionnaire to all commission members, including Urey, asking for relevant citations to include. A copy of this June 1954 request found in Urey's files, and a query to Kuiper about it, indicates that Urey indeed knew of this deadline in advance.[68]

It is difficult to avoid the conclusion, however, that Kuiper had deliberately decided to omit contributions to cosmogony, aware that a summary of such research would include works with which he disagreed strongly. In 1954, for example, as he contemplated the IAU draft, Kuiper had written Struve that "Alfvén has just published a book based on ideas which I personally think are incorrect and as you know several other books and long articles have recently been written by Urey and others that might come in for criticism." Kuiper's draft report did not refer to Alfvén's work. Moreover, Kuiper advised a colleague that he would postpone review of cosmogonal works until a "final" publication, not wanting his own work influenced by "past interpretations." Wittingly or not, Kuiper sought to use the IAU report to emphasize astrophysical contributions, particularly his own.[69]

Kuiper's draft omissions were not an isolated instance, but rather reflected a pattern of citation practice throughout his professional career. His strongest supporters in the United States, including Russell and Whipple, had previously chided him for absent references to their own work, and in the early 1950s Dutch meteorologist and astronomer Hendrik Berlage had accused Kuiper of borrowing without attribution from his cosmogonal publications. Although there is no evidence to support Berlage's claim that Kuiper had directly appropriated elements of his theory – and Urey had independently pointed to serious difficulties in Berlage's model – Kuiper angrily advised Berlage that he would cite only work that has "affected my own," and irritated other American colleagues by declaring his intention to erect standards of work in solar system astronomy. In 1954, fellow IAU official Leo Goldberg criticized Kuiper for omitting new work on atmospheric ozone by Adel, then at Arizona State College, in his draft IAU report. While agreeing with Kuiper that Adel's methodology was likely flawed, Goldberg reminded Kuiper that the reports of committee members were "not subject to 'refereeing' in the usual sense."[70]

[68] Quoted from Kuiper (1957, 257); see idem, "[To the members of IAU Commission 16]," June 3, 1954, and Urey to Kuiper, June 23, 1954, both Box 51, HCU. Kuiper's references to Urey's two works were in the final, published version; see Kuiper (1957, 257). A copy of the original draft report could not be found.

[69] Quoted from Kuiper to Struve, Aug. 6, 1954, Box 14, and idem to E. Abendanon, Feb. 20, 1956, Box 10, both GPK; see also Kuiper to Struve, Aug. 15, 1955, Box 14, GPK.

[70] Quoted from Kuiper to Berlage, Feb. 8, 1952, Box 30, and Goldberg to Kuiper, Oct. 11, 1954, Box 11, both GPK; see Kuiper to Russell, Mar. 18, 1935, Box 47, HNR (I thank David DeVorkin for calling this letter to my attention); Whipple to Kuiper, Feb. 7, 1950, Box 3, FLW–HU; Berlage to Kuiper, Feb. 3, 1952, Box 30, GPK; Urey to Berlage, Apr. 21, 1953, Box 13, HCU; and Kuiper to L. Aller, Nov. 22, 1954, Box 29, GPK.

To judge by his subsequent actions, Urey also recognized the importance of the IAU *Transactions* as astronomy's journal of record. The IAU remained astronomy's sole international organization of note. Moreover, the only comprehensive review books in solar system astronomy were those edited by Kuiper himself. Infuriated by Kuiper's draft report, Urey prepared two supplementary reports, "The Age of the Solar System" and "The Origin of the Solar System," in late July, and insisted that Kuiper include them with his official *Transactions* report. Urey then mailed these reports to at least seventy-five scientists engaged in solar system research, including all fellow members of Commission 16. (Urey had been elected to it by Kuiper in 1952.) Two weeks later, learning that Kuiper had approved his "Age" but not his "Origin" report, Urey mailed a second, embittered letter to these scientists. Accompanying it was a two-page summary of what he regarded as a pattern of improper citations by Kuiper. "If writers of review articles and books systematically regard such ideas as obvious and use them anonymously," Urey declared, "the original scientist, who risks making errors, accumulates only his mistakes, while the reviewer, who risks nothing, acquires a magnificent and completely undeserved reputation for being profound." Urey's campaign was a thinly veiled challenge to Kuiper's continued presidency of the commission, scheduled to run for three additional years.[71]

A partial resolution of the controversy came in summer 1955, when the IAU met in Dublin. Struve, president of the IAU, found it necessary to review the matter before the Executive Committee, which appointed a panel of three members of Commission 16 to review Urey's charges. These astronomers endorsed Kuiper's plan to include Urey's first report (age) but not second (origin). On the whole, astronomers closed ranks behind Kuiper. Aided by Urey's absence from Dublin, Kuiper succeeded in altering Commission 16's title from "Observations physiques des planétes et des satellites" to "Etudes physique" of the same, helped establish a new subcommission for meteorite studies, and used his authority to alter Commission 16's membership, weeding out visual planetary observers such as Lowell's Henry Giclas and French astronomer Marcel Gentili. Kuiper also gained IAU endorsement for his planned atlas of the Moon and related lunar projects. Superficially, the controversy seemed to leave his stature within the IAU undiminished.[72]

The conflict, more divisive than any within the IAU in memory, nevertheless lessened Kuiper's authority within American astronomy and made solar system astronomy a less inviting field of research. Within the IAU, Kuiper's effectiveness in dealing with fellow commission members became reduced. Wearying of his role as an intermediary between Urey and American astronomers, Struve urged Kuiper to cease his battle with Urey; yet Kuiper persisted, publishing a labored rebuttal to

[71] Quoted from Urey to commission members, Aug. 9, 1955, Box 107, HCU; see also Urey to P. Oosterhoff, July 29, 1955, Box 107, HCU.

[72] P. Oosterhoff to Kuiper, Dec. 3, 1955, and Kuiper to Oosterhoff, Dec. 6, 1955, both Box 18, GPK; and Struve to Urey, Sept. 13, 1955, Box 16, OS.

Urey's criticisms in the Academy's *Proceedings.* Reading this, Urey prepared a strident "Reply to Kuiper's Reply." Though never published, this final attempt to define their views on the Moon's thermal history hardened the enmity between them. More important, it eroded the informal network of research and support for astronomers and geochemists interested in solar system phenomena. Harvard geochemist Birch received raw, bitter letters from both Kuiper and Urey, as did Michigan astronomer and abundances expert Aller. Infuriated by Kuiper's citation practices, Urey nearly convinced NSF officials to cut support for his editorial projects. Complaining that Chandrasekhar alone understood chemical discrepancies in Kuiper's cosmogony, Urey reneged on a promised article for Whipple, declaring, "I have hated all astronomers temporarily." Above all, their conflict, trying for their colleagues, utterly exhausted both men. Neither regained much creative energy before 1956, and each seriously considered abandoning solar system research to avoid contact with the other.[73]

The controversy also proved costly to the informal research schools that Urey and Kuiper had established at Chicago. His public battle with one of the university's best known and respected Nobel laureates, Kuiper discovered, did little to boost his standing with Chicago administrators. Moreover, it affected the research strategies of both men: Kuiper steered clear of planetary geochemistry after 1955; Urey abandoned Kuiper's cosmogonal framework in order to restructure his investigations along geochemical lines, and came to favor the comet accretion model proposed by Lyttleton to explain the formation of planetesimals. Their increased emphasis on *disciplinary* criteria was well understood by graduate students of both men after 1955. One found it like being "the child of divorced parents," an unnerving and difficult experience that deterred prospective researchers from solar system topics. The lack of transient structures at Chicago to nurture solar system research had a lasting effect. More than six years after the controversy erupted (and after Kuiper had transferred to Arizona and Urey to San Diego), Urey still sent sharply critical letters to Kuiper's graduate students, chiding them for failures to cite previous literature and for ignoring the fundamentals of chemistry. The field thus lost not just its intellectual unity, but its personal cohesion as well.[74]

[73] Quoted in Urey to Whipple, Aug. 11, 1955, Box 15, FLW–HU; see Kuiper to D. Bronk, Aug. 13, 1955, idem to Struve, Aug. 15, 1955, and Struve to Kuiper, Oct. 4, 1955, all Box 18, GPK; Urey, "Urey Reply to Kuiper's Reply," draft MS, n.d. [ca. 1955], Box 16, OS; see also Kuiper (1955) and Kuiper and Johnson (1956). For the aftermath of the controversy, see Urey to T. Page, Sept. 27, 1955, Box 16, idem to Aller, Sept. 7, 1955, Box 3, and idem to Struve, Dec. 14, 1955, Box 23, all HCU; and Kuiper to Birch, Sept. 23, 1955, Box 28, GPK. Previous priority disputes within the IAU had been handled internally and privately; see, e.g., H. Shapley to H. N. Russell, Mar. 30, 1928, Box 61, HNR; W. Baade to D. H. Menzel, Jan. 29, 1953, Box 28, GPK.

[74] Quoted from Carl Sagan OHI (Ronald E. Doel, Aug. 27, 1991, at AIP); see also Urey to M. Schwarzshild, June 14, 1956, Schwarzshild Papers, Princeton, courtesy David DeVorkin); Urey to L. Slichter, Jan. 25, 1955, Box 85, and Z. Kopal to Urey, Oct. 10, 1957, Box 50, both HCU; Harmon Craig OHI (Ronald E. Doel, Nov. 27, 1990 at AIP); Urey to D. P. Cruikshank, Nov. 13, 1962, Box 24, idem to W. K. Hartmann, Nov. 13, 1962, Box 51, and idem, "On the Origin of the Planetesimals and Asteroidal Bodies," unpub. paper, n.d. [ca. 1956], Box 136, all HCU; and Urey (1959; 1960).

Such difficulties were not confined to these local research communities. By the late 1950s, despite the heightening interest in space research, the Kuiper–Urey controversy hindered efforts to promote this field within the IAU. One example occurred in 1957, when Urey sought to organize special symposiums on cosmogony and the evolution of Earth at the planned 1958 Moscow meeting. Convinced that Kuiper would "try to prevent me from having anything to do with the subject," Urey requested B. J. Levin, a meteorite researcher at the Moscow-based Shmidt Institute of Earth Physics, to convene these programs instead. Aware of their feud, not wishing to offend either man, and hampered by limited cold-war communications, Levin organized a low-keyed affair. The discussions that took place produced few results, and no collected volume ensued. The post-*Sputnik* demands on scientists able to contribute to space research certainly taxed participants, but so did the strains of interdisciplinary cooperation and the absense of a single, central personality to coordinate such a volume.[75]

The intensely personal nature of this dispute, however, should not mask the fact that it also mirrored an important shift in thinking among American researchers about the thermal history of Earth. This Urey rediscovered late in 1955 when, while preparing an article on the origin of Earth for the *Handbuch der Physik,* he solicited renewed support for his cold accretion model from geologists and astronomers. James Gilluly of the U.S. Geological Survey, an active promoter of the cool-Earth consensus forged at Rancho Santa Fe five years before, still held to this view, noting that Bowen's studies of surface rocks still made him doubt the molten Earth hypothesis; but Rubey admitted he now leaned toward a molten earth to explain the origin of the atmosphere and oceans. The thermal history of the Moon seemed to many researchers similarly unresolved. Baldwin continued to defend a warm-Moon model, whereas Whipple and Eugene M. Shoemaker, a young USGS geologist then turning to lunar studies (see Chapter 5), accepted Urey's argument that the Moon had not experienced an intense molten phase. No set of geological, geophysical, or astronomical evidence presented by Urey or Kuiper now convinced the broad community of researchers interested in solar system evolution. Shoemaker, while at odds with many details of Urey's model, criticized Kuiper for ignoring the lunar bulge, noting that the strength required to support it "appears to contravene the evidence on the very low long term rigidity of the Earth's mantle. This evidence is not to be lightly dismissed just for the sake of simplifying hypotheses about the Moon. In fact, it is one of the most damning pieces of evidence against your hypothesis." The only area of agreement left among members of these communities by the late 1950s was that the question of thermal evolution would not be settled before improved knowledge of natural abundances were in hand. This was itself a significant change, one that

[75] See Urey to Levin, Nov. 20, 1957 (quoted), and May 14, 1958, as well as Levin to Urey, Mar. 13, 1959, all Box 52, HCU.

made cooperative efforts of the kind Kuiper and Urey had attempted to pursue harder to initiate.[76]

Bitter feelings between Kuiper and Urey lasted throughout the late 1950s and the 1960s. Officials of the National Aeronautics and Space Administration, who had appointed both men to the Ranger lunar project science team in 1961, found it necessary to seat neutral individuals between them at public functions. Finally, in 1967 Urey offered an olive branch. Concluding a formal note, Urey added, "What do you think of the creek beds that [lunar] Orbiter IV turned up? I have been suggesting water on the moon, yet always I am surprised to find the evidence. . . . Do you have another explanation?" Kuiper offered no reply. Yet after Kuiper collapsed and died on a visit to Mexico City on December 24, 1973, Urey wrote his widow, the former Sarah Fuller. "I am so very sorry to hear of the death of Gerard," Urey lamented. "He was a good scientist and man." After two decades, the Kuiper–Urey controversy was over at last.[77]

Despite its intensity, the Kuiper–Urey clash should not overshadow the fact that American astronomers did succeed in forging a consensus about certain solar system phenomena that had remained puzzling in 1945. Within little more than five years, the relationship between asteroids and meteorites, like that between comets and meteors, was no longer questioned; at the same time, the distinct origin of asteroids was thereafter no longer doubted. Oort's theory on the diffusion of comets into the distant comet cloud (i.e., the Oort cloud), the distinction between old and new comets, and Whipple's icy comet model were accepted by most American astronomers; they were also accepted by Urey, Brown, Patterson, and other geochemists at Chicago and Caltech interested in solar system problems. American astronomers similarly embraced Kuiper's nebular cosmogony as a promising model through the 1950s, even though Urey abandoned it completely.

It is also important to note that these ideas were not universally shared by astronomers in other countries, underscoring the degree to which national styles in astronomy remained significant. In Great Britain, Hoyle, Bondi, Gold, and Lyttleton continued to champion Lyttleton's interstellar model of comets as a natural consequence of the steady-state cosmology, and after 1953 there was less a dialogue between the Cambridge school and American astrophysicists than acceptance that irreconcilable differences existed. Yet their influence on solar system research lessened as they focused on defending the steady-state model against the expansionist Big Bang theory. The considerable volume of data on the properties of meteors, asteroids, and mete-

[76] Quoted from Shoemaker to Kuiper, Jan. 26, 1959, Box 31, GPK; see Urey to Rubey, Nov. 23, 1954, Box 22, and idem to Baldwin, Aug. 30, 1957, Box 12, both HCU; Gilluly to Urey, Dec. 9, 1955, Box 38, Whipple to Urey, May 22, 1958, Box 102, and Shoemaker to Urey, June 14, 1960, Box 85, all HCU; and Urey to J. T. Wilson, May 21, 1954, Box 21, JTW; see also Elsasser (1963, 2), Kuiper (1955, 820), and Russell (1980, 50), Suess and Urey (1956, 53).

[77] Quoted from Urey to Kuiper, July 19, 1967, and idem to S. Kuiper, Dec. 28, 1973, both Box 51, HCU; see Eugene M. Shoemaker OHI (Ronald E. Doel, June 16–17, 1986, at AIP); Kuiper to H. Newell, July 12, 1971, Box 2, GPK; Kuiper to Urey, Jan. 9, 1965, Box 51, HCU.

orites in the United States – largely achieved through Whipple's Navy-supported meteor program, Kuiper's NSF-funded asteroid survey, and Urey's and Brown's successes in applying nuclear techniques to meteorite analyses – simultaneously heightened the confidence of American researchers in their interpretations. These results depended only partially on the availability of world-class telescopes in the United States so critical to the development of stellar and galactic astrophysics.[78]

Like other major disputes in the history of science, the Kuiper–Urey controversy over the thermal history of the planets was shaped by unique factors besides the intellectual issues at hand. Only partially was it about the Moon; it had more to do with the increasingly divergent cosmogonal models that these men championed and the problem of finding common frameworks of reference for evaluating evidence from astrophysical and geochemical sources. The failure of Kuiper and Urey to maintain an interdisciplinary framework – further damaged by Urey's confrontational style and Kuiper's citation pattern – led both men to follow more discipline-centered approaches to planetary evolution. Kuiper accused Urey of ignoring astronomical constraints, while Urey, returning to his chemistry workbench, complained that "anyone who has looked at a cloud chamber" would see the fallacy of Kuiper's accretion model. The problem for these men was not a lack of charisma or drive – factors important in creating research schools; rather, it was the inability of each scientist to tolerate the disciplinary standards of the other, demonstrating the importance of personality in maintaining scientific work outside its customary disciplinary structures.[79]

Another branch of interdisciplinary research that proved equally difficult to cultivate lay on the border between astronomy and geology. Although astronomers and geologists agreed that the evolution of Earth as an astronomical body was an important topic – a view shared by leading patrons of American science – finding mutually acceptable answers regarding the origin of lunar and terrestrial craters remained a uniquely difficult challenge for American researchers through the 1950s. It is to this important story that we now turn.

[78] Urey (1956, 623); see also Brush (1981, 94–5; 1990). On Gold's continued interest in dust as an astronomical process, particularly its role in lunar surface phenomena, see Eugene Shoemaker OHI (Ronald E. Doel, June 16–17, 1987, at AIP), Gold (1955), and Wilhelms (1993, 26–7).

[79] Urey to E. Rabe, Dec. 31, 1957, Box 77, HCU.

5

Astronomers and Geologists: Uneasy Alliances, 1920–1960

Despite the emphasis that Kuiper and Urey placed on geologic criteria in their lunar studies, cooperation between geologists and astronomers was occasional and infrequent before the mid-1950s. One cause was a lack of shared consensus over whether lunar craters derived from geological or astronomical forces – and whether terrestrial impact structures existed at all. Another was that members of both communities remained uncertain about what was needed to demonstrate the correctness of impact versus volcanic origins for enigmatic lunar and terrestrial features. Astronomers and geologists alike perceived that important theoretical stakes lay behind this debate. Knowing whether impacts were rare or commonplace was central to understanding the formation and evolution of planets; it also bore critically on what role catastrophic events had played in geologic history. Yet finding common ground and research strategies proved more difficult for astronomers than building similar relations to meteorologists, geochemists, or geophysicists. In contrast to "astro-physics," which described distinct institutions and practices by the late nineteenth century, "astro-geology" first became a distinct entity more than a half-century later.[1]

Nevertheless, astronomers and geologists worked jointly to tackle this problem in the 1920s and again in the 1950s. As in other areas of solar system astronomy, researchers created transient institutions to enable workers from both disciplines to address common problems; but crater studies proved different from other joint ventures in this respect as well. The large costs of mounting expeditions to suspected impact craters on Earth, and the still larger expense of studying high-energy and nuclear explosives that became critical to the impact–volcanic debate, were too great to be borne by universities. American research in this field came instead to be concentrated in the Dominion Observatory of Canada and the U.S. Geological Survey, national institutions of science with staff drawn from a range of academic disciplines. The consensus behind impact that emerged in the United States and Canada after World War II owed much, directly or indirectly, to military interest in the effects of chemical and atomic explosions and the conviction of a few influential in-

[1] E. M. Shoemaker, "Investigation of Craters," draft proposal, n.d. [Nov. 1959], Box 3, EMS.

dividuals in these interdisciplinary organizations that geological studies of high-energy explosions merited pursuit.

COOPERATIVE RESEARCH IN THE 1920S: THE CARNEGIE MOON COMMITTEE

In 1920, few American astronomers saw the lunar surface a promising research topic. The problem was not a lack of interest (despite irritation that astronomers felt when light-of-the-Moon times interfered with observations of faint astronomical targets), but rather that determining the cause of lunar craters seemed an insurmountable challenge. Most American astronomers then accepted the judgment of nineteenth-century lunar researchers that the circularity of lunar craters, the volume of material surrounding them, and the apparent absence of meteorite craters on Earth favored a volcanic explanation; yet few astronomers thought the matter settled. Moreover, after Russell, Campbell, and other prominent American astronomers endorsed the view that Arizona's Coon Butte (also known as Crater Mound) deserved its informal name of Meteor Crater, interest rose that other terrestrial impact craters might be found.[2]

Most American geologists, however, believed that lunar craters and Meteor Crater could be explained by geologic processes alone. In assigning Meteor Crater a geologic origin, U.S. geologists pointed to a study by Grove Karl Gilbert, Chief Scientist of the U.S. Geological Survey and the most eminent American geologist in the early twentieth century. Hoping to find evidence of a celestial collision, Gilbert had visited Coon Butte in 1891. He discovered no sign of a buried metallic mass beneath the crater, however, and declared it the result of a volcanic steam explosion. Although Gilbert defended an impact origin for lunar craters in 1893, he published this later-famous argument in the obscure *Philosophical Society of Washington Bulletin*, and by the early twentieth century few American geologists remembered it. Most agreed with Wallace W. Campbell's 1920 assessment that lunar craters are "the ordinary products to be expected from evolutionary processes purely geologic."[3]

What also encouraged geologists to adopt a volcanic explanation for craters was the principle of uniformitarianism. Articulated by the British geologist Charles Lyell in the mid-nineteenth century, *uniformitarity* – that the present is the key to the past – seemed to most geologists to exclude deus ex machina agents like cataclysmic impacts as significant geological processes. Although geologists had long sought to disentangle religious and scientific accounts of Earth's history, defending uniformitarianism gained new urgency as the Scopes trial began in the mid-1920s. In 1927 Charles Berkey, chair of Columbia's Department of Geology, declared that geolo-

[2] Hoyt (1987, 1–30), Marvin (1986, 153–4), and Todd (1897, 250).

[3] Campbell (1920, 138); see also Gilbert (1893); on Gilbert, see Hoyt (1987, 54–72), Mark (1987), and Pyne (1978; 1980).

gy's central contribution was in demonstrating the "importance of natural laws and orderly processes" for creating Earth's features. "Geology," Berkey asserted, "has done more than any other single study in destroying the spell of magic and the belief in supernatural causes for natural phenomena." Major meteorite impacts as a geologic force seemed precisely the kind of supernatural magic that geologists had worked to discount.[4]

New interest in the origin of lunar craters was nevertheless sparked by John C. Merriam, the acerbic director of the Carnegie Institution of Washington. Largely forgotten today, Merriam's lunar program merits examination. Active for over a decade, the CIW Moon Committee was the first transient institution organized between astronomy and geology in the early twentieth century. Its members were drawn from the top ranks of American science. Moreover, it resembled other "cooperative" research programs created by Merriam at the CIW in biochemistry, seismology, geodesy, nutrition, oceanography, and physiology, all designed to supersede disciplinary boundaries and to resolve outstanding scientific and applied problems, including improved methods of harvesting forests and ocean fish. This major patron's efforts to stimulate goal-oriented research thus reveals the cooperative ideal of early-twentieth-century American science in practice.[5]

Merriam's determination to build interdisciplinary research programs was shaped by his previous research experience, the CIW's institutional structure, and the political economy of American science after World War I. As a professor of paleontology at Berkeley, Merriam had not actively promoted interdisciplinary approaches; yet after taking charge of defense research in California during World War I, and then as head of the new National Research Council (NRC), Merriam had grown increasingly familiar with major research problems outside geology. He had also been tutored in interdisciplinary practice by Hale, the U.S. doyen of cooperative research, who had founded the NRC to promote interdisciplinary studies. It was natural that Merriam, who directed the CIW from 1920 to 1938, chose to stress cooperative strategies: The CIW had long been regarded as an alternative to academic departments, and its two main branches – the Washington-based Geophysical Laboratory and Hale's Mount Wilson Observatory – embraced interdisciplinary research. By the early 1920s Merriam had become a leading U.S. advocate of cooperative approaches in science, convinced that focusing on the borderlands of established disciplines would yield important results.[6]

Merriam first proposed bringing physicists, astronomers, and geologists together to study the Moon's surface in 1919. Precisely what stimulated this idea is uncertain, although several factors were involved. While serving on the California State Coun-

[4] Quoted from Berkey, [untitled survey response], Feb. 22, 1927, Central Administration files, Columbia University; see also Greene (1982), Hooykaas (1963), and Numbers (1992, 54–101).

[5] Doel (1994); see also Introduction, n. 11.

[6] DeVorkin (1994), Doel (1994), Hale (1920, 112–13), Kohler (1991, 84–7), Servos (1990, 233–8), and Stock (1947).

cil on Defense, Merriam came into close contact with Campbell, then a fellow science administrator. Campbell was deeply interested in Meteor Crater and convinced of its meteoritic origins; in 1920 Merriam joined a small group scheduled to visit the crater. His interest also seemed piqued by remarkably sharp lunar photographs made in 1917 by Francis G. Pease, a Mount Wilson astronomer, partly to demonstrate the abilities of the newly completed 100-inch telescope.[7] In any event, Merriam became convinced that finding the origin of lunar craters could bridge Mount Wilson and the Geophysical Laboratory, further demonstrating the viability of interdisciplinary approaches within the CIW.

Ironically, Merriam's lunar proposal met with opposition from his mentor in cooperative research, Hale. In a well-orchestrated effort to close debate on lunar crater origins, Hale turned a 1919 symposium to honor Yale lunar expert Ernest Brown into a dismissal of the impact theory. Campbell declared that the absence of elongated craters (expected from nonvertical impacts), and his own claim of visual discovery of tiny craters topping the central peaks of large lunar craters, demanded a volcanic origin. Hale himself insisted on morphological analogues between lunar craters and Vesuvius-type volcanoes. Although Hale proclaimed his willingness to welcome Merriam's project if "sufficiently promising," he warned that the "heavy demands of our regular program of research" would leave few moments for lunar photography. Hale's anxiety had less to do with planning, however, than patronage. In 1910 Hale had been approached by Daniel M. Barringer, the flamboyant engineer and entrepreneur who had purchased Meteor Crater to mine iron ore he believed lay below it; Barringer had requested Hale's endorsement of the scheme and a major contribution from the independently wealthy Hale. Coming just four years after the Martian canal controversy, which Hale feared would dissuade patrons from endorsing his plans for major new telescopes, the aghast Hale likely perceived the lunar crater controversy as another unresolvable quarrel that Mount Wilson staff members had best avoid.[8]

Merriam ignored Hale. In 1923, Merriam convinced the New York-based Carnegie Corporation to fund the Moon Committee. One year later he secured additional Carnegie Corporation funds to begin a geology division at Caltech, whose potential as a home for cooperative interdisciplinary science he regarded no less highly than Hale. John P. Buwalda, a Merriam student, became chair of this new graduate program in geology in 1925, one year before Merriam selected a leader to run the lunar committee.[9]

Frederick E. Wright, Merriam's choice, was seemingly an ideal candidate for the task. American-born and Heidelberg-trained, an expert in petrographic techniques,

[7] Hoyt (1987, 215).

[8] Quoted from Hale (1920, 112–13); see also Campbell (1920, 133), Pickering (1920), and Plotkin (1993a); on Barringer, see Barringer to H. Shapley, Apr. 26, 1929, Box 2, DHCOs, as well as Hoyt (1987, 146).

[9] Merriam to W. Rose, Dec. 26, 1925, JCM; see also Goodstein (1991, 136–8) and Kohler (1991, 123).

Wright had joined the CIW's Geophysical Laboratory at its founding in 1907. Although chiefly interested in rock formation, Wright contributed to gravity research, optical design, and instrument making, a not-uncommon mix within the eclectic, interdisciplinary structure of the CIW. Wright also had extensive field experience, having traveled to the enigmatic Bushveld Igneous Complex of South Africa in 1922 as part of Harvard's Shaler Memorial Expedition. Content with Merriam's insistence that all geological investigations be framed by multiple working hypotheses, Wright analyzed his Bushveld results bearing in mind Alex du Toit's defense of continental drift as one way to explain its peculiar geology. After Wright advocated multiple working hypotheses to evaluate arguments about the Moon's origin by another South African geologist, E. H. L. Schwartz, Merriam made Wright committee chair.[10]

Following familiar patterns, Merriam used the committee to link the CIW's various divisions with outside experts. He appointed to Wright's committee Arthur L. Day and Walter S. Adams, respective heads of the Geophysical Lab and Mount Wilson. He also appointed Buwalda and theoretical physicist Paul S. Epstein from Caltech, Pease, Edison Pettit, and Edwin P. Hubble from Mount Wilson, and Russell, newly appointed research associate at Mount Wilson. By appointing Russell, the unexcelled leader of theoretical astrophysics in America, Merriam seemed hopeful that "astro-geological" studies might achieve parity with the trailblazing geochemical work emerging from the Geophysical Lab and with stellar astrophysics at Mount Wilson.[11]

The first challenge Wright faced was finding a suitable research methodology. To explore possible lunar–terrestrial analogues, Wright visited Meteor Crater in the company of Merriam, Day, and Pease in September 1927. Their stay, lasting several days, made a strong impression on them and underscored the importance of this local geological feature in generating sympathy for the impact hypothesis in the United States. The volcanologist Day admitted that, before visiting the crater, he had "never felt quite satisfied with the adequacy of the proofs offered" for its impact origin, which he attributed to "conservatism in adopting a concept for which there was no directly comparable precedent." Although Geological Survey leaders, including Gilbert's former student Nelson H. Darton, continued to regard Meteor Crater as volcanic, it is significant that no member of the CIW committee questioned its impact origin.[12]

Wright, however, soon grew discouraged about employing terrestrial analogues to "study the physical features of the surface of the moon," as Merriam had urged. One reason was that all prior analogy-based arguments had failed to resolve the

[10] Wright to Merriam, July 12, 1922, Box 186, JCM; idem to W. M. Gilbert, Mar. 2, 1925, Box 1.1, FEW; see also Fleming and Piggot (1956), Mark (1987, 182), and Servos (1990, 223–8). On the significance of Bushveld research, see Oreskes (1990).

[11] Merriam to Russell, Oct. 9, 1926, Box 50, Folder 8, HNR; see also DeVorkin (1994).

[12] Quoted from Day to D. Barringer, Nov. 23, 1928, Box 2, DHCOs; see Wright to Merriam, Sept. 3, 1927, Box 186, JCM; Hoyt (1987, 143–5).

volcanic–impact controversy. More important, Wright became convinced that the Moon was too geologically alien to permit direct comparisons. Arguments that lunar craters poorly resembled terrestrial calderas, he believed, erred in ignoring the Moon's low surface gravity. He calculated that a shell fired from the Paris Gun of World War I, if placed on the Moon, would travel over two thousand miles, compared to seventy-five miles on Earth. Wright also dismissed the Campbell–W. H. Pickering claim that lunar craters would be elliptical rather than circular if caused by meteorites striking at a slant, declaring that the mechanics of ultra-high-velocity collisions were too imperfectly understood to make such judgments. Perturbed too that the Moon lacked an atmosphere and water (and hence familiar sedimentary rocks), Wright advised committee members to discard "most of the geologist's background of experience." He determined instead to build on the quantitative foundations of his own field, geochemistry. "Before we attempt to theorize too much about the formation of the lunar surface features," he advised Russell, they had to "ascertain as definitively as possible what the materials are that we see on the surface."[13]

Wright's preferred solution was to use the Mount Wilson 100-inch telescope as a massive petrographic microscope. Employing a 12-inch refracting telescope at the U.S. Naval Observatory, across town from the Geophysical Laboratory, Wright obtained a trial run of polarization measurements of the Moon's disk. He then compared these to polarization values for terrestrial samples analyzed at the lab. Wright found lunar polarization values generally lower than 20 percent, similar to volcanic pumice but unlike that of common basalts. He concluded that a careful survey of the Moon's surface could reveal its composition and thus rule out either the impact or volcanic hypotheses for lunar surface features.[14]

Although genuinely interested in the Moon, Wright nonetheless saw lunar research principally as a means to study Earth. Like Shapley, then trying to build his cosmogonal institute at Harvard, Wright regarded cooperative research as a way to apply insights from one discipline to another, and believed investigations of the Moon would aid terrestrial geochemistry. For instance, Wright argued that lunar studies could resolve the paradox of isostasy. Noting that evidence favoring a strong crust (the fact of earthquakes) existed alongside evidence of a weak crust (including isostatic rebound and local adjustments to mountains), Wright wrote in 1927 that global studies of the Moon's gravity could overcome the difficulty of observing Earth on a similar scale. He also remained committed to multiple working hypotheses, promising an Austrian colleague to search the Moon for evidence of continental drift (he found none). Far from accumulating data without regard to theory, a charge often leveled against American geologists in this period, Wright sought to

[13] Quoted from Wright to R. Daly, June 10, 1927, Box 1, FEW, and idem to Russell, Nov. 26, 1926, Box 68, HNR; see also Wright (1927, 451).

[14] Wright to Daly, June 10, 1927, Box 1, FEW; idem, "Committee on Surface Features of the Moon," draft, n.d. [ca. Dec. 1926], Box 68, HNR.

use the Moon Committee's results to test competing geological concepts at a time of theoretical pluralism within the American geological community.[15]

The committee's cooperative phase lasted from 1926 to 1928. Although Mount Wilson in this period is most remembered for its pioneering studies of stellar characteristics and Hubble's investigation of spiral nebulae redshifts, culminating in his singular discovery of the expanding universe in 1929, the Moon Committee's own activity underscores the diversity of research sustained at Hale's observatory. Buwalda used the 100-inch telescope visually to record the Moon's "physiographic features from a geologic perspective." Pease used vacation time to visit volcanic regions in California to understand volcanic landscapes better, whereas Wright began his polarization studies. The most ambitious observations, likely suggested by Russell, were radiometric temperature measurements by Pettit and his colleague Seth B. Nicholson of the Moon during lunar eclipses in 1927 and 1928, the first made since Samuel P. Langley's bolometric studies of 1885. Epstein used the Pettit–Nicholson data to calculate that the lunar surface is covered by a thin insulating layer like pumice, rather than great expanses of exposed rock masses. This computation was the first to indicate the actual surface conditions on another planet, and one of the most significant achievements of the Moon Committee.[16]

Wright, who began spending two months per year at Mount Wilson, soon extended the committee's goals. Learning of lunar ray polarization studies by French astronomer Bernard Lyot, Wright began additional measurements with this technique. He also started lunar contour maps and proposed making photographic lunar globes to eliminate the distortion of limb foreshortening. Costly undertakings, these testified to the CIW's abundant resources for favored projects. The planned lunar projections required hand-fashioned, photographically sensitized glass globes (built to order by Eastman Kodak laboratories) and construction of a "Moon House" atop Mount Wilson. Required to project lunar photographs onto the emulsion-coated globes, the Moon House, erected north of the 100-inch dome, stretched nearly a city block. To reduce distorting air currents, workers fashioned its outer walls with two layers of corrugated metal and filled its floor with sawdust to a depth of six inches. Wright's "magnified chicken coop" was merely one of several instruments constructed for the committee, including a stereoptical projector and a visual spectrophotometric eyepiece for the 100-inch reflector.[17]

15 Wright to Merriam, Jul. 23, 1927, Box 186, JCM; idem to W. A. J. M. van der Gracht, Mar. 5, 1931, idem to Daly, May 12, 1927, and idem to W. Boorman, Sept. 14, 1935, all Box 1, FEW; see also Wright (1927, 460).

16 Quoted from Buwalda to Wright, Nov. 19, 1928, Box 1, FEW; see Wright to Merriam, July 23, 1927, Box 186, JCM; Pettit to Russell, Dec. 13, 1926, Box 55, HNR; and Pease to Wright, Feb. 6, 1926, Box 1, FEW.

17 Wright to Merriam, Mar. 5, 1931, idem to E. Öpik, June 5, 1931, idem to Adams, June 8, 1931, all Box 1, FEW; Minutes of the Moon Committee Meeting, Oct. 4, 1930, and D. L. Webster to Wright, Feb. 19, 1931, both Box 1, FEW; see also Wright (1935, 175–6).

Mount Wilson astronomers welcomed Wright in their midst well into the early 1930s. One reason they did was their initial confidence that this effort *would* indeed resolve the centuries-old puzzle of lunar craters. Another reason was that Pettit and other staff members saw great promise in applying Wright's polarization techniques to more distant and enigmatic planets, including Mercury; like Adams, Russell, Nicholson, and Dunham, Wright looked favorably on solar system research. Finally, institutional support from the parent CIW was sufficiently secure so that riskier interdisciplinary programs did not threaten funding for core research programs in stellar and galactic astronomy. When Russell queried directors of major U.S. observatories about their need for extramural funds in 1927, Adams alone claimed none. Abundant patronage encouraged these astronomers to innovate and experiment more than colleagues at smaller, more autocratic institutions.[18]

Wright nevertheless found it difficult to maintain the committee's interdisciplinary composition. Growing frustration with their failure to secure a knockout defense of the volcanic theory, which most members supported, was one reason. Epstein privately advised Wright that he had completed his demanding reductions of the 1927 and 1928 lunar eclipse results solely to discredit the impact hypothesis, which he found impossible to take seriously. The lack of disciplinary rewards for committee work also became troubling. When Epstein complained about inaccessible astronomical data, Wright had no leverage to compel astronomers to assist him. Eager to complete his studies of stellar spectra, Russell refused Wright's request to recompute lunar crater diameters, citing a lack of graduate students. Even when graduate students were available, their participation in research was often limited. Arthur L. Bennett, later Russell's student, studied the brightness curve of the lunar surface for his Ph.D. He found that it diverged upward from theoretical expectations near full moon, an important result; but when Bennett was appointed to Yale in 1931, he had to devote himself full-time to its programs in stellar parallax and infrared stellar photometry, and found no opportunity to mine his investment in lunar research.[19]

Wright also ignored new research into impact craters on Earth, a growing field by the 1930s. Presumably this stemmed from Wright's determination to avoid morphological comparisons in favor of petrographic analysis; but his decision cut him off from American astronomers and geologists who favored the impact theory. Wright seemed little aware of the brief but furious debate over the fate of the meteorite that had caused Meteor Crater in 1929, precipitated by Barringer investors worried that this body had broken up on impact and thus could not be mined. This brief flap, involving Russell and Chicago mathematical astronomer Forest Ray

[18] Wright to Russell, Nov. 3, 1927, Box 68, and W. Adams to Russell, May 7, 1928, Box 31, both HNR. On Mount Wilson's management in the 1920s, see Peter R. Dear, "The Mount Wilson Observatory 1910–1918: Scientific Production and the Research Institution," unpub. MS, Sept. 1980, in my possession.

[19] Epstein to Wright, Jan. 10, 1928, Box 1, FEW; Russell to Wright, Nov. 7, 1927, Box 30, HNR; see also Bennett (1931, 1–2, 38–9) and Russell, Dugan, and Stewart ([2d ed., 1944, 179]). Pease alone seemed ambivalent about the volcanic theory; see Pease to Wright, Feb. 6, 1926, Box 1, FEW.

Moulton, led both men to conclude that the body had indeed been destroyed, an important insight into impact mechanics apparently unknown to Wright.[20]

Equally significantly, Wright seemed unaware of growing sympathy for the impact theory at Harvard, despite a letter from his Shaler Expedition colleague, Harvard geologist Reginald V. Daly, announcing it. By 1929, both Shapley and Daly accepted the meteoric origin of lunar craters, Shapley's commitment nurtured by his intense interest in meteors as cosmic debris. Both Shapley and Daly kept track of new lunar research, including polarimetric studies by Soviet astronomer Nikolai Barabashov, and considered organizing a symposium on lunar studies. However, neither published on this subject through the 1930s. Shapley's reticence may well have resulted from Russell's sharp criticism of his meteor–stellar energy speculations (see Chapter 1); but Shapley, Daly, Öpik, and other members of the Harvard Arizona Meteor Expedition encouraged important new research on terrestrial impact craters by their graduate students through the 1930s. Watson, who had studied small bodies in the solar system while writing his Ph.D. under Shapley, investigated new terrestrial impact candidates; discrediting a proposed impact origin for the Carolina Bays, he nonetheless argued that the circularity of newly found impact craters on Earth favored the impact origin of lunar craters. An even more important result was obtained by two of Daly's students. In a classic 1937 study, John D. Boon and Claude C. Albritton favored an impact origin for the fifty-mile-wide Vredefort Ring Structure, south of Johannesburg, previously surveyed by the Harvard Shaler Memorial Expedition. They also suggested an impact origin for other large, enigmatic structures like Vredefort, categorized by University of Cincinnati structural geologist Walter Bucher as "cryptovolcanic" features. Despite Harvard's being as active a group for crater studies as the Moon Committee, Wright showed little interest in its emerging results.[21]

Although Merriam continued CIW support, by the mid-1930s the project essentially became Wright's alone. Still convinced that only geochemical methods could resolve crater origins, Wright continued his solitary polarization program. As with many types of planetary observations, polarization measurements of discrete lunar regions were time consuming and difficult to make. Wright's program competed for telescope time with other Mount Wilson programs. Matching observed lunar polarizations with terrestrial materials also proved difficult. In 1935 Wright admitted that the composition of lunar surface materials could be determined only with "a fair degree of probability." Similar difficulties plagued other lunar projects. A galvanometric relay to measure lunar temperatures proved too insensitive even with

[20] Webster to Wright, Feb. 19, 1931, and Wright to Öpik, June 5, 1931, both Box 1, FEW.

[21] Daly to Wright, May 8, 1927, Wright to Daly, May 12, 1927, and idem to Öpik, June 5, 1931, all Box 1, FEW; Shapley to D. Barringer, Apr. 29, 1929, Box 2, and idem to Daly, Feb. 15, 1928, Box 4, both DHCOs; Barringer to Daly, Dec. 10, 1926, RD; see also Boon and Albritton (1937), Burke (1986, 274–5), Hoyt (1987, 259), and Watson (1936b, 11, 13–14). Geophysicist M. King Hubbert recalled that Albritton, a contemporary and fellow Southerner raised in a religious family, became interested in impact phenomena only after breaking with this tradition; see Hubbert OHI (Ronald E. Doel, Feb. 1989, SPHG at AIP).

the 100-inch telescope, and insufficient insulation within the Moon House distorted air currents, blurring projected images. Moreover, the Moon House proved vulnerable to Mount Wilson's severe climate: It collapsed in 1930 during 90-mph winds, and unusually heavy snows crushed its replacement seven years later.[22]

The Moon Committee itself finally collapsed from a paucity of results. In the summer of 1938, when most Mount Wilson astronomers traveled to Stockholm for the IAU general assembly, Wright secured four hundred photographs of the Moon, many of unusually high quality; but his polarization studies lagged. Moreover, finding it impossible to draw positive conclusions, Wright published virtually no data. Confessing that he now felt "a bit of an intruder" among Mount Wilson astronomers, unable to accept Russell's admonishment that he relax his exacting demands for precision, Wright spent increasingly less time in California and apparently did not return after Merriam retired in 1938. Instead, he turned full-time to other research programs he had maintained throughout his tenure on the Moon Committee, including gravity studies and optical research. The formal end of the Moon Committee came in 1946 when Ira S. Bowen, replacing Adams as Mount Wilson director, sought Wright's permission to tear down the Moon House. In reality, the committee had ceased functioning half a dozen years earlier.[23]

In several important respects, the CIW Moon Committee had failed to meet its objectives. Compared to Merriam's other cooperative programs in oceanography and ionospheric studies, the Moon Committee had produced no conceptual breakthroughs and led to no sustained research traditions. Except for Epstein's lunar temperature profiles, Wright's data, extracted at high cost, had been neither utilized nor published. No successful instrumental techniques had been pioneered. The failure of the committee to leave a significant mark was evident from the speed with which it was forgotten. Queried by a junior astronomer in 1942 about lunar research, Russell failed to mention the committee's work or the few publications its members had produced, as if forgetting its very existence.[24]

Dismissing the Moon Committee as insignificant, however, would miss a larger point. The problem of lunar crater origins was one of the most intractable in solar system astronomy in the early twentieth century. The difficulty of adapting astrophysical (or petrographic) instruments and techniques to solar system research led to few useful data. Daly's own reticence to publish regarding the Moon before 1940 bespoke the difficulty of marshaling data to support one's convictions, and the contingent issue of uniformitarianism was omnipresent. Merriam's CIW lunar project

[22] Wright to A. W. Hull, May 18, 1931, and Adams to Wright, Nov. 26, 1930, both Box 1, FEW; W. S. Adams to Wright, Mar. 3, 1937, Box 48, WSA; and Kuiper to N. U. Mayall, Mar. 23, 1955, Box 12, GPK; see also Wright (1935, 180).

[23] Russell to Wright, Nov. 7, 1927, Box 30, Folder 61, HNR; Wright to Merriam, July 23, 1935, Box 185, JCM; Wright to Adams, Aug. 10, 1940, Box 73, Folder 1318, WSA; and Bowen to Wright, Mar. 29, 1946, Box 1, FEW; see also Fleming and Piggot (1956, 327, 342) and Wright (1938, 70).

[24] Russell to R. B. Baldwin, Oct. 14, 1942, Box 34, Folder 26, HNR.

– the first institution-based, interdisciplinary effort to address the origin of lunar craters and the longest-running transient institution in solar system astronomy – was an ambitious, creative effort that demonstrated continuing interest in this interdisciplinary problem. What would turn the tide were technological and institutional factors that did not exist before the start of World War II. These included the atomic bomb, new research into related explosion processes and hypervelocity impact early in the cold war, and common involvement of American astronomers and geophysicists in a wartime institutional framework that made these results accessible and demonstrable.

FORGING CONSENSUS BEHIND IMPACT: THE BALDWIN–HARVARD AXIS

Although American astronomers interested in the Moon had shared no common view as to the origin of lunar craters in the late 1930s, a consensus behind the impact theory emerged ten years later. As we have seen, Kuiper (like Urey) steadfastly accepted an impact origin for craters after beginning lunar research in the early 1950s. This transformation stemmed from several factors, including the deaths of Hale and Campbell and the diminished influence of Russell, all leading detractors of the impact theory; but it owed most of all to new research by Ralph Belknap Baldwin, whose 1949 *Face of the Moon* married quantitative measurements of lunar craters to military studies of craters created by shell bursts and chemical bombs. Reinforced by new knowledge of high-energy explosives derived during World War II, and actively defended by Whipple, Daly, and other members of Harvard's experimental geophysics community, Baldwin's arguments marked the end (at least among American astronomers) of a fundamental astronomical controversy.

As with Kuiper, Baldwin's lunar research signaled a detour from earlier work in stellar astrophysics. A student of Heber D. Curtis and Dean McLaughlin at Michigan, Baldwin had earned his Ph.D. in 1937 for spectroscopic studies of novae. Between his graduation and 1942, Baldwin had accepted short-term positions at Pennsylvania and Northwestern, where he continued work on physical models of novae and unusual binary stars. His interest in the Moon began not at Pennsylvania, despite the presence of noted meteor researcher Charles P. Olivier, but rather at Northwestern. Finding his salary at Northwestern insufficient after he married in 1940, Baldwin had begun lecturing part-time at Chicago's Adler Planetarium, where in free moments he grew intrigued by large photographs of the Moon hung as planetarium exhibits. His interest was soon drawn to radial markings that cut across mountains ringing Mare Imbrium, the largest lunar sea; he became more intrigued after finding no discussion of them in publications he consulted. While it was Baldwin's curiosity that ultimately led him to this question, he was also open to fresh research topics at the time. By 1940, Baldwin had grown alarmed over the plans of Oliver J. Lee, Director of Northwestern's Dearborn Observatory, to orient observatory research around identifying faint red stars from spectroscopic survey plates; Baldwin expected little to come from the effort.[25]

Maintaining his research in stellar astrophysics, Baldwin continued to investigate the radial markings surrounding Mare Imbrium. He soon found that lines projected from these features all intersected in this mare. By 1941, Baldwin concluded that these radial markings indicated a common cause for the mountains and lava-filled mare, which he attributed to an enormous meteorite of "asteroidal dimensions." At a colloquium at nearby Yerkes Observatory, arranged by Lee, Baldwin argued that Mare Imbrium was the largest crater on the Moon, suggesting an impact origin for other maria and virtually all craters. He then published a brief summary of this argument in *Popular Astronomy* in 1942, followed by a second, somewhat more technical article in this same journal the following year.

That the first modern treatment of the impact theory of lunar craters and seas appeared in a semipopular journal and generated a lackluster reaction when presented to the Yerkes community in 1941 has been cited as evidence that American astronomers refused to consider the impact theory seriously;[26] but this is misleading, for Baldwin's argument was initially less than compelling. His review of the historical literature was cursory: His paper made no reference to Gilbert's similar 1893 suggestion that the lunar seas owed to impact – a hardly surprising omission, given its publication in an obscure periodical – but also failed to cite work by Herbert Ives, Wright, and others who had published about the Moon and the impact hypothesis in astronomical journals. Moreover, Baldwin initially did not attempt quantitative or statistical arguments to support his interpretation; the morphological relationships he presented were more familiar to geologists than astronomers. Finally, the geological features he emphasized in lunar photographs were difficult to see when reproduced. It is not surprising that Struve had rejected the paper when it was submitted to the prestigious *Astrophysical Journal.*[27]

More important, Baldwin was ill informed about contemporary work on cosmogony and solar system research, costing him the support of Russell, potentially his most significant ally among American astronomers. In the autumn of 1942, one year after his Yerkes colloquium, Baldwin requested Russell's comments on a new, more quantitative paper favoring the impact origin of the lunar seas, enclosing with it his published article in *Popular Astronomy.* Clearly, Baldwin desired Russell's help to place this second article in the *Astrophysical Journal;* but Russell raised objections to both pieces. He criticized Baldwin for ignoring Harold Jeffreys's conclusion that thick lavas would disrupt a permanent crust and sink, challenging Baldwin's calculation that the Mare Imbrium lavas might reach a depth of one mile. He also chided Baldwin for ignoring Jeffreys's conclusion that the Moon could not have

25 Ralph B. Baldwin OHI (Ronald E. Doel, Oct. 25, 1989, at AIP); see also Wilhelms (1987; 1993, 14–19).

26 Wilhelms (1993, 15).

27 Baldwin to H. N. Russell, Sept. 19, 1942, Box 34, HNR; see also Baldwin (1942).

derived from Earth through rotational fission, calling this Baldwin's "weakest point of argument." It did not matter that this point was incidental to Baldwin's thesis: By neglecting the results of Russell's friend Jeffreys, whose views had so strongly influenced his own writings on cosmogony in 1934, Baldwin had foregone support from Russell, who irritably suggested that Baldwin invest more time in war research. Russell, significantly, was not by the early 1940s wholly set against an impact origin for certain lunar features. He mused in his reply to Baldwin that "there seems then to be no great objection to assuming that small bodies were associated with the two larger ones" (Earth and the Moon) at the time of the solar system's birth. However, Russell's negative review effectively blocked the second paper from the discipline's major journals. Baldwin thus placed it, like the first, in *Popular Astronomy.*[28]

World War II soon provided Baldwin an opportunity to answer Russell's criticisms and to greatly strengthen the scope and the authority of his arguments. When Baldwin wrote Russell, he had already left Northwestern for the Applied Physics Laboratory of Johns Hopkins. Through 1946, Baldwin worked under Merle Tuve on what became known as the "proximity fuze," a device that detonated warhead explosives in the vicinity of their intended targets and an important component of the Allied response to German V-1 missiles and Japanese kamikaze raids. After Baldwin suggested that the proximity fuze would be an effective offensive weapon (since airbursts have greater destructive power than explosives detonated at surfaces), he took on new studies of bomb explosions, reviewing classified Army records of bomb, artillery shell, and mortar explosions, the diameters of craters they produced, and the shape of craters caused by explosions at, above, and below ground level.[29]

Baldwin used access to these data to explore, in his spare time, quantitative relationships among artificial explosion craters, meteorite impact craters on Earth, and lunar craters. To learn more about terrestrial meteorite craters and gather statistics on lunar craters, Baldwin hounded the nearby Library of Congress as well as the U.S. Naval Observatory library. The project was well suited for Baldwin's solitary inclination and his desire to immerse himself in geological, geophysical, and astronomical studies critical to the fuze project and his lunar interest. By the end of 1946, Baldwin had become familiar with much American research on small bodies in the solar system, including that of Watson and Whipple of Harvard. He had also read and accepted Jeffreys's arguments about the distinct origins of Earth and the Moon. Lacking access to telescopes, Baldwin employed Lick and Mount Wilson lunar photographs for his research. By 1945 Baldwin began synthesizing these studies into what became his *Face of the Moon,* as important for the field as Kuiper's *Solar System* volumes and Urey's *The Planets.* He completed the manuscript in 1947 in Grand Rapids, Michigan, having made the difficult decision to resign his junior

[28] Baldwin to Russell, Sep. 19, 1942, and Russell to Baldwin, Oct. 14, 1942 (quoted), both Box 34, Folder 26, HNR; see also Baldwin (1942, 367).

[29] Baldwin to H. C. Urey, Nov. 19, 1958, Box 12, HCU; Baldwin OHI (Ronald E. Doel, Oct. 25, 1989, at AIP, pp. 27–34); see also Baldwin (1949, 127–8, 224–6; 1980), Baxter (1946, 36), and Hoyt (1987, 358).

post at Northwestern to work for (and eventually direct) the Oliver Machinery Company, a maker of machine tools that his family had cofounded. [30]

The significance of Baldwin's book was its synthesis of research in physics, geology, geophysics, meteorology, and astronomy, all with the aim of demonstrating the explosive (impact) origin of lunar craters as well as the existence of impact scars on Earth. Baldwin devoted several chapters to evidence that, in his view, indicated the prevalence of impact features on Earth's surface. He emphasized the discovery since 1921 of seven previously unrecognized meteorite craters, a rate of one every three years. He stressed that Watson's research indicated that large asteroids struck Earth on average every hundred thousand years, and used Whipple's atmospheric work to argue that the Moon's own atmosphere at most equals Earth's at eighty miles, insufficient to protect against incoming meteorites. Baldwin also emphasized the studies of Albritton and Boon (and later Daly) that suggested the large Vredefort structure was caused by impact, and supported their argument that all cryptovolcanic features were potential impact structures.[31] Extending these lines of evidence, Baldwin concluded that most lunar craters had formed early in the Moon's history, suggesting that Earth's were similarly ancient. Most terrestrial impact features, he declared, had fallen victim to "rising and falling lands, advances and retreats of vast epeiric [epicontinental] seas, mountain building, faulting, folding, erosion, and deposition"; most would lie below the oceans. Baldwin's synthesis also included extensive discussions of atmospheric phenomena and their effect on infalling meteorites, reflecting a familiarity with aeronomy and ballistics honed by long association with Tuve's APL group.[32]

Most important, Baldwin argued that energy–diameter relations experimentally derived from chemical explosions could be scaled up to craters of lunar size. In part this argument rested on Baldwin's compilation of depth–diameter values for more than three hundred craters, obtained from published studies by the nineteenth-century astronomers Wilhelm Beer and Johann Mädler, by the contemporary British astronomer Thomas L. MacDonald, and from his own measurements. He placed still greater emphasis, however, on precise measures of four meteor craters on Earth and the Army's abundant data on chemical explosion craters. By simultaneously plotting these data, Baldwin found that explosion craters, terrestrial impact craters, and lunar craters all fell along a single logarithmic curve "too startling, too positive, to be fortuitous." Although Ives had previously called attention to morphological similarities between chemical explosion craters and lunar craters, Baldwin was the first to demonstrate a *quantitative* relationship between (both impact- and bomb-created) terrestrial craters and lunar craters. Baldwin underscored this relation with

[30] Baldwin, "Biographical Information," OHI files, AIP, and Baldwin OHI (Ronald E. Doel, Oct. 25, 1989, at AIP); see also Baldwin (1949, 11).

[31] Baldwin (1949, 95) and Watson (1941, 170–1).

[32] Quoted from Baldwin (1949, 93–4 [cf. 172–3]); see Baldwin OHI (Ronald E. Doel, Oct. 25, 1989, at AIP, p. 40); see also Hoyt (1987, 355).

aerial images of bomb craters at Regensburg and Marienburg in Germany, fresh memories for many readers.[33]

The publication of *Face of the Moon* in 1949 persuaded a majority of American astronomers to shift allegiance to the impact theory. One reason was that Baldwin was by then better known to his American colleagues than in 1942: Throughout the early 1940s Baldwin had continued publishing new articles on stellar astrophysics, including eight papers in the *Astrophysical Journal.* Another reason was that Baldwin's book, unlike his wartime articles, was buttressed by quantitative arguments, including his logarithmic graph that linked lunar with terrestrial craters and existing data to theory. In a laudatory review, Whipple singled out as particularly convincing the "continuous smooth curve" linking craters of various energies and diameters. Urey also cited this quantitative relationship, declaring that it showed "in a very convincing way that the lunar craters were mostly explosion pits" caused by the collisional energy of high-velocity bodies.[34]

Other factors also encouraged American astronomers to accept Baldwin's results. One was that a number of American geologists and geophysicists had, independently of Baldwin, begun publishing similar conclusions in the mid-1940s. American geophysicist Robert S. Dietz, unaware of Baldwin's work, had defended an impact origin for Mare Imbrium in 1946. One year later, Daly had endorsed the meteoric origin of lunar craters and the Vredefort Ring Structure, reinforcing the 1937 Boon–Albritton argument. Like Dietz, Daly emphasized that his ideas matched those advanced in 1893 by Gilbert, whose classic study of lunar craters had also escaped Baldwin's attention. That the astronomer Baldwin had reached conclusions similar to those of Gilbert, Daly, and Dietz suggested to Urey and others that an interdisciplinary consensus was forming.[35]

A second factor that aided Baldwin's argument was that many American astronomers, including Whipple, Leland Cunningham, and J. Allen Hynek, were familiar with high-energy explosions and ballistics through their wartime research. This experience increased their confidence about applying chemical explosion data to the problem of crater formation. Hynek declared that Baldwin had shown "a very close connection" between the lunar surface and the world wars. Whipple lamented the "world's increasing use of high explosives" but similarly argued that explosion craters were a legitimate stepping-stone to lunar craters. The advent of the atomic age also made vast explosions much less incomprehensible than they had seemed in 1942. Baldwin himself made explicit analogies between the Hiroshima, Nagasaki, Alamogordo, and Bikini Island nuclear explosions and the great Tunguska impact in Siberia in 1908, arguing that a major asteroid impact would have an effect "far larger than any battlefield holocaust, atomic bomb or otherwise." Scientists who backed

[33] Baldwin (1949, 131 [quoted], 132, 162, 129) and Ives (1919).

[34] Quoted from Whipple (1949c, 258) and Urey, revised draft, "The Planets," chap. 2, p. 22, Box 136, HCU.

[35] Daly to Baldwin, Apr. 5, 1945, RD; Baldwin (1949, 63), Daly (1943; 1947), Dietz (1946a,b), and Urey (1956).

Baldwin's reasoning typically had close connections with atomic research as well. Urey, a high-ranking member of the Manhattan Project, had even witnessed the Crossroads nuclear test on Bikini Island in 1946 as a guest of General Leslie R. Groves, the project's wartime director.[36]

A final factor was firm endorsement of Baldwin's results by Whipple, then the leading American researcher on small bodies of the solar system. With his Navy-funded meteor study well underway and his stature at Harvard rising rapidly, Whipple was among the younger generation of astronomers, including Kuiper and Brouwer, who gained new authority within the American astronomical community in the late 1940s. In a widely read 1949 review, Whipple declared that Baldwin had "constructed an almost unassailable argument for the explosive origin of major lunar craters" from which "the meteoric theory follows as a natural consequence." Whipple praised the accuracy of Baldwin's quantitative computations, a strong endorsement given Whipple's recognized skills in mathematics. Behind the scenes, Whipple privately championed Baldwin's conclusions to influential astronomers who initially hesitated to accept them, including Struve, and smoothed the feathers of individuals ruffled by Baldwin's strident tone and bluntly argued passages. Whipple, in short, became for Baldwin the champion that Russell had not.[37]

Whipple's efforts were particularly significant since he, like Russell, had rejected the impact theory less than five years before. His change of mind owed less to Baldwin's 1949 monograph than to extended discussions within the Harvard experimental geophysics community. In the first edition of his semipopular *Earth, Moon, and Planets,* published in 1941, Whipple had not ruled out meteoric processes in creating lunar craters, although he cited the low contemporary influx of meteors and the lack of large craters within the maria as evidence that volcanic processes had predominated. In private, however, Whipple had dismissed the impact theory for all except tiny ray craters. Writing the respected economic geologist Josiah E. Spurr, who in the early 1940s had suggested an igneous origin of all lunar features, Whipple had declared himself in wholehearted agreement. Allowing that he had "a fairly good proof" of the volcanic theory "based on the rate of capture of meteoritic bodies by the planets," Whipple had declared his intention to complete this proof after the war.[38]

His faith in the volcanic theory had nonetheless been undermined by Daly. During 1943–7, Daly, a longtime adherent of Alfred Wegener's theories of continental drift and the impact origin of lunar craters, had taken up anew the problem of terrestrial and lunar evolution. Shortly after Spurr's manuscript reached Whipple

[36] Quoted from Hynek (1949, 257), Whipple (1949c, 258), and Baldwin (1949, 112); see Whipple to T. Sterne, Dec. 27, 1946, Box 15, FLW–HU; Groves to Urey, n.d. [1946], Box 94, Folder 7, HCU; see also Merrill (1950).

[37] Quoted from Whipple (1949c, 258); see Whipple to Struve, Mar. 20, 1950, Box 4, idem to R. D. Hemens, Feb. 7, 1949, and Struve to Whipple, Mar. 23, 1950, Box 5, all FLW–HU.

[38] Whipple to Spurr, Jan. 26, 1943 (quoted), and Spurr to Whipple, Mar. 10, 1943, both HUG 4876.806, FLW–HU; see also Hoyt (1987, 356), Spurr (1944, 106–10), and Whipple (1941, 141–5, 147–9, 245).

early in 1943, Daly had inconclusively debated Whipple on the impact theory. After further discussions with Watson, Shapley, and local geophysicists, however, Daly had prepared three significant papers on impact studies. In the first, published in the *Bulletin of the Geological Society of America,* Daly had argued that Earth was composed of meteoric material accreted through collisions. Three years later, he had published an extended speculation about the Moon's early history, arguing that recent geophysical studies of heat flow and subocean structures was "forcing students of geological dynamics into the field of cosmogony." Here Daly amplified an unpublished suggestion by Russell that the Moon had been formed by a glancing collision between Earth and a massive object early in the solar system's history – perhaps the first time this fertile idea (widely adopted by American astronomers forty years later) was raised in print. His third paper, in 1947, had accepted that the Vredefort Ring Structure and cryptovolcanic structures were all likely meteorite scars. These articles reflected the twenty-five-year-long dialogue between Daly and Shapley regarding meteors and the lunar surface. Furthermore, they reflected Shapley's conviction, shared by members of Harvard's Committee on Experimental Geology and Geophysics, that borderland research between astronomy and geology was extremely promising.[39]

By 1947, Whipple's views toward impact had begun shifting toward Daly's. He became more firmly convinced while reading Baldwin's manuscript late that year for the University of Chicago Press (a task urged by Shapley, who saw impact studies as a possible new research line for the observatory). Whipple accepted Daly's claim that the Vredefort structure was formed by impact, and hence demonstrated that Earth too had a geologic record of major impacts. Indeed, most factors that Whipple cited in Baldwin's defense were geophysical rather than astronomical. He now argued that the mountains ringing Mare Imbrium were inexplicable except by impact, citing the convective theory of orogeny promoted by Griggs, his former Harvard colleague. Whipple also hoped that the meteor theory might reveal clues about the origin of ocean deeps – a central interest of Daly's in the early 1940s – and the stability of the continents. Writing Baldwin, Whipple urged him to find evidence that ocean deeps were also caused by impacts, noting that such proof would challenge the theory of continental drift. "I've always been repulsed by the Wegener hypothesis of the moving continent," Whipple confided to Baldwin, underscoring a remaining source of tension between him and Daly.[40]

The growing consensus behind the impact theory in the United States was not shared in Britain or in Europe; nor was it accepted by astronomers or geologists in

[39] Daly (1943; 1946, 104 [quoted]; 1947); Daly to L. T. Nel, Mar. 30, 1949, RD; Baldwin and Wilhelms (1992).

[40] Quoted from Whipple to Baldwin, Sept. 1, 1949, Box 1, FLW–HU; Whipple to Struve, Mar. 20, 1950, Box 4, and Baldwin to Whipple, Aug. 29, 1949, Box 1, both FLW–HU; Ralph B. Baldwin OHI (Ronald E. Doel, Oct. 25, 1989, at AIP, p. 38). See also Daly (1942, esp. 3–47; 1946, 105, 109, 118) and Le-Grand (1988, 116–17).

the Soviet Union. In part this reflected unfamiliarity with Baldwin's work and reputation outside America, and the more limited involvement of astronomers there in ballistics and weapons research during World War II. Other factors also played a role, however. For instance, in Great Britain Jeffreys remained firmly opposed to the impact theory, declaring his faith in the sufficiency of the volcanic theory alone. As the dean of British mathematical geophysicists, his views were influential. Moreover, Jeffreys's equally strong support for the contraction theory of mountain building – then supported by relatively few U.S. researchers – blunted the significance Baldwin, Whipple, and Daly attributed to the mountains ringing Mare Imbrium as signifiers of impact. Thomas L. MacDonald, a lunar researcher and longtime director of the Lunar Section of the British Astronomical Association, declared in 1949 that Baldwin had only succeeded in making the impact theory more plausible. In reviewing Baldwin's work, MacDonald argued that lunar and terrestrial mountains were adequately explained by horizontal compression caused by contraction, drawing on arguments Jeffreys had advanced through subsequent editions of his monumental *The Earth*.[41]

Within the United States and Canada, interest in the problem of meteorite impacts began taking on a distinctly disciplinary cast. With few exceptions, American astronomers came to accept Baldwin's arguments about terrestrial and lunar craters. Whipple and Kuiper regarded the controversy over lunar crater origins as essentially solved; neither engaged in debates with geologists over the merits of the impact theory. By contrast, most American geologists joined Bucher in assigning volcanic origins to cryptovolcanic structures and to lunar craters.[42] Although geologists ardently defended the volcanic theory in the late 1950s and early 1960s, these debates no longer influenced research by American astronomers.

It is also important to observe that institutional support for impact studies also bifurcated along disciplinary lines: Funds for researchers who wished to study craters as explosive phenomena far outweighed that for those who saw them as examples of unusual volcanic activity. Institutional backing soon became a leading influence on this field. Baldwin's *Face of the Moon* was among the last major contributions that an isolated individual would make to terrestrial impact studies. Investigating Earth's impact scars and the physics of impact required logistical support that no individuals and very few universities could provide. Neither the aging Daly nor Whipple, despite their interest in cosmogony, mustered new Harvard efforts to test Baldwin's claims. While Baldwin himself continued part-time research, remaining at Oliver Machinery, his main contribution remained synthesis. In the 1950s, the two institutions that did mount extensive studies of terrestrial impact craters were large, well-financed centers with extensive experience in mounting expeditions and con-

[41] Jeffreys to Urey, May 31, 1951, Box 47, HCU; see also Hoyt (1987, 358), LeGrand (1988, 104–5), and MacDonald (1949, 59).

[42] Bucher (1963). Among American astronomers, only Dinsmore Alter continued to defend Spurr's ideas; see Doel (1992, 246–7).

ducting field measurement: the Dominion Observatory in Canada and the U.S. Geological Survey. Their programs were led, respectively, by astronomer Carlyle S. Beals and geologist-geophysicist Eugene M. Shoemaker.

SEARCHING FOR CANADIAN IMPACT CRATERS

American astronomers who accepted Baldwin's arguments for lunar impacts agreed that large bodies had also struck Earth; but not all planetary researchers accepted Baldwin's assertion that a "painstaking scrutiny of the earth's surface in search of faint traces of ancient impacts" was justified. Although at odds over the Moon's thermal history, both Kuiper and Urey believed that virtually all lunar craters had formed shortly after the Moon's birth, a view Whipple also shared. This suggested to them little reward in searching for additional meteorite craters on Earth, since most would have been formed before the start of geologic time and erased by tectonic forces.[43]

Despite this pessimism, a research program to address these terrestrial impact craters was organized at the Dominion Observatory, located in Ottawa. The project, headed by observatory director Carlyle S. Beals, aimed to discover and investigate impact sites within the Canadian Shield, and to assess Baldwin's claims that Earth's impact history could be read from the geological record. Beals's program was significant not only in that it was the first institutional response to Baldwin's book, and did in fact discover by the late 1950s impact craters that provided the first indication of impact rates in the Precambrian era, but also because it was the first major interdisciplinary project since the CIW Moon Committee to straddle the border between astronomy and the geological sciences. No less interesting is the fact that it was one of the few major, post-1945, North American research programs in solar system astronomy that did not depend on military patronage.

Beals seemed an unlikely leader of this effort, judging solely from his earlier research interests. By 1946, when he was appointed Dominion Astronomer (making him director of Canada's national observatory), Beals had worked entirely in stellar astronomy. Born in Nova Scotia, Beals had studied spectroscopy with Alfred Fowler at Imperial College in London, where he had earned his Ph.D. in 1926. Returning to Canada, Beals had worked at the Dominion Astrophysical Observatory in British Columbia before being called to Ottawa. More comfortable with laboratory spectroscopy than observing, Beals had tackled such problems as the spectra of Wolf–Rayet and P Cygni stars and the distribution of interstellar gas clouds. In the process he had gained a solid record as an instrumentalist. This – plus his considerable skills in negotiation and personal diplomacy, noted by Richard Jarrell – explains why

[43] Quoted from Baldwin (1949, 127); see Kuiper, "Surface of the Moon and Early Development of the Solar System," draft, n.d. [1958], Box 35, GPK; see also Kuiper (1954), Urey (1955b), and Whipple (1959a).

Beals was chosen to lead the older, urban-centered Dominion Observatory, which had declined in research output and resources since the late 1930s.[44]

Several factors influenced Beals to search for fossil meteorite craters. One was Dominion staff member Peter B. Millman. A Harvard astronomy Ph.D. (1933) and product of Shapley's meteor research school, Millman had continued studying meteors and their spectra at Toronto during the 1930s and 1940s. After World War II, where he had served in the Research and Development Division of the Royal Canadian Air Force (RCAF), Millman had convinced the Canadian National Research Council to take part in Whipple's Navy-funded meteor project, a task aided by concurrent radar meteor research by another Canadian NRC scientist, Donald W. R. McKinley. Millman, indeed, had been the first to recommend testing Baldwin's arguments by searching for terrestrial impact craters. Beals came to share Millman's enthusiasm for meteor research, and in 1949 successfully championed his idea to organize a symposium on this field at the Ottawa meeting of the American Astronomical Society.[45]

A second factor, underscoring the importance of regional geology to studies of terrestrial and lunar craters, was the proximity of the Dominion Observatory to the Canadian Shield. Covering nearly two million square miles in northern Manitoba and Québec provinces, the shield was judged by North American geologists to be one of the oldest and best-exposed regions of the planet. By the early 1950s, geologists regarded the shield's age as 1.5–2 billion years, close to the 2–3-billion-year age then accepted for Earth itself. Moreover, except for episodes of glaciation, the shield's exposed rocky surface was judged essentially undisturbed for at least half a billion years. Despite its remoteness, geologists had mapped the shield's properties after the discovery of rich ore deposits there in the nineteenth century. Beals sensed that the shield's great age and undisturbed state would facilitate the discovery of any impact features there, and determine their frequency over geologic time as well. Baldwin had not explicitly cited the Canadian Shield in his 1949 book, but he had predicted that Archeozoic-age rocks were possibly old enough to have formed simultaneously with the lunar surface.[46]

Yet another reason why Beals felt attracted to this problem was the institutional structure of the Dominion Observatory. The Ottawa-based observatory was a disciplinary hybrid that uniquely reflected Canada's transformation from a British colony to a distinct nation. From its founding in 1905, the observatory's service role had expanded from traditional time service and positional determinations to providing geodetic, seismic, and gravity measurements, particularly of Canada's largely unexplored northern and northwest territories. As a result, the Dominion Observatory's

[44] Ralph Baldwin OHI (Ronald E. Doel, Oct. 25, 1989, at AIP, pp. 42–5); see also Jarrell (1988, 119–21) and Locke (1979).

[45] Ralph Baldwin OHI (Ronald E. Doel, Oct. 25, 1989, at AIP, p. 43); Beals to F. Whipple, Oct. 8, 1947, Box 1, and Whipple to O. Struve, Dec. 2, 1948, Box 4, both FLW–HU; see also Chant (1949, 139), Halliday (1991, 69–70), Jarrell (1988, 155), and Millman and McKinley (1967, 281–3).

[46] Baldwin (1949, 194) and Beals, Ferguson, and Landau (1956, 210).

scientific staff included geophysicists as well as astronomers, a far different disciplinary mix than that found at the U.S. Naval Observatory and other comparable national observatories. Beals recognized that impact crater research – which he revealingly termed a "branch of astronomical-geophysical science" – could unite his staff around a common research problem, bringing together divisions that previously had functioned as autonomous units. Besides this competitive advantage, Beals perceived a further advantage in launching the impact crater search. The growing strength of the Geological Survey of Canada and the Canadian Department of Mines and Technical Surveys, also headquartered in Ottawa, left the geophysical divisions of the Dominion Observatory vulnerable to annexation by these fellow civil service agencies. Pursuing a genuinely interdisciplinary program, he believed, could reinforce the observatory's autonomy.[47]

Beals was also aware by 1950 that many geophysicists had become interested in using aerial photographs for geological mapping, a technique pioneered in Canada owing to its vast and remote northern territories. Because of its strategic economic value and large exposures of barren rock, the Canadian Shield had become a major target of aerial surveys after World War II. As a civil service agency, Dominion Observatory enjoyed direct access to the swiftly expanding collection of aerial photographs of the shield's terrain assembled by the RCAF. Beals also had authority to request high-resolution aerial photographs of any suspected impact features detected in lower-quality images, as well as geological samples of these structures. This was a unique institutional resource, one that no other private or federally supported astronomical observatory could muster. Beals perceived this as crucial for identifying actual impact craters.[48]

Beals's decision to launch the impact crater search was further stimulated by the discovery of a fossil impact crater on the shield in 1950. Using aerial photographs, Fred Chubb, a prospector, had become interested in the late 1940s in a circular lake 1,100 miles north of Montréal in the Ungava Peninsula. Chubb thought the lake marked the top of a kimberlite pipe, a narrow volcanic structure frequently associated with diamonds. After staking a claim, Chubb had sought advice from geologists. Seeing potential in Chubb's story, the *Toronto Mail and Globe* had sponsored an expedition to the crater in 1950 by geophysicist V. B. Meen of the Royal Ontario Museum. After finding no evidence of volcanism, Meen declared it meteoritic. A second expedition, cosponsored by the U.S. National Geographic Society, reached the crater the following summer. Resultant publicity inspired pilots crisscrossing the Canadian Shield to scan this territory for other circular features. In 1951, Ottawa's Spartan Air Services Limited reported to the Dominion Observatory a crater-

47 Quoted from Beals, Innes, and Rottenberg (1963, 382); see Chant (1949, 139), Jarrell (1988, 88, 109–11, 157, 170), Ross (1957, 341), and Zaslow (1975, 404).

48 J. Tuzo Wilson OHI (Ronald E. Doel, Feb. 16, 1993, at AIP); George Garland OHI (Ronald E. Doel, Feb. 18, 1993, at AIP); Beals to M. Boyer, June 11, 1953, Box 15, CSB; see also Harrison (1954, 17), Ross (1957, 344–5), Zaslow (1975, 404, 409).

like structure near Brent, Ontario. These sightings convinced Beals that aerial photographs could indeed identify candidate impact sites.[49]

Despite these advantages, the Canadian Shield impact program emerged slowly during 1950–5. Beals instead concentrated on building new research programs within each of the observatory's main branches, stressing, for instance, its Super-Schmidt meteor camera survey. Two factors underlay his caution. One was the difficulty of finding diagnostic tools for distinguishing impact craters from those created by volcanic forces. Although Baldwin had confidently predicted that his logarithmic depth–diameter curve for explosion and impact craters could identify ancient impact structures despite a billion years of "folding, faulting, metamorphism, and erosion," in practice the geologic record admitted more ambiguity. Beals learned that deviations from this empirical relationship could result from the strength of the target rocks, the velocity and angle of impact, and subsequent distortions of the bedrock below the collision site.[50] It was not until 1955 that Morris Innes, chief of the Observatory's Geophysics and Gravity Branch, finally completed gravitational, seismic, and magnetic observations of the newly discovered Brent feature. A Toronto Ph.D. in geophysics, Innes had absorbed J. Tuzo Wilson's fascination with the Canadian Shield and his interest in tectonic problems. He and his staff found that the Brent crater's underlying structure generally conformed to the relation that Baldwin's equation predicted, a relief for Beals.[51]

Beals also proceeded cautiously because of a heated mid-1950s controversy over the Chubb crater, by then variously known as Ungava or New Quebec. In 1953, Paul Healy and Frederick C. Leonard of Los Angeles joined director Lincoln LaPaz of the New Mexico Institute of Meteoritics in a stinging dissent of V. B. Meen's conclusions about its impact origin. Meen had employed geologic evidence in declaring the crater meteoritic, noting its clear morphological resemblance to Meteor Crater in Arizona and the absence of volcanic activity in the region. Healy, LaPaz, and Leonard dismissed Meen's evaluation as scientifically "quite unjustifiable." Instead, they declared that only two identification criteria were valid indicators of impact: actual discovery of meteorites within craters, or actual observations of the fall of large meteorites. These criteria stemmed from the older geochemical tradition of meteorite analysis, where meteorites were the central analytical focus; LaPaz and his colleagues paid no attention to the quantitative structural relationships that Baldwin had proposed. The sharpness of their attack reflected continued divisions within the badly fractured, interdisciplinary community interested in meteoritics.[52]

[49] Beals (1958b, 39) and Jarrell (1988, 156–7).

[50] Baldwin (1949, 94), Beals, Ferguson, and Landau (1956, 205–7), Beals, Innes, and Rottenberg (1963, 246–7), Halliday (1991, 70), Jarrell (1988, 156–7, 160), Millman and McKinley (1967, 289), and "Notes from Observatories" (1954).

[51] J. T. Wilson to Kuiper, July 8, 1952, Box 32, GPK; J. Tuzo Wilson OHI (Ronald E. Doel, Feb. 16, 1993, at AIP); see also Beals, Ferguson, and Landau (1956, 207).

[52] Quoted from Healy, LaPaz, and Leonard (1953, 160); see Beals, Ferguson, and Landau (1956, 206) and Beals, Innes, and Rottenberg (1963, 277).

American astronomers interested in meteor research showed no interest in the arguments by Healy, LaPaz, and Leonard. As we saw in Chapter 3, LaPaz was known as a controversialist, and neither LaPaz's New Mexico institute nor the Meteoritical Society that he directed were highly regarded by Whipple, Brown, Urey, Wildt, or other American leaders of astronomy and "astro-chemistry." Beals recognized, however, that the Dominion Observatory was a branch of the Canadian Department of Mines, whose geologists tended to support volcanic interpretations of lunar craters and proposed terrestrial analogues.[53] To undercut the criticisms of the LaPaz group, Beals persuaded geologist James M. Harrison to visit New Quebec Crater in 1953. An expert in economic geology at the Geological Survey of Canada and a specialist in the Precambrian geology of the shield, Harrison was a senior and respected Canadian geologist known for his interest in applying new analytical tools to geology. After camping beside New Quebec Crater for several days, mapping its structure and surroundings, Harrison issued a new interpretation of the crater's geologic history. He agreed with Meen about its impact origin; but whereas Meen had interpreted immense boulders ringing the crater as emplaced by the blast, making the absence of associated meteorites disturbing, Harrison argued that these boulders had been positioned by glaciers, indicating that original debris within the crater had been displaced by flowing ice. Although not referring to Baldwin's work, Harrison's report explicitly rejected the LaPaz group's proposed criteria for identifying impacts. Harrison's statement, prominently published in the *Journal of the Royal Astronomical Society of Canada,* cleared the way for Beals to undertake a comprehensive survey.[54]

Beals launched the Dominion Observatory impact crater survey in two stages. First, he arranged to review nearly four hundred thousand aerial images of the Canadian Shield, maintained locally at Ottawa's Air Photographic Survey Library. This task occupied Beals and several staff members throughout the summer of 1955. (Despite his interest, Millman, who had reluctantly left the Dominion Observatory to take over McKinley's large radio meteor project at the Canadian NRC, did not participate.) Their search strategy was conservative: Rather than compiling a full census of possible fossil craters, they focused solely on best cases, discarding suspected ancient volcanoes, sinkholes, shallow lakes in boggy ground, or circular lakes with no evident impact structures. Using a stereoscope to heighten topographic relief of candidate features, Beals's team winnowed the final list to under ten. Theirs was the first systematic attempt to search for impact craters over a bounded geographic region. Responding to critics who argued that ground based geological surveys were better for identifying suspect craters, Beals argued that large circular structures were

[53] On the attitudes of Canadian geologists, see Halliday (unrecorded OHI by Ronald E. Doel, Jan. 22, 1992, AIP); Harold Masursky OHI (Ronald E. Doel, June 22, 1987, at AIP); and Mark (1987). LaPaz accepted the volcanic theory for lunar craters (except ray craters); see LaPaz to E. Shoemaker, July 4, 1960, Box 3, EMS.

[54] Beals (1958b, 245), Harrison (1954, 16–20), and Zaslow (1975, 416).

more easily glimpsed from the air, pointing out that lumbering and mining operations had gone on for years around the Brent crater before aerial views had identified this structure.[55]

The second and more difficult step involved direct geophysical studies of suspected craters to search for evidence of impact origin. Beginning in 1955, Beals authorized several expeditions to these features, permitting staff researchers to employ various tests for Baldwin's depth–diameter relations for explosion craters and associated geological structures. Gravity measurements, seismic studies of shallow explosive charges, and magnetic surveys were employed to assess general characteristics. In the winter of 1955, observatory staff members added a new technique: diamond drilling of core samples both within and around candidate craters. Since genuine impact structures were expected to show distinct layers of crushed breccias and undisturbed bedrock, in contrast to typical volcanic formations, Beals's staff viewed this approach as particularly promising. The observatory's preexistent multidisciplinary character eased expedition planning. Greater hindrances were weather and logistics. Extracting cores from the bottoms of water-filled craters could proceed only when their surfaces were frozen solid, typically late December to March, and equipment failures marred early attempts. The shortness of the summer season at suspected craters and cloudy weather delayed efforts by the RCAF and Geological Survey of Canada to obtain additional aerial photographs or direct samples.[56]

Within two years, nevertheless, Beals's team announced new results. Although several candidates identified in 1955 were eliminated after geophysical measurements suggested volcanic causes, other features showed underlying geological structures that adhered to Baldwin's formula. Besides yielding additional evidence about the New Quebec and Brent events, Beals's group found two larger features (Holleford and Deep Bay) whose structures matched Baldwin's depth–diameter relations. Beals used these results to argue two points. Deep Bay Crater, discovered and surveyed by Innes in 1956, was seven miles across – ten times the diameter of Meteor Crater. Its discovery closed the gap between smaller visible lunar craters and the largest known terrestrial impact scars. For Beals, however, the survey's cosmogonic implications were most important. Declaring the frequency and distribution of Earth's meteor craters greater than astronomers had supposed, Beals stressed Baldwin's argument that cratering was a geologic force that had affected the entire solar system. Though his initial 1957 and 1958 reports contained no precise calculations of Earth's impact rate in geological time – Beals noted only that Earth still seemed to have fewer recent impacts than the Moon – he declared that these craters "*may* represent the final stages in a process of accretion that formed the earth." The project's results were the

[55] Beals (1958b, 39), Beals, Ferguson, and Landau (1956, 208), Beals, Innes, Rottenberg (1963, 271), Mark (1987, 133), Ross (1957, 344), and Zaslow (1975, 431, 438); a helpful overview is Krinov (1963).

[56] Beals to M. Boyer, June 11, 1953, Box 15, CSB; I. Halliday (unrecorded OHI by Ronald E. Doel, Jan. 22, 1992, AIP); G. Garland OHI (Ronald E. Doel, Feb. 18, 1993, at AIP); see also Beals, Ferguson, and Landau (1956, 258), Beals, Innes, Rottenberg (1963, 277).

first terrestrial data to suggest the importance of celestial impacts as a recurrent geological process, rather than occasional or rare occurrences.[57]

Despite their novelty, Beals's stature within Canadian astronomy enabled him to disseminate his team's results quickly. Beals used the observatory's own *Publications* to communicate lengthy investigations of individual craters, employing the more widely read *Journal of the Royal Astronomical Society of Canada* to report more general findings. Mindful of the multidisciplinary audiences for these results, Beals also published notices of their survey results in such periodicals as *Nature* and *Scientific American.* He used these articles not only to address the significance of impacts for Earth's evolution, but also to defend the program's reliance on aerial surveillance and promote similar studies elsewhere. A majority of the ten impact craters discovered since 1950 (adding to the seventeen already known), Beals declared, had been found from the air.[58]

By the late 1950s, Beals's impact crater program was a well-established branch of the Dominion Observatory. One reason why it enjoyed remarkably steady growth, aside from the observatory's uncommon astronomical-geophysical orientation, was that the program encountered little reaction from geologists committed to volcanic interpretations. Although intense opposition later came from researchers within the Geological Survey of Canada, particularly geologist K. L. Currie, virtually no attacks on the Canadian Shield program were made before 1960 in the major geological journals of either Canada or the United States. The absence of early debate on what would soon become one of the largest and most significant interdisciplinary conflicts of the twentieth century is a poignant reminder of the intellectual distance still separating the astronomical and geological communities. However, another reason that Beals's program prospered was the continuing identity of meteor research with Canadian astronomy. Nearly 40 percent of articles in the Canadian-centered *Journal of the Royal Astronomical Society of Canada* in 1950 addressed meteor astronomy, a figure that remained over 25 percent throughout the decade. With strong support from Kuiper, Whipple, and Urey, who regarded the Dominion project as contributing to but not competing with their own lunar and meteor studies, Beals's program occupied a distinct, secure niche within Canadian science.[59]

Nevertheless, despite rapid expansion of U.S. solar system astronomy after the launch of *Sputnik,* Beals's efforts enjoyed only modest growth through the early 1960s. This remained so even as team members argued an impact origin for ever larger features, including the highly eroded, sixty-mile-wide Manicouagan structure in eastern Québec. An important distinction between Canadian and U.S. astronomy in the post-*Sputnik* era was that U.S. researchers experienced tremendous pressure

[57] Quoted from Beals (1957, 526); see also Beals (1958b, 39) and Beals, Ferguson, and Landau (1956, 211).

[58] Beals (1958a,b); see also Beals to G. Kuiper, Nov. 28, 1958, Box 18, GPK, and Jarrell (1988, 156).

[59] Baldwin OHI (Ronald E. Doel, Oct. 25, 1989, at AIP, pp. 42–5); Kuiper to Beals, Oct. 20, 1955, Box 10, GPK; Beals to Urey, Nov. 17, 1959, Box 12, HCU; see also Currie (1965), Jarrell (1988, 157), Mark (1987, 156), and Whipple (1959a).

to apply new information to their burgeoning defense and space programs. These pressures had much to do with the growth of terrestrial impact studies in the United States.

IMPACT RESEARCH AT THE U.S. GEOLOGICAL SURVEY

While Beals attempted to develop interdisciplinary studies of terrestrial impact structures within an astronomical institution, a number of geologists and physicists, particularly in the United States, sought to address the origin of craters on both Earth and the Moon. The most significant research of this type – which greatly influenced astronomical studies of the Moon and small bodies as well as investigations of impact phenomena – was done by Eugene M. Shoemaker, a young geologist at the U.S. Geological Survey (USGS). Shoemaker's attempts to combine geological and astronomical studies are important not only for illuminating continued methodological and professional disagreements between these communities, but also the strong military influence on Shoemaker's research characteristic of so many fields of American science after World War II.

Like Beals, Shoemaker had become attracted to the impact theory of lunar and terrestrial craters by reading Baldwin's 1949 monograph. A Caltech B.A. (1947) and M.A. (1948) in geology, Shoemaker had first been exposed to field geology and mineralogy at the Museum of Science in Buffalo (where his mother taught public school) and by trips into the mountains of Wyoming (where his father had found work in a Civilian Conservation Corps camp). After working two years with the Geological Survey, taking part in its extensive postwar project to search for new domestic sources of uranium, Shoemaker had taken leave to begin Ph.D. research at Princeton, where he worked with Hess. He had interrupted his graduate education in 1952 to take on a new assignment within the survey. By then, Shoemaker was interested in determining the origin of lunar craters and, more broadly, in understanding the Moon's geology. He also saw the V-2 program as a sign that direct exploration of the Moon could occur within his lifetime.[60]

Shoemaker recognized the institutional advantages that the Geological Survey could offer someone interested in studying terrestrial phenomena possibly related to lunar craters. At the same time, Shoemaker well understood the traditional service role that the USGS filled for other government agencies, which limited the opportunities of employees to pursue research projects of their own choosing. As a means of staying close to lunar studies while retaining his position within the survey, Shoemaker had accepted an opportunity to begin a large study of *diatremes,* volcanic vents formed by violent eruptions. The diatreme investigation stemmed directly

[60] Shoemaker to S. W. Hobbs, Dec. 30, 1958, Box 3, EMS; Shoemaker OHI (Ronald E. Doel, Jan. 30, 1986, and June 16, 1987, both at AIP); see also Shoemaker (1981, 51–3).

from geochemical research in which Shoemaker had participated at the survey's office in Grand Junction, Colorado, which coordinated studies of uranium deposits in the American West on behalf of the Atomic Energy Commission. Early in the 1950s, Shoemaker had demonstrated that certain diatremes contained elevated concentrations of uranium. By then Shoemaker had been intrigued by the close morphological similarity between diatremes and small lunar craters, particularly since Baldwin in 1949 had suggested that "chains" of small craters near the Moon's largest craters were likely volcanic. Knowing that diatremes often occurred along linear faults, Shoemaker believed that he had found a terrestrial analogue to a particular lunar feature. The diatreme project, backed by ample funding from the survey and the AEC, had therefore been much to Shoemaker's liking.[61]

Yet Shoemaker also accepted Baldwin's claim that most lunar craters were caused by impact, and sought to study high-energy explosion processes related to geology. One year after initiating the diatreme project, Shoemaker had welcomed an invitation from survey officials to join a larger, secret project under AEC direction. Worried that U.S. plutonium production could no longer meet military needs after the Korean War broke out in 1950, AEC officials had endorsed a proposal by Los Alamos weapons designer Theodore Turner. Turner's plan envisioned the quick manufacture of plutonium by the detonation of a nuclear device inside a thick blanket of uranium. To contain the blast, Turner suggested exploding the device in ice, then filtering plutonium from the resultant meltwater cavity. Turner's planned Megaton Ice-Contained Explosion project (MICE), beset with practical difficulties, was never attempted; but AEC officials soon proposed exploding such a device within a natural rock cavity, such as a salt dome, in the western United States. While allowing this highly classified project to continue, the change complicated the task of recovering plutonium from the anticipated jumble of fragments, a problem for geology and geochemistry as well as geophysics. Aware of Shoemaker's geochemical studies at Princeton, his training in physics, and his achievements within the Grand Junction geochemical project, Shoemaker's mentors within the USGS, including future survey director Vincent McKelvey, had appointed their 25-year-old candidate to MICE in 1953.[62]

Shoemaker's research in the mid-1950s came to focus on natural underground explosions. Using the geologic structure of diatremes as his source of evidence, Shoemaker sought to ascertain whether debris from a nuclear blast inside an underground cavity would reach the surface or remain confined below ground. He pursued this by studying diatreme formations in the Hopi Buttes of northern Arizona, a

[61] R. Jahns to Shoemaker, July 20, 1956, Box 7, and Shoemaker, "Investigation of Craters," n.d., accompanying Shoemaker to S. Hobbs, Nov. 25, 1959, Box 3, both EMS; Shoemaker OHI (Ronald E. Doel, June 16, 1987, at AIP); see also Baldwin (1949, 37, 216), Shoemaker (1981, 52–3), and Shoemaker, Byers, and Roach (1958).

[62] Shoemaker OHI (Ronald E. Doel, June 16, 1987, at AIP); and E. M. Shoemaker, "An Account of Some of the Circumstances Surrounding the Recognition of the Impact Origin of the Ries Crater," draft MS, n.d. [ca. 1986], in my possession. On MICE and related projects, see Thomas R. Nolan to Secretary of the Interior, Apr. 21, 1960, Box 260, USGS; also McPhee (1974, 113) and Shoemaker (1981, 52).

task that occupied him during 1953–7. He also saw this work, however, as a means for discovering new ways to distinguish impact from volcanic craters. Partly to carry out his MICE assignment, and partly to learn more about the physical process of high-energy shock, Shoemaker began an intensive study of shock wave theory, and arranged to teach geochemistry and geophysics at the University of Colorado in 1956. His course notes reveal considerable familiarity with Griggs's arguments about mountain building and Francis Birch's studies of high-pressure phase transitions in minerals, reflecting his close association with Hess at Princeton as well as his own far-ranging, systematic reading.[63]

The most critical step in Shoemaker's diatreme and nuclear explosion research was his decision to investigate Meteor Crater in Arizona, together with two craters produced by nuclear explosions. In late 1956, Shoemaker began intensive mapping of this best-known impact crater, focusing on relationships between its structure and shock wave mechanics. He then investigated craters formed by underground nuclear weapons tests at the Nevada Test Site (NTS). Created by the AEC in 1950, after the Korean War had eroded domestic resistance to atomic tests within the United States, the NTS conducted its controversial tests of atomic blasts involving buildings, armored vehicles, and the morale of nearby troops throughout the 1950s. Although many of these detonations were aerial bursts, a few were exploded in shallow ground. A 1951 test shot code-named Jangle U and the similar 1955 Teapot Ess shot both created craters in the Nevada desert. Shoemaker's study of the structure and geology of the Jangle U and Teapot Ess craters soon convinced him that the larger Meteor Crater was "an eerily close scaled-up" version of the atomic craters.[64] His explosion crater investigation turned out to have greater relevance to his MICE work than the diatreme study, which pleased his AEC patrons; but Shoemaker also believed these results would allow him to extend "certain generalizations based primarily on NTS studies to meteorite craters as a class." What Shoemaker had in mind was Bucher's cryptovolcanic structures and the circular features on the Canadian Shield that Beals had tentatively identified as meteoric.[65]

In time not devoted to the diatreme project or MICE, Shoemaker worked to reinforce his conviction that Meteor Crater was "the bridge between shock wave phenomenology and the surface features of the Moon." Drawing on theoretical and experimental results familiar to him through his AEC-sponsored research – including Samuel Glasstone's research on the distribution of energy in atomic explosions, the Birch–Bridgman studies of rock volume under strong shock, and L. V. Altshuler's work on the compressibility of iron – Shoemaker produced several studies in

[63] Shoemaker to C. Anderson, Nov. 1, 1955, Box 7, and "course notes," loose-leaf sheets, n.d. [ca. 1955], EMS; see also Shoemaker, Byers, and Roach (1958).

[64] Shoemaker (1959; 1962, 315; 1963; 1981, 52 [quoted]). On the early history of the NTS, see Eckel (1968), Hewlett and Duncan (1969), Melosh (1989, 228–32), and Miller (1986).

[65] Shoemaker to E. Hobbs, Dec. 30, 1958 [quoted], and idem, "Geological Exploration of the Moon," draft MS, Oct. 1959, both Box 3, EMS; idem to W. Pratt, Nov. 19, 1959, and idem to E. Chao, June 16, 1960, both Box 7, EMS.

rapid succession in the late 1950s. Using the Nevada bomb crater data, Shoemaker computed the total shock energy of Meteor Crater as equal to 1.7 megatons. Assuming 15 kps velocity (which Whipple had derived for meteorites of presumed asteroidal origin), he computed a mass for the impacting body of 63,000 tons; but he primarily sought this result to support the impact origin of lunar craters. Noting that the pattern of ejecta surrounding diatremes was typically circular, whereas that around Meteor Crater (and the Nevada craters) irregular and hummocky, he contended that this irregularity favored the impact theory for lunar craters. Using a single, remarkably sharp Mount Wilson photograph of the crater Copernicus, Shoemaker plotted precise positions of 975 tiny craters surrounding this large structure. Arguing that these features were secondary impacts – a position opposite that of most lunar researchers, and his own initial assumptions – he computed their ballistic trajectories. These distributions matched the observed trajectories that caused hummocky ejecta patterns around nuclear craters. On this basis, Shoemaker declared that Copernicus was produced by an impact generating 1,800 megatons of energy. The target's calculated velocity of 4.8–9.6 kps was within the range of Earth-crossing asteroids. It was the first time that the impact history of any lunar feature had been derived in quantitative, geophysical terms, a singularly important result.[66]

Shoemaker's working-out of the shock mechanics in Meteor Crater and the Nevada Test Site craters – a core contribution to impact studies that extended Baldwin's work into new theoretical realms – was made possible by several factors. Certainly the most critical was Shoemaker's intense interest in the problem of lunar geology, and his insight that the theory of shock wave mechanics could be applied from nuclear explosion craters to impact features. Another factor was his persistence and personal diplomacy. Shoemaker gained permission to map the privately owned Meteor Crater only after cultivating relations with members of the Barringer family, aware that an intemperate rebuttal of its impact origins by a survey geologist in the mid-1940s had soured family members to the USGS. However, Shoemaker's research was also aided by his high-level security clearance, his access to classified data, and his contacts with scientists in the nuclear physics community also working on shock wave theory and its applications. Shoemaker was well aware of both classified and public investigations by researchers at the RAND Corporation and the Lawrence Radiation Laboratory, including that of Griggs, Teller, and R. L. Bjork; he also knew of other attempts to find correlations between meteor craters and nuclear craters by RAND physicists John J. Gilvarry and Jerald E. Hill. More important, Shoemaker had access to classified data from NTS nuclear shots and theoretical studies from the Stanford Research Institute. An insider within the community of scientists engaged in applied shock wave studies, Shoemaker was sufficiently well

[66] Hoyt (1987, 320–1) and Shoemaker (1960a, 432; 1962, 316–19, 332). Baldwin and Shoemaker disagreed on defining impact events as "explosions"; see R. B. Baldwin OHI (Ronald E. Doel, Oct. 25, 1989, at AIP), and Shoemaker (1960b; 1962a, 327).

placed by the late 1950s to resist successfully efforts to have his findings classified in the name of national security.[67]

By then Shoemaker was planning a lunar program within the Geological Survey. In 1956, he suggested to USGS Director Thomas B. Nolan a lunar mapping project to compare impact with diatreme and maar-type volcanic craters. Although Shoemaker communicated his interest in lunar studies to survey leaders – including William W. Rubey, a confederate of Kuiper and Urey through his investigation of the origin of seawater – Nolan took no action. After the 1957 launch of *Sputnik,* Shoemaker learned that John A. O'Keefe, a Chicago-trained astronomer and geodesist in the Army Corp of Engineers, had commissioned a two-year lunar mapping program within the survey's Military Geology Branch. Reflecting heightened military anxiety about space, the project, directed by senior USGS geologist Arnold Mason, assessed the lunar "terrain" as a staging base for strategic operations.[68] In early 1959, anxious to get into lunar work, Shoemaker proposed a much larger program, one designed with the survey's traditional support role in mind, and urged survey leaders to seek direct congressional appropriations for this work. When this appeal failed, Shoemaker persuaded USGS leaders to allow individual members interested in lunar problems to pursue them, time permitting, as extensions of ongoing survey research – thus linking Lorin Stieff's studies of isotopes, William Fischer's work on aerial photogeologic mapping, and George Neuerberg's petrographic research with his own. Through these efforts, Shoemaker built a constituency within the survey favorable to studying lunar geology.[69]

That a large research program devoted to lunar or impact studies did not emerge within the Geological Survey in the 1950s reveals important constraints within this institution and within the broader American geological community. One major issue was the survey's troubled relationship with external patrons, particularly the AEC, following World War II. Anxious to replenish uranium supplies virtually depleted during the Manhattan Project, the AEC had directed the USGS to revitalize uranium and radium mining industries in the United States, many of which had gone out of existence when mineral markets crashed after World War I. The uranium program had become the survey's largest project during the first decade of the cold war, requiring a new field office, the hiring of hundreds of additional geologists and geochemists, and proficiency in the use of new tools such as airborne

[67] Shoemaker to S. W. Hobbs, Aug. 11, 1959, and idem to Chao, June 10 and 16, 1960, all Box 7, EMS; V. Wilmarth to E. Ruppel, Feb. 26, 1959, H. James to S. W. Hobbs, Mar. 13, 1959, and V. McKelvey to Hobbs, Mar. 13, 1959, all Box 7, EMS; see also Gilvarry and Hill (1956), Shoemaker (1960a), and Wilhelms (1993, 21). For RAND Corporation interest in nuclear and meteor crater studies, see *Index of Selected Publications . . .* (1962, A-43, A-55, P-72, P-75).

[68] Rubey to Shoemaker, June 14, 1957, W. Fisher to P. Orkild, Oct. 8, 1958, and M. Elias to Shoemaker, Dec. 10, 1958, all Box 3, EMS; O'Keefe to G. P. Kuiper, Feb. 27, 1959, Box 30, GPK; "Lunar Studies" folder, Box 61, and Nolan to T. K. Glennan, May 5, 1960, Box 90, both USGS; see also John A. O'Keefe OHI (Ronald E. Doel and Joseph N. Tatarewicz, Feb. 2, 1993, at AIP) and Doel (1992, 239–40).

[69] Shoemaker to M. Elias, Jan. 22, 1959, idem to A. C. Mason, May 29, 1959, McKelvey to Shoemaker, May 29, 1959, and Shoemaker to G. Neuerberg, Nov. 24, 1959, all Box 3, EMS.

magnetometers. Shoemaker's initial assignment with the survey, like his later diatreme and MICE studies, was funded through this initiative. Between 1949 and 1953 the USGS Geologic Division budget had risen from about $7 million to almost $16 million, the AEC underwriting nearly half this total. In June 1958, however, the AEC declared that sufficient uranium reserves had been found and abruptly canceled much of the uranium program. It also sharply reduced the autonomy of survey geologists in related undertakings such as Project Plowshares, designed to justify nonmilitary uses for nuclear explosions, such as the creation of ship canals. (Shoemaker was himself transferred from the now defunct Grand Junction field office to the survey's branch office in Menlo Park, California.) Bluntly advising AEC Commissioner Willard Libby that he would rather terminate existing AEC contracts than perform "a role contrary to the primary mission of the Geological Survey," Nolan hesitated to link his agency to any other unfamiliar governmental programs, a view shared by many senior survey administrators.[70]

A second factor was that the emerging federal patron for lunar research, the National Aeronautics and Space Administration, was itself uncertain about the role it wanted geologists to play in lunar exploration. NASA's emerging lunar program was largely formulated by physicist Robert Jastrow, head of this agency's nascent Goddard Space Flight Center, who himself relied on Urey for advice on organizing lunar research. Although Urey was familiar with Shoemaker's research, and came to accept Shoemaker's claim that a recent asteroidal collision had created Copernicus – challenging his own assertion that all lunar craters were formed before geologic time – he remained dubious about the significance of lunar geology. For Urey, whose influence in lunar science rose following his appointment to the Space Sciences Board in 1958, geochemical sampling remained the best method to solve the problem of the Moon's origin. Moreover, he continued to entertain the claim of Thomas Gold, then at Harvard, that many lunar features were agglomerations of dust (hence implying that geologic principles derived from Earth had limited applicability elsewhere). Despite new Shoemaker proposals for lunar research – and a plea by Manfred Eimer, a Jet Propulsion Laboratory (JPL) scientist and former Caltech classmate, to begin lunar photogeology and detailed geologic studies of terrestrial craters – NASA limited funds for geologic studies in the late 1950s.[71]

[70] On the AEC's uranium program, see Helmreich (1985, 225), Sylves (1987), and Taylor (1979, 29–30). On postwar USGS relations with the AEC, see Nolan to Libby, n.d. [ca. Oct. 1958], and J. E. Reeves to Nolan, Oct. 16, 1958, both Box 81, USGS; A. Goldenberg, "Geologic Division Initial Financial Operating Plan, FY 1959," June 19, 1958, Box 181, USGS; and Rabbitt (1974, 24–9). Many survey geologists and geochemists displaced by the AEC withdrawal ultimately found work in a large mapping program begun by Kentucky state officials in 1960; see Cressman and Noger (1981).

[71] Urey to Gold, July 17, 1955, Box 39, idem to Whipple, May 14, 1958, Box 102, and idem to Jastrow, Sept. 14, 1960, Box 47, all HCU. Urey nevertheless favored appointing a "hard rock geologist" to the Space Science Board; see Urey to L. Berkner, Aug. 1, 1958, and idem to Berkner, Dec. 2, 1958, both Box 74, LVB. On NASA's early lunar policy, see Nolan to T. Keith Glennan, May 5, 1960, Box 90, USGS; also Tatarewicz (1990a, 27–8) and Wilhelms (1993, 19–20, 35–7).

What further limited survey participation in impact-related studies was widespread skepticism over the importance of impacts in geologic history. As we have seen, most American geologists in the 1950s, just as earlier in the century, regarded meteorite impacts as a rare rather than significant geologic force, and hence found themselves disturbed by what amounted to a catastrophist challenge to the foundations of modern geology. Their intellectual leaders on this issue were not upstarts like Shoemaker or disciplinary intruders like Beals but senior geologists like Bucher. Bucher continued to oppose impact explanations for cryptovolcanic structures and other craters, arguing that their apparent alignment with tectonic features and new insights into violent volcanic processes made impact explanations unnecessary. Anxiety about catastrophic impacts was suggested by the virtual absence of references in American geology texts to Baldwin, Beals, or other impact advocates. Senior survey geologist James Gilluly, preparing the 1959 edition of his well-known textbook *Principles of Geology,* affirmed the uniformitarian principle by declaring that "geological study extending over many generations has failed to find evidence of ancient conditions totally unlike those existing today"; he made no reference to the impact hypothesis. An even more revealing dismissal came from geologist Gerhardt Amstutz, who complained that space research had created a "wave of meteorite-impact belief," leading to a "revival of creationistic beliefs." Such theories, he declared, threatened geologists with a "relapse into patterns which belong to the Dark Middle Ages."[72]

Too much emphasis can nevertheless be placed on opposition to the impact hypothesis within the USGS as causing its limited investment in lunar studies in the 1950s. A few survey leaders, particularly Rubey, were well versed in the impact theory for lunar craters through contacts with Kuiper and Urey. Nolan and William Pecora, an economic geologist and future survey director, were willing to accept that certain lunar features were meteoric. Moreover, several American geologists independently of Shoemaker published articles in the 1950s favoring the impact theory for terrestrial structures, suggesting opposition was far from uniform among American geologists. Instead, it is important to understand that pragmatic career issues also influenced survey work in this field. Lunar geology was far from a secure career pathway in the late 1950s, and NASA remained an unknown, possibly fickle patron. Nolan did move quickly to begin new geochemical studies of meteorites and tektites in 1958, largely with NASA funds, finding this work an unambiguous extension of the survey's mission and a means of employing the geochemists displaced by AEC cutbacks. Moreover, Shoemaker's backlog of uncompleted reports for his diatreme and AEC projects also limited his ability to forge a new lunar program. S. Warren Hobbs, chief of the survey's Branch of Mineral Deposits, hesitated to permit Shoemaker to "take on any other work" before these commitments were ful-

[72] Quoted from Gilluly (1959, 20–1); Amstutz is quoted in Mark (1987, 156). See Bradley (1963), Bucher (1963, 643), Dunbar (1960, 58–9), Elston (1990), and Mark (1987, 148–51).

filled. Living with this backlog was by then familiar to Shoemaker, who did not formally complete his Princeton Ph.D. requirements (by submitting a version of his Meteor Crater paper) until 1959.[73]

It is also clear that Shoemaker hesitated to become entangled with interdisciplinary collaborations that promised few geological awards, or might limit his access to the Geological Survey's unparalleled resources for geological fieldwork, a central thread in his research strategy. Hoping to initiate new investigations of cryptovolcanic structures, Shoemaker established ties with physicist and ballistic expert Alex C. Chambers of the nearby Ames Research Laboratory in Mountain View, California. Shoemaker judged that Chambers's work, then focused on ultra-high-velocity micrometeorite impacts on spacecraft and intercontinental missiles, would give him access to high-pressure facilities the USGS lacked.[74]

At the same time, Shoemaker shied away from establishing more than cordial relations with the lunar research team that Kuiper had begun to assemble at Yerkes (see Chapter 6). Both he and Kuiper were interested in understanding lunar geologic evolution; yet Shoemaker also sensed that critical issues divided them. In part these issues were scientific and technical. He disagreed with Kuiper over the value of polarization studies (which Kuiper favored), the thermal history of the Moon, and the origin of major lunar craters (Kuiper continued to accept that virtually all had formed when the Moon moved through a debris ring just after its birth). Shoemaker alone wanted to correlate the Moon's impact history with that of Earth's, and to establish the validity of applying geologic principles to the Moon.[75] However, his hesitation to ally himself with Kuiper also stemmed from disciplinary and personal factors, as always a critical issue in shaping potential transdisciplinary collaborations. Shoemaker believed that an Air Force lunar mapping project Kuiper initiated at Yerkes would fall short, in part because Kuiper paid limited attention to cartographic issues. Moreover, Shoemaker remained disturbed by Kuiper's still smoldering controversy with Urey, his reluctance to share promised data, and his domineering personality. Shoemaker advised survey colleagues to maintain cordial but wary relations with Kuiper's group, since "we are dependent, for the time being, upon astronomers" for photographs and access to large telescopes. Although he considered

[73] Challenges to impact interpretations include Hager (1953), Hardy (1954), and Wilson (1953); on survey interest in meteorite geochemistry, and Shoemaker's conflicting commitments, see Thomas B. Nolan, "U.S. Geological investigations related to the Lunar Science Program [proposal]," n.d., cover letter to K. Glennan, Dec. 16, 1959, Box 90, USGS; Hobbs to Shoemaker, Dec. 15, 1958, and L. Page to Shoemaker, Jan. 20, 1959, Box 7, both EMS; and Shoemaker to Kuiper, Dec. 2, 1958, Box 14, GPK.

[74] Shoemaker to Chao, June 16, 1960, Box 7, EMS; see also Charters (1960, 136–40) and Muenger (1985, 66–7, 73–4).

[75] Urey to Z. Kopal, Nov. 9, 1955, Box 50, and idem to R. S. Dietz, Apr. 10, 1959, Box 26, both HCU; Kuiper, [untitled lunar notes], Dec. 17, 1956, Box 11, GPK; see also Kuiper (1954) and Shoemaker (1962, 332). Shoemaker insisted, e.g., that "[t]he geological law of superposition is as valid for the moon as it is for the earth" (Shoemaker 1962, 346–7). By the early 1960s his stratigraphic system, subsequently modified, became the basis for lunar geology; see Shoemaker and Hackman (1960), and Wilhelms (1993, 40–2).

beginning a cooperative photographic lunar program elsewhere, Shoemaker conceded that this "would amount essentially to entering into competition" with Kuiper team members, who "now have a head start of about 6 years." Shoemaker frequently attended the Lunar and Planetary Exploration Colloquia, sponsored by North American Aviation's Missile Division, to learn from astronomers about lunar conditions, but avoided Kuiper's occasional invitations to make extended stays at Yerkes. Shoemaker seems not to have even entertained the idea of collaborating with Whipple, likely because Whipple remained focused on artificial satellite studies and felt convinced that most impact craters were formed before geologic time.[76]

What actually boosted Shoemaker's lunar research program was not his contacts with astronomers, but related work in physics and geophysics. By the spring of 1960, he was interested in a dense form of silica named "coesite," which Massachusetts industrial scientist Loring Coes had produced several years earlier using a pressure of 20,000 atmospheres at roughly 1,200° C. Aware that coesite had not been found in nature, Shoemaker thought to search for it in rock samples he had from Meteor Crater and the nuclear craters at the NTS. Before he inspected these, however, he learned that Edward Chao, a geochemist from the survey's headquarters in Washington, D.C., had in fact discovered coesite in other samples from Meteor Crater. Although disappointed by his loss of priority, Shoemaker hurriedly mapped the distribution of coesite surrounding Meteor Crater from his own samples, announcing these results jointly with Chao in *Science.*[77] Within two months, while traveling in Germany before the start of the Twenty-first International Geological Congress in Copenhagen, Shoemaker retrieved rock samples from the sixty-mile-wide Ries Basin, a large cryptovolcanic structure. He sent these to Chao for analysis, soon learning that they, too, contained coesite. This discovery led to a second joint paper with Chao, wherein Shoemaker claimed coesite was an unambiguous diagnostic for impact and that all cryptovolcanic structures were potentially of impact origin.[78]

Publication of these papers led to greatly increased NASA funding for "astrogeological" studies, as well as heightened enthusiasm for impact studies within the survey. Overcoming anxiety about taking on another AEC-style project, Pecora, then USGS chief scientist, candidly termed the coesite discovery a significant institutional achievement "of great value at a time when funds and outside support was

[76] Quoted from Shoemaker to M. Elias, Jan. 22, 1959, Box 3, EMS; see Shoemaker to Kuiper, Jan. 26, 1959, Box 14, and Kuiper to Alperin, Oct. 4, 1958, Box 18, both GPK; "Lunar Studies" folder, Box 61, USGS; Shoemaker OHI (Ronald E. Doel, June 17, 1987, at AIP); Wilhelms OHI (Ronald E. Doel, June 22, 1987, at AIP); see also Wilhelms (1993, 40–2). For Kuiper, see "Notes on conversation with Harry Hess," n.d. [ca. 1958], Box 12, GPK.

[77] At Shoemaker's urging a third author on this paper was his collaborator in x-ray crystallography, B. M. Madsen; see Shoemaker to Chao, June 7 and 16, 1960, both Box 7, EMS; also Chao, Shoemaker, and Madsen (1960), Mark (1987, 114–17, 166), O'Keefe (1963, 52–5), Shoemaker (1961).

[78] Shoemaker, "An Account of Some of the Circumstances Surrounding the Recognition of the Impact Origin of the Ries Crater," undated draft [ca. 1986], in my possession; see also Dietz (1959), Engelhardt (1982), Shoemaker and Chao (1961).

needed for the Survey." (Pecora seemed less convinced concerning the geological significance of Shoemaker's impact research, noting in a popular geology journal that it simply permitted "belief that some prominent lunar craters may well be craters induced by falls of meteorites.")[79] NASA officials overcame their hesitation to fund lunar geological research following these discoveries, in part because scientists outside geology – including *Science* editor Philip Abelson – testified that these results had far-reaching implications for physics, geochemistry, and astronomy. By late August, despite unfilled obligations to other projects, Shoemaker became head of a three-person effort to investigate possible impact craters on Earth and to initiate studies of lunar geology. Relieved of having to collaborate with Kuiper after gaining access to the 36-inch refractor at nearby Lick, Shoemaker recruited to his staff individuals with backgrounds in field geology, geophysics, or geochemistry. His Branch of Astrogeologic Studies, with an initial annual budget of $200,000, became the first permanent institution of lunar science established within the United States, and soon its most significant.[80]

In 1960, much of Shoemaker's work on impact studies and lunar geology lay before him, as NASA began funding what became its intensive decade-long focus on manned and unmanned lunar exploration. (Even after his formal retirement from the Geological Survey thirty-four years later, Shoemaker would gain new fame as a codiscoverer of Comet Shoemaker–Levy 9, which crashed spectacularly into Jupiter in the summer of 1994.) However, it is important to observe that military emphasis on explosion processes, the success of Shoemaker's shock wave research in the diatreme and MICE programs, and his own entrepreneurial skills in promoting these studies within and outside the survey, effectively hindered institutional backing for advocates of volcanic theories for terrestrial and lunar craters during the late 1950s and thereafter. Jack Green, a former Bucher student at Columbia who accepted Bucher's volcanic arguments for cryptovolcanic structures and all craterlike features on Earth and the Moon, became the most enthusiastic U.S. defender of the volcanic theory during this time. Despite gaining a position in 1957 at North American Aviation, itself anxious to secure advice on the lunar environment as it bid for anticipated lunar spacecraft contracts, Green was criticized and subsequently ignored by Urey, Shoemaker, and others who influenced funding decisions within NASA and the USGS. Shoemaker's research – shaped by Baldwin's arguments and the context of postwar American science – helped forge an institutional commitment to impact

[79] Quoted from Pecora to Shoemaker, June 13, 1960, Box 7, EMS, and Pecora (1960, 19); see also Pecora to Shoemaker, June 17, 1960, Box 7, EMS.

[80] Shoemaker OHI (Ronald E. Doel, June 17, 1987, at AIP); A. R. Baker to T. K. Glennan, Dec. 2, 1960, "Budget Estimate for Geological Survey Lunar Science Investigations, FY 1961," P. H. Abelson to N. W. Cunningham, May 31, 1961, and T. Nolan to A. T. Waterman, Dec. 29, 1961, all Box 90, USGS; see also Shoemaker (1981, 53). Total NASA appropriations to the USGS rose from $ 205,000 in FY 1960 to $554,500 in 1961; see Baker to Golden, Sept. 12, 1960, and Aug. 25, 1961, both Box 90, USGS.

studies at this leading agency of American geology long before impact theory gained similar standing within comparable bodies in Europe and the Soviet Union.[81]

Studies of the origin of lunar craters thus proved to be one of the best-funded yet most difficult fields of solar system astronomy prior to 1960. The problem was not, as Cambridge geophysicist Edward C. (Teddy) Bullard declared to Kuiper in 1952, that astronomers expected "a few broad general principles worked out in detail to cover most of the subject," whereas geologists were "so overwhelmed by detail that [they are] apt to be very impatient of any sort of theory and lapse into merely describing things." Theory testing clearly concerned geologists such as Wright, Daly, and Shoemaker no less than astronomers Baldwin, Beals, and Kuiper. The problem was rather one of finding a demonstration of either the impact or volcanic hypothesis that satisfied leading disciplinary participants in this issue. Not until Beals found evidence of impact craters in the ancient rocks of the Canadian Shield did Kuiper, Urey, and other lunar researchers accept that impact collisions had continued through geologic time. Similarly, not until Shoemaker demonstrated the equivalence of nuclear explosion craters and impact structures in the late 1950s did geologists begin to accept the troubling implication that a cataclysmic process rarely witnessed in human history was a significant geological force.[82]

Several factors proved important in bringing about this emerging consensus. One was the local geological landscape, particularly the accessible Meteor Crater in the U.S. Southwest and the large expanse of the Canadian Shield that the Dominion Observatory staff was assigned to explore. Familiarity with the geology of these regions made it more difficult for geologists to disregard craters within them; Day's willingness to entertain the possibility of impact collisions following his visit to Meteor Crater in 1927 was an example of the power of visual demonstration in science. Economic and military factors, however, particularly aided research into (and, importantly, acceptance of) the impact theory among American astronomers and geophysicists. Beals's desire to search for impact sites on the Canadian Shield was aided by the Canadian government's desire to explore and exploit the rich resources of this region, and by his institution's strategic place to carry out this mission. Shoemaker's own investigations of diatremes, nuclear craters, Meteor Crater, and shock wave mechanics, while reflecting the creativity and intellectual resourcefulness of this gifted researcher, were made possible only by the broader military context in which his work was carried out. Even Baldwin's original defense of the impact origin of lunar craters, although written after he had left his wartime assignment with the OSRD proximity fuze project, benefited from his access to Army records of

[81] Urey to Z. Kopal, Sept. 20, 1960, Box 50, HCU; Baldwin OHI (Ronald E. Doel, Oct., 25, 1989, at AIP); see also Doel (1992), Elston (1990), and Markov (1962).

[82] Quoted from Bullard to Kuiper, Sept. 3, 1952, Box 11, GPK; see also Albritton (1989) and Simpson (1963).

bomb explosion craters and his evident close familiarity with these data. No comparable institutional or programmatic support went to researchers attracted to volcanic explanations for lunar craters or terrestrial analogues.

Transient institutions forged between astronomers and geologists played a role in resolving this controversy and bringing about cooperative research programs, as they did in other fields of solar system astronomy; but the importance of their contributions was mixed. Despite its resources and bold agenda, the CIW Moon Committee, backed by one of the largest patrons of American science in the early twentieth century, failed to yield new clues about the nature of the Moon and by the late 1930s sank from memory. Shapley, Daly, and other members of Harvard's own informal astronomical–geophysical group, although unmentioned in previous histories of impact studies, clearly played a more significant role in resolving the origin of lunar craters. Partly this demonstrated the difficulty of determining fruitful investigatory pathways: Wright's petrographic technique turned out to be an ill-fated gamble, despite the advice of leading experts in astronomy and geology within the CIW. However, the enormous expense of mounting expeditions to suspected impact structures and studying the shock wave mechanics under extreme pressures also made it easier for permanent national institutions, rather than transient organizations or university communities, to pursue this field. The tools and techniques commonly perceived as most relevant to impact crater studies were largely in the hands of geologists, geochemists, and geologists; by the late 1950s, as the NASA era began, astronomers became less central to lunar research.

As we have seen, Beals and Shoemaker frequently met difficulties in creating an institutional context for solar system research within their home facilities. Although Whipple, Kuiper, and others did not perceive this problem in the early 1950s, and felt confident that their field would continue its advance, rapid changes in the discipline placed them by the decade's end in increased competition with stellar and galactic researchers for telescope time and patronage. The splintering of American astronomy had a profound influence on the fate and self-perceptions of American astronomers interested in the solar system. It is to this important development that we now turn.

6

Patronage and American Astronomy: Turmoil and Transition, 1952–1960

Compared to their colleagues in physics, American astronomers sought little federal aid between 1945 and 1955. The Brookhaven, Los Alamos, and Lawrence national laboratories, symbolic of the emerging Big Science age in physics (with their emphasis on money, manpower, and machines) and the domination of physics in the early cold war, had no immediate counterparts for U.S. astronomers. Most government grants for astronomy ranged from hundreds to just several thousand dollars; the U.S. Naval Observatory's continued investment in celestial mechanics remained a low-keyed affair. Even Whipple's Navy-sponsored meteor–upper-atmosphere project, one of the largest in U.S. astronomy after World War II, consumed only about $100,000 per year, a tiny fraction of the budgets for particle accelerators and federal research institutes. Although most research programs in solar system astronomy after 1945 were initiated by federal or military funds, much of the cost of doing astronomy in America – the salaries of principal investigators, the maintenance of observatory telescopes, the upkeep of academic departments – continued to be borne by private philanthropic organizations or by general university budgets, much as in the years between the world wars.

The 1950s nevertheless became a period of intense change for American astronomy. *Sputnik I*'s launch in October 1957, and the torrent of federal funds for astronomy and space-related research released in its wake, are the best-known aspects of this change. Even before *Sputnik* soared into orbit, however, astronomy had been growing rapidly in size, in its diversity of patrons, and in its base of instrumentation. Important changes were also taking place on the periphery of this discipline, including the rise of geophysics and the International Geophysical Year (IGY), an undertaking shaped by cold-war anxieties and geopolitical strategies.[1] The rise of radio astronomy, a migration of researchers from physics, and the influence of federal patrons intent on building national observatories further transformed American astronomy, as did sharply increased competition for telescope time. Attendance at American Astronomical Society meetings rose from 150 in 1950 to over 500 in 1960, and U.S. as-

[1] LeGrand (1988, 170–5), McDougall (1985), and Needell (1987).

tronomy began expanding 20 percent per year, faster than any other physical science discipline. Big Science also took root in American astronomy at this time, defined not simply by instrument cost and size but by a style of research distinct from the traditional "lone furrow" approach still common among astronomers at university and private observatories. New programs involved specialized workers and multidisciplinary teams; their great size was accompanied by a loss of autonomy and greater responsiveness to political factors.[2] U.S. astronomy was radically transformed, not during World War II – often seen as the watershed of twentieth-century American science – but in the 1950s.

No field of American astronomy felt these changes more than solar system research. The 1950s were a time of unprecedented expansion for planetary science, particularly as science budgets mushroomed after *Sputnik*. New instruments and new agencies, including NASA, were created in these years, and research programs developed earlier in the 1950s by Kuiper, Whipple, and others expanded dramatically. Although the first permanent institutions in planetary science (including Kuiper's Lunar and Planetary Laboratory) were formed during this decade, its senior leaders found this period difficult. Familiar ways of doing research were disrupted. Patronage, once steady but small-scale, became large and uncertain, with funding levels and priorities no longer reflecting the judgments of disciplinary peers but national policies. Whipple, who in 1947 had cheerfully advised Harrison Brown that a "new era is opening up" in meteor and cometary studies, dourly noted in 1956 that solar system astronomy now seemed a "neglected" field of research. In 1953 Kuiper had similarly invited readers to compare his edited *Solar System* volumes with the *Handbuch der Astrophysik* of 1929 to see "the advance made in the last two decades," declaring that "equally productive" times would continue through the 1970s; but in 1956 Kuiper too voiced concern about what he termed the virtual "abandonment of planetary studies" since Percival Lowell's heyday.[3] The chief anxiety for both men was that American astronomy, until then a unified discipline, had begun splitting into competing specialties, placing their field at comparative disadvantage.

Since 1960, most planetary astronomers have labeled the pre-*Sputnik* era a time when stellar astronomers ridiculed and ignored solar system research. For instance, Gérard de Vaucouleurs declared that planetary physics and astrophysics were "neglected since the turn of the century in favor of the deeper vistas opened by the large reflectors applied to the exploration of galactic and extragalactic space."[4] Given the rich and complicated history of solar system astronomy through the twentieth century, it might be tempting to dismiss these practitioners' contemporary statements stressing neglect as little more than misperceptions, exaggerations, or fund-

[2] Berendzen and Moslen (1972, 49), Smith (1989; 1992, 186–7). Smith nevertheless points out that, by the late nineteenth century, major national observatories such as Greenwich already operated with divisions of labor characteristic of modern Big Science.

[3] Quoted from Whipple to H. S. Brown, Sept. 24, 1947, Box 1, FLW–HU, Whipple (1956c, vii), and Kuiper (1953d, vii; 1956b, 89); Kuiper was referring to Abetti et al. (1929).

[4] De Vaucouleurs (1960, 27).

raising tactics. Such statements are themselves important, however, for they reveal how astronomers responded to the eclipse of earlier patronage systems in astronomy, and fundamental changes in the ground rules by which one obtained funds, sought instrument access, and did science. The anguish of solar system researchers was genuine, even if the villains at whom they pointed were not. These accounts illuminate the fragmentation of American astronomy, and indeed American science, in the 1950s, stimulated in large part because of shifts in patronage that had begun long before *Sputnik I* reached its launching pad.

FRAGMENTING THE DISCIPLINE: THE NSF AND POSTWAR ASTRONOMY

During the early 1950s, as throughout the first half of the twentieth century, American astronomy remained a unified discipline. To be sure, astronomy in 1950 embraced more distant fields than in 1920, including galactic structure, interstellar matter, and far-infrared research; but several factors served to protect the discipline from the rise of specialization that had already fragmented physics and biology in America. One was its limited size: In 1950 fewer than three hundred individuals called themselves astronomers, a smaller number than for any other physical science discipline. Another was its maintenance of discipline-wide professional journals, such as the *Astronomical Journal* and the *Astrophysical Journal,* and its reliance on common graduate training for all fields of astronomy. A third factor was that most astronomers worked at academic institutions. Nearly 82 percent listed universities or private research organizations as their employers, a higher ratio than for any other science discipline in America. Since most universities would not hire two astronomers with the same specialty, and few hired more than one, astronomers were encouraged to maintain their generalist outlooks.[5]

An even more important factor that preserved the coherence of astronomy in the United States was its patronage structure – particularly the reluctance of astronomers to accept major federal contracts. The experience of astronomers in this respect was quite different from that of other physical scientists, and refutes the notion that members of all scientific communities rushed to embrace government and military patronage after World War II. Throughout the early 1950s, for example, leading members of the astronomy advisory committee to the Office of Naval Research requested just $80,000 per year for small-grant research, a figure that paled besides million-dollar annual allocations for physics and geophysics. Moreover, astronomers generally showed little interest in competing for large individual grants from military or federal agencies. Although Whipple welcomed his million-dollar Navy contract for meteor research, few American astronomers felt this way; in the early

[5] Berendzen and Moslen (1972, 49), Smith (1989, 22–3), Struve (1960, 18–19), and Struve and Zebergs (1962, 141).

1950s, a virtually isolated Whipple repeatedly criticized his colleagues for what to him seemed excessive timidity over federal funds.[6]

The reluctance of U.S. astronomers to accept government aid largely derived from the instrumental traditions of astronomers and the unique place of astronomy within the political economy of American science. In contrast to physicists, accustomed to building costly apparatus applicable only to a limited number of experiments, the central instrument for astronomers remained the large, multipurpose reflecting telescope, with lifetimes spanning many decades; the cost of constructing auxiliary detectors for these telescopes, such as spectrographs, was relatively minor. Moreover, astronomers had long secured substantial support from such powerful patrons as the Carnegie and Rockefeller foundations, and enjoyed modest but stable support from their home universities; few astronomers were willing to risk this security to pursue wealthier but untested patrons and unfamiliar instrumentation. In 1949, Lick's C. Donald Shane privately warned colleagues that accepting federal patronage invited "the danger of unwise dictation of the astronomical program" by individuals outside the discipline. Although few leading U.S. astronomers saw matters so sharply, they did share his concern. When Kuiper urged the construction of a large reflector at McDonald in 1953, he eyed the Ford Foundation and the state of Texas as possible patrons, ignoring federal and military agencies.[7]

The coherence of American astronomy meant that solar system studies remained an integrated component of the discipline. A majority of U.S. astronomers continued to concentrate on the discipline's core research problems, particularly the evolution of stars and structure and evolution of galaxies; but celestial mechanics and solar system astrophysics were identifiable branches of the discipline. For instance, at Cincinnati, which remained the home of the IAU's Minor Planet Center, mathematical astronomers Herget and Rabe focused their attention on asteroids; at Yale, Brouwer continued to investigate planetary and asteroid orbits; and in Washington, D.C., Clemence continued his work on celestial mechanics of the solar system. Aided by other researchers at Indiana, asteroid research remained a strong field of American astronomy. Leland Cunningham, a Whipple student who earned his Ph.D. in 1946, similarly launched new investigations of cometary orbits at Berkeley in the early 1950s, employing comet observations provided by Lick astronomer Harold Jeffers. Although none of these efforts was large, they were identifiable programs of mathematical astronomy whose results were utilized by Whipple, Kuiper, and other leaders of the field.[8]

[6] Whipple to Kuiper, May 18, 1948, Box 3, FLW–HU; DeVorkin (in press).

[7] Quoted from "Draft of Proposal for the Support of Astronomy by The National Science Foundation," n.d. [1949], annotated copy in Box 5, JLG; and Needell (1991). Kuiper expected such a telescope would be part of a university consortium; see Kuiper, "[untitled memo on large telescope]," Mar. 26, 1953, Meinel to Kuiper, Apr. 2, 1953, and Kuiper to Struve, May 29, 1953, all Box 17, GPK.

[8] Cunningham to D. Brouwer, Apr. 20, 1946, DB; Jeffers to Pettit, Nov. 5, 1953, Box 13, and Kuiper to D. Brouwer, Jan. 26, 1952, Box 10, both GPK; Rabe to W. W. Morgan, Sept. 13, 1959, Herget file, WWM; see also Cuffey (1956) and Rabe (1950).

A number of astrophysicists also contributed to solar system studies part-time during the early 1950s, continuing the pattern of field crossing practiced by Russell, Shapley, and Menzel in the 1920s and 1930s. Although the difficulty of securing astrophysical data for planetary bodies limited such contributions, several astrophysicists did turn to solar system research whenever instrumental advances or new theoretical developments allowed. At Michigan, for example, stellar astronomer Dean B. McLaughlin studied observations of seasonal albedo variations on Mars, arguing that their pattern seemed less likely the result of vegetation than of windblown dust. In 1951, Wildt aided a graduate student, Wendell C. de Marcus, to produce a thesis on the properties of hydrogen at pressures he calculated to exist in Jupiter's core, seeing this work closely tied to his own studies of planetary and stellar interiors. Wylie at Iowa and Bobrovnikoff at Ohio Wesleyan continued their respective work on meteor characteristics and comet spectra, while Lowell astronomers observed planetary atmospheres. Related planetary studies were launched by Robert S. Richardson (Mount Wilson–Palomar) and John D. Strong (Johns Hopkins). All considered themselves to be full professional members of the American astronomical community, despite the diversity of their research interests.[9]

Additional evidence supports the view that solar system research remained a wholly integrated branch of American astronomy through the early 1950s. Roughly 10 percent of all articles published in the *Astronomical Journal,* the *Astrophysical Journal,* and the *Publications of the Astronomical Society of the Pacific,* the major periodicals of U.S. astronomy, reported results in this field. Colloquia series in astronomy at Caltech and at Berkeley for 1951 and 1952 similarly reveal that one in ten, on average, addressed solar system topics. Equally important, astronomers like Brouwer, Kuiper, Whipple, and Clemence, whose expertise lay in solar system astronomy, were elected to leadership posts in the 1950s within the American Astronomical Society, the discipline's main professional organization. Whipple and Kuiper became members of the influential AAS Advisory Council, and Clemence, a mathematical astronomer at the U.S. Naval Observatory, became its vice-president in 1952, after Brouwer's term expired.[10]

Perhaps the most significant indication of the field's solid standing within the larger discipline, however, comes from careful analysis of funding patterns. Between 1948 and 1953, the ONR became the preeminent general patron for astronomy and other physical sciences, filling a void caused by repeated delays of the proposed National Science Foundation. Its astronomy advisory committee, composed of elite

[9] L. Aller to G. Kuiper, July 19, 1955, Box 10, and Kuiper to B. Strömgren, Jan. 24, 1952, Box 30, GPK; Wildt to M. Schwarzshild, Oct. 5, 1955, RW; Strong to R. Putnam, June 4, 1953, and idem to E. C. Slipher and A. Wilson, Aug. 13, 1953, both LOW; see also Bobrovnikov (1951, 341), McLaughlin (1955), Osterbrock (1986), Pettit and Richardson (1955), and Wildt (1947; 1958). Wildt also retained his membership in the IAU's Commission 16 (planetary surfaces and interiors) during the 1950s.

[10] "Colloquium Speakers," 1951 and 1952 lists, and Struve to J. Oort, Feb. 20, 1952, both Box 11, UCA[os]; also colloquium lists, ca. 1951–3, Box 32, GPK. See also Appendix Table A.3.

members elected by the AAS leadership, was assigned to select competitive proposals submitted to the ONR. During these years, between one-eighth and one-quarter of total grant dollars for astronomy funded solar system research, supporting proposals from Wylie, Cunningham, Wildt, and Kuiper. This meant that solar system astronomy fared better than other fields, since it produced 10 percent of the discipline's output. Letters from Kuiper, Whipple, Brouwer, and their colleagues before 1953 voiced little anxiety about their status within this community. Indeed, Kuiper's main worry was that Lowell astronomers would inadvertently reopen the Martian canal controversy, drawing unfavorable publicity to their field.[11]

By 1953, however, American astronomers realized that important changes were occurring within the discipline, driven by shifts in funding. Within several years these developments began to heighten tensions among component members of this community. One such factor was the growth of new, specialized research instruments. Although large, multipurpose reflecting telescopes remained the discipline's principal tools, specialized instruments became increasingly common in the mid-1950s. The Baker Super-Schmidt cameras constructed for the Harvard meteor program were one example. Radio telescopes were another, including the 25-ft dish installed at Harvard in 1952. Because radio astronomers were largely trained in electrical engineers or in physics, and because radio astronomy attracted few U.S. participants prior to the late 1950s, the American astronomical community retained its orientation toward optical techniques through the early 1960s. However, the proliferation of specialized instruments lessened the significance of large reflecting telescopes as the discipline's central research instruments. Those who sought access to these new instruments were required to learn techniques particular to them, eliminating the common heritage of research practices and tacit knowledge formerly shared by all optical astronomers.[12]

A second factor was the rapid proliferation of patrons for American astronomy after 1950. Although government support for space research mushroomed after *Sputnik* and accelerated in the 1960s, its *range* of patrons had begun diversifying much earlier, a development with tremendous implications for this community. Until the immediate postwar period, American astronomy had remained a tightly organized, centrally controlled discipline. When the ill-fated National Research Fund was created in 1928, for example, Russell was the sole astronomer selected to review funding proposals submitted to this body. Twenty years later, just a half-dozen astronomers in the ONR's astronomy advisory committee had similar author-

[11] No central figures for ONR grants were published, although astronomy funding can be determined from the private records of advisory committee members; see, e.g., Kuiper to B. Strömgren, Jan. 24, 1952, Box 30, GPK. Of ten projects in solar system astronomy approved for funding in 1951–2, six ranked above the median score of evaluated proposals. On the canal controversy, see Kuiper to G. de Vaucouleurs, Sept. 6, 1950, Box 18, GPK; A. Dollfus to de Vaucouleurs, Feb. 23, 1948, BL; Audouin Dollfus OHI (Ronald E. Doel, Jan. 14, 1987, SHMA at AIP); and de Vaucouleurs (1955, 197).

[12] Needell (1991) and Sullivan (1984).

ity, keeping the discipline tightly focused; but thereafter astronomy's patronage grew fragmented. In part because of delays in creating the National Science Foundation, military agencies eager to promote research in their domains emerged as major patrons of science by the late 1940s. Air Force contracts supported Menzel's multi-million-dollar studies of correlations between solar flares and Earth's ionosphere, critical for military communications, and, as we have seen (Chapter 2), the Navy's Bureau of Ordnance underwrote Whipple's investigations of the upper atmosphere via meteor investigations. By the early 1950s military agencies also signed contracts with Brouwer for studies of planetary orbits and with Clyde Tombaugh for a search for small natural satellites of Earth. These projects were far larger than the small-grant funds administered by astronomers through the ONR. Many elite American astronomers grew alarmed by this trend. For example, in 1953 Kuiper complained that projects like Tombaugh's were ill-advised. Despite their potential contribution to solar system studies, he declared, they limited the autonomy of astronomers to allocate disciplinary resources.[13]

These tensions intensified when the National Science Foundation (NSF), finally established in 1952, became the new chief patron of U.S. astronomy. Initially American astronomers expected the NSF to maintain the hands-off policy and modest support levels characteristic of the ONR. This expectation was partially realized: In 1953 the ONR's Astronomy Advisory Committee was transferred to the NSF, and astronomers in this body continued their periodic review of small-grant proposals as under the ONR. Yet American astronomers soon discovered that NSF officials were eager to promote large new research facilities. Several factors influenced this policy, including the outbreak of the Korean War in 1950, which made government officials eager to remobilize U.S. scientific capacity for scientific and military needs. However, this policy also reflected the vision of Raymond J. Seeger, a Yale-trained theoretical physicist and former George Washington University professor second in command under Alan Waterman, the former ONR chief scientist who became the first NSF head. As director of the foundation's Mathematics, Physics, and Engineering Division, Seeger had considerable leverage in translating his ideas into policy. Whereas Struve, Clemence, and other senior U.S. astronomers had requested a one-time grant of $150,000 for capital improvements to existing telescopes in 1950, Seeger and Waterman soon proposed vastly greater undertakings. In 1953, the NSF convened a conference at Lowell Observatory, at which U.S. astronomers were instructed to discuss what kinds of telescopes were most useful for the discipline. Seeger and Waterman then announced a multi-million-dollar plan to build the first national optical facility, soon adding funds for a separate national radio telescope.[14]

[13] DeVorkin (in press). Brouwer's research was also supported by the Watson Scientific Computing Bureau of IBM; see Brouwer (1956). On Tombaugh, see Kuiper to L. Goldberg, Mar. 4, 1953, Box 18, GPK; Whipple to R. C. Gibbs, May 26, 1953, Box 11, FLW–HU; and Tombaugh (1956).

[14] Kuiper to B. Strömgren, Jan. 24, 1952, Box 31, GPK; Edmondson (1991, 70), England (1982), Kevles (1977, 344–7, 356–60), and Needell (1987).

Attempts by Waterman and Seeger to create new instrumental facilities drew mixed reactions from American astronomers. Many remained anxious about the federal government as a sponsor of major instruments, and few wished a dramatic expansion of the scale of astronomical research. A typical response came from Kuiper, who in 1953 urged NSF officials to reexamine their commitment to building new telescopes, declaring that American researchers were already competitive with their European and Soviet counterparts and that existing U.S. telescopes were not oversubscribed.[15] Kuiper also criticized a 1953 plan by Spitzer, then chair of Princeton's Department of Astronomy, to create an orbiting astronomical observatory. Although Spitzer's approach reflected his postwar involvement in large-scale fusion research, and gained sympathy among a few East Coast astronomers who were familiar with Big Science research and lacked competitive optical telescopes at their home institutions, strong opposition came from Struve, Greenstein, and other optical astronomers worried about its long gestation, uncertain future, and probable cost. Kuiper recommended making no official mention of Spitzer's project to government patrons "[u]nless there should be some advantage in getting government services used to the notion that astronomers are capable of asking for a lot of money." Kuiper was not opposed to the *idea* of such a telescope; rather, he worried that such ventures would jeopardize the modest but stable resources of the discipline and distort its research priorities.[16]

The NSF's determination to build national optical and radio telescopes nevertheless eroded the intellectual cohesiveness of American astronomy. One reason was that the planning groups for each of these new facilities came to comprise distinct sets of astronomers, with little common overlap. Leading the national optical telescope committee were directors of the major U.S. optical observatories, such as Bowen of Mount Wilson–Palomar and Shane of Lick; the radio telescope committee, by contrast, was primarily composed of East Coast theoretical astrophysicists such as Menzel and Leo Goldberg of Michigan, active participants in IGY planning groups and comfortable with the prospect of federal funding for science. Different attitudes toward federal funding for science were fortified within these committees. Bowen stressed that "government grants are normally made for short periods only and their renewal is always subject to many very uncertain factors, political or otherwise," whereas Goldberg and Menzel berated fellow astronomers for not recognizing that the United States could "provide an enormously expanded effort in basic research." Members of the radio astronomy panel, including Menzel, pressed for large-scale developments at this second national observatory. Radio astronomy's rapid growth in the mid-1950s, and the overlap between radio astronomers and IGY

[15] Kuiper to Seeger, June 18, 1952, idem to H. C. Kelly, Feb. 10, 1953, Seeger to Kuiper, Feb. 16, 1953, Kuiper to L. Goldberg, Mar. 10, 1953, all Box 30, GPK; and Whitaker (1985, 17).

[16] Quoted from Kuiper to Struve, May 29, 1953, Box 10, GPK; see also idem to L. Goldberg, Mar. 4, 1953, Box 18, GPK; and Smith (1989, 27). This project eventually became the Hubble Space Telescope, launched in 1990.

scientists, further distinguished astronomers who preferred small-science methods from those who envied the growth of large-scale physics research.[17]

A more important issue for solar system research, however, was that astronomers planning the national optical telescope decided to emphasize stellar and galactic astronomy over other fields. In mid-1954 members of this advisory panel, including Bowen, Struve, Albert Whitford, Bengt Strömgren, and Robert McMath (the latter three respectively representing the Lick, Yerkes, and University of Michigan observatories) met to discuss the design of this new instrument. Despite Struve's insistence that the group recall that lunar and planetary studies required high atmospheric *steadiness,* not simply transparency, the full committee ignored this point in later site deliberations. Moreover, although committee members called for a "general-purpose" reflecting telescope of use for all American astronomers, it also endorsed a lengthy memorandum submitted by Bowen that advocated designing a "stellar and nebular" telescope for faint-object spectroscopy, and also backed construction of a separate telescope for solar astrophysics. Not surprisingly, the Bowen Report, as it was later called, irritated and discouraged solar system researchers, particularly Kuiper. "Planning, when implemented, leads to financial support, [which] brings with it enormous pressures and prejudices," Kuiper grumbled to a close associate, complaining that small-scale projects had much greater positive influence on American astronomy than multi-million-dollar programs like Menzel's solar studies. Bowen's priorities simply reflected the research interests of a majority of U.S. astronomers, whose fields had received less government support than solar system studies since 1945; nevertheless, his report signaled that competition between astronomy's component fields was on the rise.[18]

Outwardly the discipline remained unitary and intact through the mid-1950s. Its journals remained inclusive, departments of astronomy experienced little curriculum change, and no splinter groups pressured the AAS for structural reforms. Senior American astronomers saw the handwriting on the wall, however, especially as NSF funds for astronomy leaped from $148,000 in 1954 to $363,000 in 1955. In that year a poignant alarm came from Struve. By then president of the International Astronomical Union and the most influential astronomer in the United States, Struve worried that funding conflicts would fragment the discipline. In a widely read article in *Popular Astronomy,* he declared that the time had come for U.S. astronomers to embrace national planning of research goals. Anticipating strong distrust of national planning, Struve endorsed it as a way for astronomers to assert greater auton-

[17] Quoted from Bowen Report, Aug. 14, 1954, Box 31, OS, and Goldberg and Menzel (1956, 103). On U.S. radio astronomy, see Needell (1987), Smith (1992, 186–7), and Sullivan (1984).

[18] Quoted from "Record of the First Meeting, NSF Advisory Panel for National Astronomical Observatory," Nov. 4–5, 1954, Box 31, UCA[os], and from Kuiper to P. van de Kamp, Jan. 22, 1955, Box 32, GPK; see also Struve, "The Educational Value of the National Astronomical Observatory," memorandum, Apr. 15, 1955, Box 31, UCA[os].

omy and control over NSF leaders, and to resolve conflicts among the discipline's ever more diverse communities.[19]

Struve continued to regard the discipline as a coherent whole. It was perhaps the last time that he or other senior U.S. astronomers would do so, for after 1955 conflicts among instrumental styles and specialties rose sharply. This trend was ironically illustrated by a planning conference on the general needs of astronomy, organized by Whipple at Struve's urging. Whipple's "New Horizons" panel, which met at Princeton in April 1955, attempted to identify domains of astronomy less likely than stellar and galactic astrophysics to benefit from new NSF funding. Two areas considered were "collatoral sciences," including interdisciplinary work in geochemistry, and "solar system astronomy," within which Whipple located thirteen distinct subfields, including studies of meteors, asteroids, comets, interplanetary matter, the Moon, planets, satellites, and celestial mechanics. Although Kuiper, Whipple, Brouwer and others interested in solar system astronomy wrote for the "New Horizons" volume, pointing out new research opportunities, the project did little to improve disciplinary unity. "New Horizons" authors sharpened rivalries by pressing claims from their particular fields, and neither Waterman nor Seeger, for whom these assessments were intended, took the effort seriously. Asked his views, Harvard's Bart Bok agreed that the NSF needed to invest more in border-crossing science; but he thought that the national observatories campaign would prevail. Praising interdisciplinary research, he grumbled to Whipple, was "like coming out against sin."[20]

It is unlikely that any actions could have averted the growing fragmentation of American astronomy by this time. After 1955, U.S. astronomers increasingly identified themselves as adherents of distinct instrumental approaches and as members of specific fields. Astronomers like Spitzer and Whipple, advocates of large-scale research programs backed by major federal grants, were opposed by astronomers who preferred to emphasize small science with traditional optical techniques, heightening distinctions between East Coast research centers and the major optical observatories in the American Southwest. Friction also arose among members of the discipline's major intellectual branches, however. Rapid growth, specialized patronage, and increased demands on instruments left senior researchers like Kuiper worried that solar system astronomy was becoming a *competitor* of stellar and galactic astrophysics. These tensions mushroomed after *Sputnik,* although they had originated in NSF funding policies.

[19] Struve to R. Seeger, Oct. 31, 1955, Box 13, FLW–SI; Whipple to Struve, Dec. 3, 1954, Box 31, UCA[os]; see also DeVorkin (in press, 48–9), Struve (1955), and Tatarewicz (1990a, 114).

[20] Quoted from Bok to Whipple, June 17, 1955, Box 15, FLW–HU; see also minutes, "The Committee on Needs in Astronomy," Apr. 3, 1955, Box 15, FLW–HU; Whipple to Struve, Dec. 3, 1954, Box 31, UCA[os]; and DeVorkin (in press). Final articles appeared in vol. 1 of *Smithsonian Contributions to Astrophysics,* begun after Whipple became director of the Smithsonian Astrophysical Observatory in 1955. Whipple also placed auroral studies and earth magnetism under "solar system astronomy," reflecting Harvard's long embrace of "astro-geophysical" research.

In the remainder of this chapter, the growth of U.S. solar system astronomy on the eve of the space age is examined within two institutional contexts: first at Harvard, where proponents of Big Science approaches to astronomy dominated, then at Chicago, where astronomers maintained their emphasis on traditional optical astronomy despite bitter quarrels over their future research mission. Previous accounts have stressed that conflicts among astronomy's component fields were largely based on intellectual prejudices, fueled by stellar researchers who derided the intrinsic value of solar system research.[21] It is argued here that these tensions were more the consequences of institutional and social factors, including the emergence of differentiated patrons and the political pressures of the space race. These made astronomy more like physics and other disciplines already buffeted by the pressures, opportunities, and discontinuities of cold-war science policy.

ASTRO-GEOPHYSICS: CREATING BIG SCIENCE AT HARVARD

During the 1950s, Whipple joined Menzel in launching new solar system research at Harvard. This was not a departure for Harvard astrophysicists: As we have seen, Harlow Shapley had stimulated interest in meteors, impact craters, and small bodies of the solar system during the 1920s and 1930s (Chapter 1). Whipple and Menzel continued to endorse the interdisciplinary approach that Shapley had nurtured through coordinated studies of the age of Earth and the origin of meteors; in 1954, for example, Menzel declared that "the interesting zone that lies between geophysics and astrophysics" remained Harvard's research mandate.[22] Whipple's meteor and upper-atmospheric programs grew remarkably swiftly; but the 1950s brought dramatic changes to astronomy at Harvard. In contrast to most U.S. departments of astronomy, Harvard astronomers embraced Big Science styles of instrumentation and management, drawing criticism from optical astronomers like Kuiper and Greenstein. Whipple and Menzel invested heavily in balloon- and satellite-based instruments, encouraged IGY research, and made Harvard the largest U.S. center for research on meteors and related phenomena. Dramatic gains in funding for Harvard astronomy strained traditional budget channels. The Harvard College Observatory resolved these tensions through a partial merger with a federal research facility, illuminating one means by which U.S. universities adapted to the bigness of Big Science.

As we have seen (Chapter 2), Whipple's Harvard-based meteor program grew rapidly after World War II. Encouraged by Navy patrons, by Baker's success in designing fast Super-Schmidt telescope-cameras, and by the abundance of information

[21] See, e.g., Tatarewicz (1990a, 116–17).

[22] Quoted from Menzel to G. Vaeth, Dec. 15, 1952, copy in Box 18, GPK; Menzel, *ARHCO*, 1954–5: 1–2.

about the origin of cometary and meteoric particles as well as the upper atmosphere, Whipple recruited ever greater numbers of astronomers to take part in this study. By the early 1950s, it had become one of the largest projects in U.S. astronomy; only Menzel's solar research program, supported by military patrons interested in the relation between solar activity and long-range communications, exceeded it.

Another reason that Whipple's meteor work blossomed, however, was that Shapley, still observatory director, faced continued difficulties in recruiting funds for other branches of astronomy. A major problem was the poor optical and mechanical quality of Harvard's 61-inch telescope at nearby Agassiz (Oak Ridge) Station, which rendered it useless for quality spectroscopic or photoelectric research. Shapley by the late 1940s was confronting unusual difficulties. The Rockefeller-funded philanthropies that had endorsed his broad interdisciplinary projects in the 1920s and 1930s no longer underwote the physical sciences. Moreover, as a political liberal who supported Progressive candidate Henry A. Wallace's 1948 presidential campaign and championed left-wing causes, Shapley became a target of McCarthyite attacks and congressional inquiries as the cold war deepened. Although Harvard president Conant publicly supported Shapley's right to free speech, privately he abhorred Shapley's politics, chilling relations between them. Nevertheless, it seems indisputable that the poor condition of Harvard's older telescopes had resulted from Shapley's feckless management of these facilities and his reluctance to fund routine maintenance. Even Bok, Shapley's closest Harvard associate, privately conceded that Harvard's optical telescopes were no match for those at competing centers of American astronomy.[23]

By the early 1950s, conflict began to rise among Harvard astronomers over the merits of large military-funded research efforts versus traditional optical studies backed by nonmilitary patrons. These tensions, which pitted Shapley and Bok against Whipple and Menzel, had deep roots. Political differences were evident – Shapley had refused to accept military funding for his research after World War II, whereas Whipple and Menzel believed that national security was properly served through university-based undertakings like the meteor project – but a more important dimension involved the scale and style of scientific research. Whipple and Menzel, comfortable with large military-sponsored projects and their hierarchical ordering of specialists, yearned to investigate atmospheric phenomena and other interdisciplinary problems with combined experimental and observational approaches; they were dissatisfied with attempts to *coordinate* independent research, as Shapley wanted to do by assembling climatologists, astronomers, and paleobotanists at his 1946

[23] On Shapley's post-1945 career, see Bok, "Confidential Report to the Harvard Observatory Council on the Summer's Activities and Cambridge and Oak Ridge," Sept. 22, 1948, pp. 2–3, as well as idem to Buck, Oct. 6, 1952, and Menzel to Provost Paul H. Buck, Feb. 24, 1953, all Box 7, FLW–SI; also DeVorkin (1984, 49, 55), Diamond (1992, 117–19), Hershberg (1993, 416, 446–8), and Kidwell (1992). For the perceptions of non-Harvard astronomers, see Struve to M. L. Oliphant, Dec. 1, 1955, Box 17, OS, Kuiper to J. Oort, Feb. 11, 1953, Box 18, and R. C. Gibbs to Kuiper, Feb. 16, 1953, Box 11, both GPK.

conference on global climatology. For his part, Shapley worried that military patronage for astronomy distorted research priorities. While admitting that Whipple's meteor project operated without overt military interference, he complained that military funding emphasized certain fields and certain questions to the detriment of the discipline as a whole.[24]

This conflict came to a head in late 1952, with major consequences for Harvard astronomy. In that year, Harvard administrators discovered that the Harvard College Observatory lacked funds to maintain both the Agassiz Station and its Southern Hemisphere observatory, the Boyden Station in Bloemfontein, South Africa. Conant convened a special HCO Visiting Committee to review Harvard astronomy; to it he appointed observatory leaders Bowen, Shane, and Strömgren, and theoretical physicist J. Robert Oppenheimer, a close associate, as chair. Committee members toured the Agassiz Station facilities that November. Shapley and Bok remained optimistic that Oppenheimer's committee would vindicate them by recommending heightened funding for traditional optical astronomy; but committee members instead voted to recast Boyden Station as a consortium-operated facility, to limit graduate training to Whipple's and Menzel's fields, and to invest just $300,000 to restore the 61-inch telescope only enough for instructional use. Apart from its Air Force- and Navy-funded observing stations, they declared, Harvard's optical equipment "by reasonable contemporary standards . . . is not adequate either for a real program of observational research, or for the proper training of graduate and advanced students in observational astronomy." Bok launched an intense rebuttal; but after a stormy private discussion among astronomers at the eighty-eighth meeting of the AAS in Amherst, Massachusetts, in December 1952, Shapley resigned the observatory directorship, the post he had filled since 1921. In early 1953 Conant appointed Menzel acting head of the HCO. Two years later, learning that Menzel had become permanent director, Bok abruptly resigned. His departure troubled Struve and others sympathetic to Bok's dedication to general-purpose astronomical instruments, but by then Menzel and Whipple were firmly in charge of Harvard's astronomy program.[25]

Stellar and galactic research at Harvard did not cease with Menzel's appointment. Shapley, Bok, and other Harvard researchers continued work in these fields, maintaining Harvard's standing in optical stellar astronomy; yet Menzel clearly saw solar astrophysics and solar system research as Harvard's most promising fields. Early in 1953, Menzel enabled Whipple to hire associate researchers, including Thomas, and brought back former Harvard astronomer Luiga Jacchia, then at the MIT

[24] Shapley, *ARHCO,* 1945–6: 1, 1946–7: 1, and 1949–50: 2; see also DeVorkin (1992, 95–6) and Kevles (1977, 355).

[25] Quoted in Conant, "Memorandum to The [Harvard] Corporation," Nov. 11, 1952, Box 7, FLW–SI; see also Bok to O. Struve, Dec. 9, 1952, and Struve to Bok, Apr. 1, 1952, both Box 10, OS; Bok to Whipple, June 17, 1955, Box 15, FLW–HU; Menzel to P. Buck, Feb. 24, 1953, Box 7, FLW–SI; Bart Bok OHI (David H. DeVorkin, May 15, 17, 19, and June 14, 1978, SHMA at AIP); and Needell (1987). Bok later joined the newly created Australian National University.

Computing Bureau, to continue work on Whipple's program. Although Menzel and Whipple collaborated only occasionally in scientific work (an important exception being their speculative 1955 paper on the atmosphere of Venus, which predicted a global ocean of hydrocarbons), they saw mutual advantage in supporting one another's programs.[26]

The fickle nature of military research contracts was one difficulty that Whipple's increased collaboration with Menzel helped to address. In 1954, as we saw in Chapter 2, Whipple received notice from the Navy Bureau of Ordnance that his Super-Schmidt meteor project was canceled. The problem was not a lack of results: The program had yielded an impressive quantity of new information on the sizes of cometary and meteoric particles, allowing Whipple to address such issues as the formation of the zodiacal light and the amount of meteoric material striking Earth. The problem, rather, was that the meteor project had failed to reveal accurately the *density* of the upper atmosphere, the datum that interested his Navy patrons; when alternative methods proved better for this task, Ordnance officials ended the study and requested return of the Super-Schmidts. Menzel secured a new mission for the Super-Schmidts from his Air Force patrons, who contracted with Whipple to study upper-atmospheric winds by photographing the lingering trains of brilliant meteors. Though this transfer caused minimal disruptions, it rekindled Whipple's frustration over his inability to study meteor ballistics experimentally at Harvard, as uncertainty over particle sizes had affected his atmospheric density estimates. It also underscored the vulnerability of this Harvard project, by 1954 operating with an annual budget of $120,000 and supporting twenty-three individuals part- or full-time.[27]

The difficulty of sustaining large-scale contract programs within Harvard's small department of astronomy continued to trouble Menzel and Whipple. Menzel soon seized on a novel solution to this dilemma. In early 1955 he convinced Leonard Carmichael, Secretary of the Smithsonian Institution, to move the Smithsonian Astrophysical Observatory (SAO) to Harvard. Carmichael had conceded the necessity of moving this facility after the Smithsonian failed to lure to Washington a qualified researcher to direct its decades-long study of the Sun's energy output following the retirement of its longtime head, Loyal B. Aldrich. Menzel persuaded Carmichael that Harvard had competency in solar physics and the capacity to train graduates in

[26] Menzel to C. Payne-Gaposchkin and Whipple, Feb. 26, 1953, Box 7, FLW–SI; idem, *ARHCO*, 1952–3: 8; and Menzel and Whipple (1955). Their Venus study was influenced by dissertation research by William Sinton, Urey's geochemistry, and Whipple's distrust of radiometric measurements suggesting high surface temperatures; see Whipple to John Strong, Feb. 18, 1954, Box 10, FLW–HU; idem to Urey, Dec. 6, 1955, Box 15, and Aug. 12, 1957, Box 102, both HCU. As noted in Chapter 1, Menzel had first studied planetary atmospheres as Russell's graduate student in the 1920s.

[27] Whipple to J. Keller, Mar. 2, 1954, Box 10, and idem to D. R. Bates, Aug. 25, 1950, Box 1, both FLW–HU; Menzel, *ARHCO*, 1953–4: 17 and 1955–6: 9; and untitled budget notes, n.d. [ca. 1957], Box 32, DHCOm; see also Jacchia and Whipple (1956), Kerr and Whipple (1954, 570), and Whipple (1955a, 384).

this expanding field, two factors important to Smithsonian officials. No less significantly, Carmichael followed Menzel's recommendation to appoint Whipple as the SAO's new director. Because Whipple was one of Harvard's most well-respected astrophysicists, and had studied solar heating on the structure of the upper atmosphere, such an offer was entirely appropriate; but Menzel and Whipple were also aware that government facilities could accommodate large federal grants better than university departments, and clearly saw advantages in bringing the SAO to Cambridge beyond the boost it gave to solar astrophysics at Harvard.[28]

Under Whipple's direction, the SAO quickly became a major center for solar system research as well as for stellar atmospheres and theoretical astrophysics. In his first year as director, Whipple (who retained his Harvard appointment) worked to establish large-scale interdisciplinary projects on several fronts. Transferring his meteor grants from Harvard, Whipple created the experimental program in meteor ballistics that he had long desired, soon arranging to use an Air Force Aerobee rocket to spray the upper atmosphere with artificial aluminum meteors. He also began studying impact craters and the micrometeoroid bombardment of vehicles in space, acknowledging the military's "extremely great interest . . . in these present days of extra-atmospheric missiles." Furthermore, Whipple developed a large program in radar meteor studies cooperatively with the Lincoln Laboratory of MIT, and hired researchers trained in fields other than astronomy, including John S. Rinehart, the former leader of ballistics research at the Navy's Ordnance Test Station at Inyokern. What limited Whipple's expansion was not money but Smithsonian fears that Whipple was trying to build a competing Smithsonian on the banks of the Charles. Defending his hiring of John A. Wood, a meteorite geochemist, Whipple promised Carmichael's office that he had no intention of duplicating meteorite studies already underway in its Museum of Natural History.[29]

Whipple's largest initial achievement as SAO director, one that profoundly influenced this observatory's future, was linking the SAO to the International Geophysical Year, scheduled for July 1957–December 1958. In early 1956, stressing the SAO's infrastructure and his experience in meteor–upper-atmosphere studies, Whipple won the IGY contract for artificial satellite tracking, one of the largest awarded to any astronomical center at that time. As a member of the IGY Technical Panel on Rocketry, Whipple understood the satellite's potential value for geodesy and upper-atmosphere studies; he also understood its perceived importance to national security concerns. Whipple used the million dollars this contract provided to finance an advanced generation of telescope-cameras, the Baker–Nunn Super-Schmidts. To facilitate computations of satellite orbits, Whipple hired as consultants two experts in

[28] For a detailed analysis, see Doel (1990b).

[29] Whipple to J. L. Keddy, June 26, 1957, Box 9, Menzel to McG. Bundy, Jan. 24, 1955, Box 6, and "Minutes of the HCOC, July 8, 1955," Box 7, all FLW–SI; Whipple and Thomas to Carmichael, May 25, 1956, Box 4, LC[si]; Menzel, *ARHCO,* 1953–4: 32, 1955–6: 9, and 1956–7: 11–12; see also Doel (1990b, 144–7), Dupree (1990), Jacchia and Whipple (1956, 351), and Whipple (1956b, 83–5).

celestial mechanics, Samuel Herrick of UCLA and Leland Cunningham of Berkeley – anticipating what, after *Sputnik,* became a dramatic revival of this field. More important, he hired nearly a dozen researchers experienced in solar system astronomy, including former Yerkes–McDonald astronomer George A. van Biesbroeck. The SAO soon grew by several orders of magnitude, its budget burgeoning from $70,000 in 1955 to $3.2 million in 1956 while its staff swelled from fifteen to roughly a hundred by 1957. Well before the IGY and the launch of *Sputnik,* Whipple had transformed the SAO partly into a center for solar system research.[30]

Whipple and Menzel were nevertheless aware that most American astronomers, including those interested in the solar system, were uncomfortable with Big Science "astro-geophysics" emerging at the Harvard–Smithsonian observatories. Optical astronomers by the mid-1950s perceived several problems with this approach. The tremendous outpouring of IGY results threatened to overwhelm astronomical journals and professional meetings; adopting a defensive posture, Struve argued that "astro-geophysics" lay beyond the boundaries of the *Astrophysical Journal,* and endorsed a plan to emphasize "purely astronomical problems" at the 1958 meeting of the IAU by limiting discussions "of a geophysical character."[31] Even more troubling to optical astronomers was the emphasis Menzel and Whipple placed on nontraditional instruments. At Mount Wilson–Palomar, Greenstein refused to join the IGY Technical Panel on Earth Satellites, declaring that Mount Wilson–Palomar staff

> must be concerned with problems which can be handled by the large telescopes, and in our good climate. I do not feel that it is probable that we will ever take an active part in the upper atmosphere or geophysical research, since such topics can be very well handled by scientists at other institutions for which these problems may be a major responsibility.[32]

Greenstein's attitude toward instrumentation troubled Menzel and Whipple, who increasingly regarded U.S. optical astronomers as obsessed with traditional approaches. In sharp contrast to Greenstein and Struve, Whipple termed the prospect of satellite astronomy "the most marvelous scientific development that has happened during my life."[33]

Despite newly established links with the SAO, Menzel nevertheless found it difficult to maintain HCO's competitiveness prior to *Sputnik*'s launch. One reason was the restrictive nature of federal and military grants, whose brick-and-mortar exclusions prevented applying funds to new buildings or capital expenses. Aware that the observatory's finances remained limited, Menzel launched a new $1.5-million en-

[30] Whipple to Herrick, June 29, 1956, and idem to W. von Braun, Nov. 13, 1956, both Box 8, FLW–SI; and Cunningham to Struve, Nov. 20, 1957, Box 8, OS; see also Doel (1990b, 137), McDougall (1985), and Sullivan (1961). Another factor that aided the SAO's selection for satellite tracking was its institutional experience in managing solar observing stations in foreign countries, as a similar global network was required for satellite tracking; see DeVorkin (1990, 129–30).

[31] Struve to P. Byerly, July 16, 1957 [quoted], and Oosterhoff to Struve, Dec. 18, 1957, both Box 23, OS.

[32] Greenstein to J. Kaplan, Oct. 7, 1956, Rocketry file, Drawer 13, IGY.

[33] Quoted from Whipple (1956a, 109); see also Smith et al. (1989, 27–34).

dowment drive in 1956. Following the advice of Marion Eppley, an eminent chemist and Visiting Committee member, Menzel promoted radio astronomy and artificial satellite studies as Harvard's "two romantic new developments." Yet Harvard's reliance on grants alone to support new research ventures left many of them on shaky ground, and annoyed reviewers at U.S. astronomy's general-purpose patron, the NSF. In early 1957 members of the foundation's astronomy advisory panel balked at fully funding Harvard proposals for radio astronomy, declaring they lacked confidence over Harvard's history of demanding a "high level of continual support" from outside sources.[34]

A more serious problem for Menzel, despite his public talk about the advantages of nontraditional instruments, was the lack of adequate telescope resources for optical astronomers at his observatory. The extent to which this troubled him was revealed by confidential negotiations he opened with Lowell administrators in November of 1956. Menzel offered to Roger Lowell Putnam, Lowell's sole trustee, a plan to link the SAO, HCO, and Lowell observatories into a common entity. As part of this restructuring, Menzel proposed appointing Lowell staff members as Harvard research associates, providing Lowell astronomers access to Harvard plate stacks and other data, and moving the 61-inch Agassiz reflector to Flagstaff – which, he argued, would permit studies of solar-induced phenomena, including aurorae, on other planets. In Menzel's view, such an arrangement would profitably merge IGY-style research with traditional optical astronomy. Putnam was intrigued, feeling it could make Lowell competitive with the national optical observatory; but John Hall, a Yale-trained astronomer and director-designate of Lowell, convinced Putnam to reject the proposal. In part, Hall raised pragmatic questions about control, including Lowell's two-thousand-mile distance from the far more powerful Harvard. However, Hall's main objections were instrumental and professional: Declaring that Harvard had neither staff nor students "actively engaged in traditional astronomy," and instead emphasized meteor astronomy and nonoptical studies, Hall argued that their "conspicuous lack of active interest in the Lowell type of astronomy means that there is little or no common spiritual ground on which a successful association must rest." For Hall, who later declared that satellites would provide only "a small fraction of one percent" of worthwhile astronomy, the Harvard–Smithsonian observatories no longer seemed close to the core of U.S. astronomy.[35]

34 Quoted from Eppley to Menzel, Jan. 20, 1956, Box 28, DHCOm, and from Minutes of the 7th Annual meeting of advisors, Advisory Panel for Astronomy, Aug. 17–18, 1957, NAS; "Notes for the Report of the Visiting Committee," Feb. 7, 1957, "Contracts & Grants at HCO during the Period July 1957–Dec. 1958," Harvard College Observatory Council, [Report on Harvard astronomy program], Dec. 3, 1957, and Menzel to Eppley, Jan. 26, 1956, all Box 32, DHCOm; Whipple to A. Loomis, Jr., Mar. 29, 1958, Box 11, FLW–SI; and Struve to M. L. Oliphant, Dec. 1, 1955, Box 17, OS.

35 Quoted from Hall to Putnam, Feb. 6, 1957, and idem to W. Sinton, Mar. 15, 1958, LOW; see also Menzel to Putnam, Nov. 13, 1956, LOW; "Proposal for Collaboration in Astronomy Between Harvard College Observatory, Lowell Observatory, and the Smithsonian Astrophysical Observatory," undated draft [ca. Nov. 1956], Box 28, DHCOm; and Whipple to L. Carmichael, Nov. 23, 1956, Box 9, FLW–SI. Whipple's interest in Lowell Observatory was further sparked by William Sinton, a former student of John Strong at Johns Hopkins and SAO visitor who accepted a Lowell appointment in 1956; see Sinton,

Having lost this bid, Menzel and Whipple turned defect into virtue by proclaiming their institutions as centers of nontraditional astronomy. In late 1956, for the first time, Menzel announced that graduate training would begin in ionospheric, geomagnetic, and upper-atmospheric research, which he collectively termed "astro-geophysics";[36] but it was the flood of funding that followed the launch of *Sputnik 1* that convinced Menzel to focus particularly on "astro-geophysics" and planetary phenomena. In addition to creating a new Harvard Ph.D. specialization in "space research," Menzel hired Thomas Gold – the British theoretical astrophysicist who, with Bondi and Hoyle, had developed steady-state cosmology – to direct research in radio astronomy and the magnetic properties of planetary bodies. Throwing himself wholeheartedly into administration, Menzel secured Air Force contracts to study planetary surfaces and atmospheres, to create a map of Mars, and to design planetary telescopes for use in manned or unmanned high-altitude balloons. These contracts enabled him to bring additional astronomers to Harvard, including the French-born Gérard de Vaucouleurs, an expert in galactic and planetary astronomy. Menzel thereby created a team of solar system scientists as Whipple had done at the SAO, announcing to interested patrons their availability to pursue pressing problems relating to space.[37]

In February 1958 Menzel and Whipple also established the Harvard Committee on Space Research, which began meeting regularly that month. Although a less visible development than the relocation of the SAO to Harvard, the creation of this committee was quite significant, for with it Menzel and Whipple announced their intention not simply to *aid* space-based Big Science, but to become full participants. The committee's initial mandate – to help establish the Harvard–Smithsonian as an independent center for satellite construction – lasted just four months, at which time the Eisenhower administration proclaimed the establishment of a civilian space agency, NASA; but the committee stimulated Harvard–Smithsonian researchers to think further about large-scale projects. By that summer Menzel and Whipple announced plans to launch major new investigations of meteorites, cosmic rays, and the upper atmosphere, and Whipple began designing a satellite-based telescope system, later termed Project Celescope. The committee's energies were a reminder that the HCO, five years after Shapley's resignation, had turned sharply toward military-dominated space research. Although a number of U.S. optical astronomers talked enthusiastically about Spitzer's proposed space telescope after *Sputnik*'s launch, their commitment to ground-based optical astronomy was undiminished.

Footnote 35 *(cont.)*
"Request for a National Science Foundation Grant-in-Aid to Support a Study of the Infrared Spectrum of the Planets and the Moon," Nov. 2, 1956, LOW, and Strong and Sinton (1956).

36 A former Menzel student, Walter Orr Roberts, also taught graduate classes in "astro-geophysics" at Colorado in the mid-1950s; see Menzel, *ARHCO,* 1955–6: 19, and "Observatory Report . . ." (1958).

37 Menzel to H. Campaigne, June 15, 1956, idem to B. Bronk, Apr. 22, 1958, and idem to L. Berkner, July 11, 1958, all Box 32, DHCOm; idem to O. Struve, June 29, 1956, and Struve to Menzel, July 9, 1956, both Box 9, OS; Urey to Whipple, May 14, 1958, Box 102, HCU; de Vaucouleurs OHI (Ronald E. Doel, Nov. 20 and 23, 1991, at AIP); and Wilhelms (1993, 26–7).

Indeed, listing likely institutional competitors, Menzel and Whipple noted the Naval Research Laboratory, the Air Force Cambridge Research Center, and James Van Allen's magnetospheric research team at Iowa, geophysical centers at best remotely familiar to most U.S. astronomers.[38]

Nevertheless, Menzel and Whipple continued to regard themselves primarily as astronomers, not geophysicists, and portrayed the emerging controversy with optical colleagues simply as one of instrumental style. In 1959, Menzel boasted that "[u]nlike astronomers in the observatories of the western United States, we are not committed to years of future research in terrestrial astronomy, with heavy equipment designed to study the skies from the earth's surface. The poor atmospheric conditions of the eastern seaboard, which hamper visual observations, form no barrier to the telescope in space."[39] Although technically correct, Menzel's declaration glossed over other problems that followed Harvard's reorientation toward Big Science "astro-geophysics." By the standards of American astronomy, funding at the Harvard–Smithsonian observatories remained high – in 1958, Whipple and Menzel managed planetary and meteor research contracts totalling over $2.6 million – but these leaders found military patrons less interested in creating stable research programs than in solving immediate technical problems, such as determining the bearing weight of the lunar surface. Among Menzel's headaches was the tendency of military patrons to cancel or alter research contracts when political priorities shifted in the emerging space program, thereby lessening his institution's autonomy. A revealing case was de Vaucouleurs's efforts to analyze Venus's atmosphere. In July 1959, de Vaucouleurs organized an expedition to study an occultation of the bright star Regulus by Venus, establishing seven observing stations along the occultation path in Spain, France, Italy, and Lebanon. Although de Vaucouleurs was delighted by the quality of data obtained, and believed these data could resolve important questions about the density and temperature of Venus's atmosphere, Air Force officials wanted him first to complete his Mars mapping, a project that promised little professional reward. Although de Vaucouleurs prevailed, the conflict was difficult. Similarly, Harvard efforts to design balloon-based studies of planetary atmospheres ceased when the relevant Air Force contract was abruptly canceled. Frustrated by these dead ends, Menzel confessed to being "acutely conscious of the fact that the

[38] Draft, first meeting, Harvard Committee on Space Research, Feb. 4, 1958, and second meeting, Feb. 25, 1958, both Box 32, DHCOm; Menzel, "Research at Harvard College Observatory," draft, Nov. 30, 1959, H. Odishaw to Menzel, Mar. 30, 1959, and Menzel to E. Dyer, Jr., Dec. 2, 1959, all Box 32, DHCOm; Whipple to H. C. Urey, Dec. 4, 1959, Box 102, HCU; Menzel, *ARHCO,* 1957–8: 2, 1958–9: 3; and McDougall (1985, 198) and Smith (1989). Caltech scientists similarly attempted to build a spacecraft capability in the brief interregnum between *Sputnik* and NASA; see, e.g., H. Brown to G. Schilling, Sept. 4, 1959, Box 27, LAD.

[39] Quoted from Menzel, "Space Research at Harvard College Observatory," Nov. 30, 1959; see also Alfred Whitford OHI (David DeVorkin, July 17, 1978, SHMA at AIP, p. 21); Nancy Roman OHI (Joseph N. Tatarewicz, Jan. 28, 1983, SAOHP, p. 7); and Nancy Roman OHI (David DeVorkin, Aug. 19, 1980, SHMA at AIP, p. 24).

Observatory now depends so heavily upon government contracts and grants that we can make only short-term appointments."[40]

Harvard's increased emphasis on nonoptical astronomy also caused increased discord within the HCO. In 1959, Gold and Cecilia Payne-Gaposchkin rose to criticize what they perceived as the menacing relation between the overarching, ballooning SAO and Harvard's smaller academic department and observatory. Payne-Gaposhkin, a former Shapley Ph.D. who had beome chair of Harvard's department of astronomy in 1956, joined Gold in blasting Menzel and Whipple for investing in satellites and other Big Science projects, claiming they undermined Harvard's teaching and research mission. Whipple's orbiting observatory plans, they charged, caused SAO "secretaries, travel red-tapists and other administrative staff" to burgeon, enabling the Smithsonian tail to wag the Harvard dog. What most perturbed them was Whipple's and Menzel's determination to build a permanent SAO headquarters next to the HCO's Garden Street location, thus making "a high church wedding out of the present common-law marriage." Matters grew fractious enough that Harvard officials appointed an internal review committee, chaired by Harvard physicist and Nobel laureate John Van Vleck. Van Vleck's review placed some blame on Whipple, declaring that he had not "cultivated the emotional support of some of his colleagues like Wilson vis-à-vis his senators in 1918." But Van Vleck was only partially correct in attributing these tensions to personality: As Big Science projects such as accelerators, rockets, and military-funded laboratories proliferated at U.S. universities in the 1950s, debates over patronage and appropriate scales for research had increased as well. What was new was that such debates, long present in American centers of physics, now spilled into astronomy. Although the SAO building was constructed, tensions did not abate. The following spring, Gold departed Harvard for Cornell.[41]

Perhaps the most important transformation that Menzel and Whipple experienced in the 1950s was finding their institutions increasingly at the periphery of American astronomy. In contrast with most U.S. astronomers, as we have seen, both men had sought large sums directly from military agencies soon after World War II; Whipple's meteor project, generously supported by Navy Ordnance, had been a far cry from the small ONR grants-in-aid for which many American astronomers had competed in the early 1950s. During this time, nevertheless, their research had

[40] Quoted from Menzel, *ARHCO*, 1958–9: 3–4; see Menzel and de Vaucouleurs to E. R. Dyer, Jr., Dec. 2, 1959, and de Vaucouleurs et al., "Research Directed Toward the Production of a Map of the Visual Features of Mars," HCO Contract AF19(604)–7461, 1964, both Box 32, DHCOm; de Vaucouleurs OHI (Ronald E. Doel, Nov. 23, 1991, at AIP); and de Vaucouleurs and Menzel (1960); see also Whipple (1959b). The Mars project also raised concerns about the unresolved canal controversy; see de Vaucouleurs to Urey, Nov. 19, 1958, Box 24, HCU, and de Vaucouleurs (1955).

[41] Gold and Payne-Gaposchkin quoted in Van Vleck, H. Brooks, and E. Purcell to G. McBundy, Jan. 6, 1959, Series II, Folder 34, JVV (I thank Spencer Weart for calling this memo to my attention); and see Leo Goldberg to members of the HCO advisory committee, Jan. 19, 1971, Box 51, JLG. Galison (1985) and Smith (1989; 1992) provide poignant analyses of the growth of large-scale research projects.

remained within the compass of American astronomy, and the HCO had remained an active (if inadequately equipped) disciplinary center. Major changes had occurred only in the mid-1950s, when Menzel and Whipple had increasingly moved toward IGY-related research, and NSF officials began defining astronomy in terms of optical and radar techniques. Their heightened feeling of exclusion were apparent as early as 1954. In that year Whipple had complained to NSF astronomy representative Peter van de Kamp that funds for meteor and upper-air research had come "almost entirely" from "defense contracts in the three armed forces," rather than from stable, disciplinary patrons such as the NSF. Yet Waterman and Seeger continued to emphasize funding for fields not already supported by military agencies, requiring Whipple to deal with less stable patrons. When he again commented on patronage in 1960, Whipple stressed the issue of abandonment, declaring that astronomers interested in solar system research had "received practically no support, financial or moral, until after World War II, when the military became quite interested in the upper atmosphere and mildly interested in the space beyond it."[42]

Menzel also sounded the theme of abandonment in his own pleas to patrons. In 1959, pitching Harvard's plans for expanding solar system astronomy to Edward R. Dyer, Jr., executive secretary of the influential NAS Space Science Board, Menzel raised the specter of possible Soviet competition in solar system research, calling attention to what he termed limited American work in this field. "We wish to point out," he wrote, "that very great activity is being shown in the Soviet Union to advance planetary studies on a national scale while in the United States . . . less than half a dozen astronomers are at present active in the field, mainly on a part-time basis and with very limited means." Although U.S. astronomers were in fact divided over the strength of Soviet work in this field – de Vaucouleurs and Kuiper estimated that Soviet advances roughly equaled those in the United States – Menzel's portrait of American activity was absurdly low, considering the Harvard–Smithsonian observatories in particular and the discipline in general.[43] His concern with neglect nevertheless indicated that he, like Whipple, understood how much the political economy of American science had changed during the 1950s, particularly after the launch of *Sputnik.* Where Menzel had once sought funds for solar system astronomy from traditional disciplinary patrons or specific military agencies, he now found himself competing on a national level, against other scientific priorities, to fund Big Science ventures. For Menzel, taking NASA as his new standard of measurement, *all* of American astronomy seemed small and underfunded; but his rhetoric also suggested that trading stable disciplinary patronage for large-scale grants was painful as well as exhilarating. In suggesting to Dyer that solar system astronomy had been neglected by its parent discipline, Menzel conveyed his frustration that Greenstein, Struve,

[42] Quoted from Whipple to van de Kamp, June 18, 1954, Box 13, FLW–HU, and in Tatarewicz (1990a, 48).

[43] Quoted from Menzel and de Vaucouleurs to Dyer, Dec. 2, 1959, Box 32, DHCOm; see also Doel (1992; in press-c).

and other optical astronomers now placed "astro-geophysics" outside the common boundaries of American astronomy.[44]

It also seems apparent that Whipple and Menzel – who, more than their colleagues in traditional astronomy, had deep commitments to their military patrons – found themselves embarrassed at being unable to respond to demands by these patrons once the era of space exploration abruptly began. Anxiety about service roles appeared in the statements of several leading Harvard–Smithsonian astronomers after the launch of *Sputnik.* In a candid assessment, one with autobiographical overtones, de Vaucouleurs observed that astronomers with experience in solar system research "have suddenly found themselves under great pressure to provide immediate and definite answers to the many questions that the technologist must raise in his planning of interplanetary craft and direct planetary exploration." For Whipple, supported by military patrons since the end of World War II, service role anxieties were even more poignant. Writing a close colleague, Whipple lamented that astronomers "have neglected" planetary studies, adding that he felt "somewhat apologetic to our space engineers for lack of knowledge that we could have obtained." Yet it was impossible for astronomers to have anticipated later, practical demands; the major research problems of solar system astronomy, as Kuiper repeatedly stressed, had less to do with the determination of decimal points than with finding correlations between fields of knowledge. Military patronage, however, clearly influenced how Whipple and his colleagues conceptualized problems in solar system studies. It shaped the pursuit of research at the Harvard–Smithsonian observatories, and ultimately influenced how Whipple perceived the history of this field.[45]

SPECIALIZATION AND CONFLICT AT YERKES–McDONALD

In many ways, the differences in research philosophies between Whipple and Kuiper by 1955 could not have been stronger. As we have seen, Whipple strongly favored large-scale research programs that were locally concentrated and federally funded; Kuiper preferred small-grant-funded projects dispersed among numerous beneficiaries. Whipple saw balloon- and satellite-mounted instruments as the future of astronomy; Kuiper staked his hopes on ground-based optical astronomy. Whipple

[44] It is important to note that de Vaucouleurs's often-quoted assertion that planetary physics and astrophysics "had been neglected since the turn of the century" was written just after Air Force officials canceled Harvard's balloon-based planetary spectroscopy project, in which he had participated; see de Vaucouleurs (1960, 27).

[45] Quoted from de Vaucouleurs (1960, 27) and Whipple to Swings, Nov. 7, 1960, Box 13, FLW–SI; see also Whipple to Urey, July 8, 1959, Box 47, HCU, and Kuiper (1956b, 89–90). Struve similarly declared that the "situation in astronomy with regard to space exploration is almost catastrophic" when quizzed by politicians about the readiness of his discipline to aid space missions, emphasizing that astronomers had largely worked to answer basic, not applied, research questions; see Struve to G. J. Feldman, May 6, 1958, Box 16, OS, and Struve (1960).

urged his fellow astronomers to endorse bold federal planning; Kuiper argued against seeking more than modest amounts from federal patrons. Whipple expected satellite-based research to transform the field; Kuiper had opposed Spitzer's proposed space telescope as nonastronomical. Whipple wanted to explore meteoric phenomena with ever-broader geophysical techniques; Kuiper continued to rely on optical telescopes to study the Moon, asteroids, and planetary atmospheres. In the early 1950s, Whipple's style of research was uncommon among American astronomers; Kuiper's, by contrast, was widely shared.

Nevertheless, inexorable pressures also came to influence Kuiper and the Yerkes–McDonald observatories during the late 1950s. Kuiper continued to favor ground-based optical techniques for solar system research, but found it increasingly difficult to maintain the informal style of collaborative, interdisciplinary research that he had nurtured since 1945. Against his will, seeking competitive advantage, Kuiper sought more specialized instruments and to build a permanent, structured institute of solar system studies, barricaded from stellar and galactic astrophysicists at his facility. In the end, even these compromises did not prevent a calamitous upheaval at Yerkes–McDonald, curtailing his term as director in 1960 and thrusting Kuiper into a new and quite different academic environment. The transformation of U.S. astronomy in the 1950s occurred on a national scale; conflicts at Yerkes–McDonald after 1957 help illuminate how shifts in patronage were largely responsible.

Kuiper's decision not to embrace Big Science approaches to solar system research, as we have seen, stemmed from several factors. One was his uninterrupted access to world-class optical telescopes. Since his first appointment at Lick in 1933, Kuiper had worked with state-of-the-art instruments, particularly the 82-inch McDonald reflector, until 1948 the world's second largest. His principal astronomical achievements, including his infrared studies of Mars in 1947 and his prewar studies of faint stars, had been observational in nature. He received a month's time annually on the 82-inch telescope and, like other senior American astronomers, counted on guest observer privileges at the Mount Wilson–Palomar 200-inch for measurements beyond reach of his own instrument. Kuiper also devised research programs that required relatively inexpensive auxiliary detectors for their completion, such as the Cashman-cell infrared detector and binocular eyepiece that he used for visual assessments of the Moon's thermal history. Moreover, he found it relatively easy to obtain small-grant aid to fund these projects. Between 1953 and 1956, Kuiper had received nearly $40,000 from the ONR and NSF to study the physical properties of asteroids, the atmospheres of terrestrial planets, and the lunar surface; these funds supported part-time collaborators, assistants, and graduate students. Kuiper did not share Whipple's anxiety about securing large contracts to maintain competitive research.[46]

[46] Kuiper, "Grant Proposal [Louis Block Fund for Basic Research and Advanced Study," Mar. 8, 1958, and "Final Report to the National Science Foundation, Grant G-4140, 'Solar System Studies,'" June 20, 1961, both Box 33, GPK; "Contracts and Grants for Which Dr. G. P. Kuiper Is Designated as Principal Investigator or Director," n.d. [ca. 1961], Box 29, GPK; Groeneveld and Kuiper (1954), Kuiper et al. (1958).

Moreover, like fellow optical astronomers, Kuiper saw a threat to his style of research from the grander operating scale of geophysics. It is important to note that Kuiper was not unfamiliar with geophysical instruments and methods: He had flirted with the possibility of placing instruments on board an Air Force B-29 to search for evidence of water vapor in the atmosphere of Venus in 1949, just before the Korean War abruptly curtailed this possibility. In that same year, Kuiper had arranged for astronomer and geophysicist Aden Meinel to come to Yerkes to make spectroscopic measurements of faint night-sky emissions. Furthermore, in 1954, on the strength of his edited *Atmospheres of Earth and Planets* and *The Earth as a Planet* volumes, he had been invited to become technical director of the IGY.[47] Kuiper, however, continued to see geophysics as an unpromising detour: Few geophysicists shared his basic interest in the evolution of asteroids and the origin of the solar system, and the enormous scale of geophysical investigations seemed antithetical to the simpler, smaller instrumental culture of astronomy. Although Meinel's auroral study impressed him, Kuiper privately fretted over its swelling size, which reached $40,000 per year by 1954 and a support staff of eight. "We should guard against a too-enormous expansion in geophysics," Kuiper later confided, "because our manpower in the shops and among the graduate students . . . is very limited, and Meinel's projects could easily absorb half or more of all we have." In declaring that "[o]ur main objective must remain to use our large telescopes," Kuiper affirmed the widespread conviction among optical astronomers that traditional instruments were more versatile tools than their geophysical competitors.[48]

Finally, Kuiper preferred an associative style of interdisciplinary research, one that differed fundamentally from Whipple's. In part, Whipple sought a Big Science approach to meteor research because scientists in ballistics laboratories had not made controlled experiments that he desired to assess meteoric properties; coordinating this work through the SAO suited his needs. Kuiper, by contrast, faced fewer worries of this kind studying asteroids and the Moon. Their respective choices were also deliberate, however. During the early 1950s Kuiper showed little enthusiasm for the permanent institutional structure that Whipple eagerly promoted at the SAO, preferring instead to stimulate interdisciplinary studies through personal contacts. The clearest expression of his faith in coordinated studies was his edited *Solar System* volumes, a project he had begun in the late 1940s and by far his most ambitious and influential. The four volumes of this series, published between 1953 and 1963, brought together an impressive range of contributions on oceanography and meteo-

[47] Kuiper had also negotiated with the Army's Flight Determination Office and the Air Force's Project Skyhook to make balloon-based studies; see G. Vaeth to Director, Yerkes Observatory, Nov. 3, 1952, Kuiper to Vaeth, Nov. 21, 1952, idem to Vaeth, Jan. 22, 1953, E. P. Martz to Kuiper, Nov. 11, 1953, and idem to Kuiper, Nov. 20, 1953, all Box 30, GPK; on the IGY, see L. Berkner to J. Kaplan, Dec. 14, 1954, Box 284, MT.

[48] Quoted from Kuiper to W. G. Whaley, Aug. 19, 1960, Box 14, GPK; see also idem to F. E. Roach, Nov. 13, 1950, and idem to Meinel, Feb. 22, 1956, both Box 30, GPK.

rology to the nature of interplanetary space and the structure of planetary atmospheres. Kuiper had intended these books to inspire colleagues in neighboring disciplines to think about interconnections among their fields. He had cajoled a chapter from the eminent oceanographer Harald Sverdrup by declaring that "the study of the circulation on Jupiter and Saturn will induce astronomers to take advantage of the methods already in use by oceanographers," and similarly had requested meteorologist Horace Byers, directing Chicago's well-funded Thunderstorm Project, to alert astronomers "to the power of meteorology" for studying solar–terrestrial relations.[49]

To a significant degree, Kuiper had regarded these volumes not merely as a way to report results, but as a vehicle to stimulate new work and to reach consensus on outstanding problems, further underscoring his commitment to small-scale, collaborative projects. During the early 1950s, as his editorial work on the *Solar System* began, Kuiper had spent nearly a third of his research time with collaborators, often in person, finding the experience a highlight of his professional career. He had urged graduate students to make new observations to refine critical values, as Russell had done editing *Astronomy* a quarter century before, and had focused on research distant from his specialty. In 1952, Kuiper had quizzed Sverdrup, Roger Revelle, Brouwer, and de Vaucouleurs as to the merits of a theory, advanced by the Serbian engineer Mulatin Milankovich, that ice ages were linked to periodic wobbles in Earth's rotational axis. He had then derided this later-heralded idea when de Vaucouleurs alone had championed it, but constant interactions with scientists outside his specialty kept Kuiper stimulated and productive. Drawing on his positive encounters at the Yerkes symposium on planetary atmospheres and his initial contacts with Urey, Kuiper urged loosely confederated research programs as a promising path for American astronomy. When NSF officials in 1954 had touted the national optical facility, Kuiper had huffed to a colleague, "coordination of knowledge in a broad field of science is just as important as building a new observatory."[50]

Within two years, however, Kuiper felt far less optimistic about this research style, arguing that to remain viable, solar system investigations needed infusions of new funding beyond what the discipline could provide. Kuiper's diminished confidence was not entirely caused by the tumultuous changes affecting American astronomy on the whole. One factor was painfully local: his intense controversy in 1955 with a fellow Chicagoan, Urey (see Chapter 4). The importance of this conflict on Kuiper's career and his attitudes toward interdisciplinary research cannot be underestimated. Although astronomers had supported him as the conflict climaxed at the IAU meet-

[49] Quoted from Kuiper to Sverdrup, Aug. 31, 1950, and idem to H. Byers, June 3, 1952, both Box 11, GPK; see also Struve to R. Hemens, Mar. 28, 1946, Box 31, GPK, and Kuiper (1953c). Kuiper's *Solar System* series included a volume on the Sun, not part of solar system astronomy as then understood, but judged important for resolving questions of cosmogony and natural abundances.

[50] Quoted from Kuiper to van de Kamp, Apr. 26, 1954, Box 31, GPK; see Kuiper to Revelle, May 22, 1954, idem to Sverdrup, May 28, 1952, Bullard to Kuiper, June 12, 1952, and de Vaucouleurs to Kuiper, July 18, 1952, all Box 11, GPK.

ing in Dublin, Kuiper soon perceived that his fight with a popular Nobel laureate at his home campus had troubled Chicago administrators and weakened his ability to forge new interdisciplinary ventures. The controversy underscored the importance of compatible personalities for sustaining interdisciplinary collaborations, particularly in the small-science communities where Kuiper and Urey worked. Not until the space race began were permanent institutions for "astro-chemistry" built; it took the needs of planetary mission specialists and the enormous operating funds of NASA to create institutions sufficiently strong to overcome disciplinary and methodological tensions between these fields.[51]

The loss of Urey's intellectual stimulation also reinforced Kuiper's sense of isolation at Yerkes–McDonald; yet the extent of his professional solitude at Chicago in the mid-1950s was later exaggerated by Kuiper and others who wrote on this period. Kuiper studied planetary atmospheres with Joseph W. Chamberlain, a young Michigan-trained spectroscopist hired to replace Meinel, who took extended leave from Yerkes in 1956 to direct planning for the national optical observatory.[52] Kuiper worked also with visiting astronomers hired to do the asteroid survey at McDonald, among them the Heidelberg-trained Ingrid Groeneveld and her future husband, C. J. van Houten; in addition, he had the satisfaction of training two new graduate students, Tom Gehrels and Carl Sagan. Still, Kuiper deeply missed his nurturing contacts with Chicago scientists of equal stature on problems of common interest, including Gerhard Herzberg and especially Urey. It is hardly an overstatement to say that Kuiper's informal partnership with Urey had inspired his faith in small-scale collaborative research. As we saw in Chapter 4, much of Kuiper's work in the early 1950s in cosmogony and the thermal properties of small bodies and the Moon had been synergistic with Urey's geochemical models of planetary formation and evolution. Kuiper's loss was deepened when his first graduate student in solar system research, Daniel L. Harris III, lost his bid for tenure at Chicago in 1956 and accepted a professorship at Northwestern.[53] The extent to which Kuiper missed his local interdisciplinary contacts showed in his longing to recreate them. He tried to launch new lunar research programs with Chicago geologist J. Harlen Bretz, later heralded for his interpretation of the scablands floodplains in Washington State, and with Douglas Allan, a Toronto Ph.D. trained by Wilson who shared Kuiper's interest in the Moon. Kuiper invited both men to visit with him and to observe the Moon at Yerkes and McDonald, hoping to inspire the intense dialogue that he had

[51] Kuiper to A. Meinel, Feb. 22, 1956, Box 17, GPK; Urey to Waterman, Feb. 9, 1955, and Waterman to Urey, Feb. 21, 1955, both Box 11, NSF; see also Tatarewicz (1990a).

[52] Sagan (1974), for example, stresses Kuiper's isolation at Yerkes–McDonald. On Kuiper's interactions with Chamberlain, see L. Berkner to J. Kaplan, Dec. 14, 1954, Box 284, MT; Chamberlain to B. Strömgren, Jan. 6, 1956, 1955–6 correspondence folder, YO; idem to Kuiper, Aug. 26, 1957, Box 3, JWC; and Chamberlain and Kuiper (1956).

[53] Harris's limited publication record and cool personal relations with Chandrasekhar were factors that Kuiper believed led to his negative tenure decision; see Kuiper to B. Strömgren, Feb. 11, 1953, YO, and idem to O. Struve, Dec. 19, 1956, Box 11, OS.

shared with Urey a half-dozen years earlier. These new collaborations were forced and shallow, however: Bretz, in his mid-70s, was retired from active research, and Allan was a recent Ph.D. who ceased writing to Kuiper after accepting a Cambridge University fellowship in 1956. Neither were substitutes for Kuiper's close institutional and intellectual bond with Urey.[54]

Yet by 1956 what worried Kuiper more than these local difficulties were broad shifts in the discipline: the continued insistence of Waterman and Seeger to build national observatories, their emphasis on aiding stellar and galactic astronomy, the proliferation of military patrons particularly interested in solar system research, the heightened competition for telescope time on major instruments, and the accelerating growth of the discipline. Several of these issues especially concerned him. The NSF's determination to aid astronomy's core research programs left Kuiper concerned that this critical patron would no longer meet the particular needs of solar system astronomers. Writing Swarthmore astronomer Peter van de Kamp – the NSF's astronomy program officer and a longtime personal friend – Kuiper complained that interdisciplinary fields were being slighted. For the first time, he called attention to the distinctive character of his field, arguing that the Moon and planets had "a definite potential interest" analogous to that of Earth's arctic regions, since "sooner or later they will become of more direct concern to us."[55] Kuiper may have had the International Geophysical Year in mind in drawing this analogy, for he soon pressed individuals connected with the IGY to consider funding solar system research as a way to place "certain broad geophysical problems in proper context." What accentuated the issue of patronage for Kuiper were new frustrations with existing detectors on the McDonald 82-inch. Despite sensitivity-enhancing modifications to his far-infrared spectrograph of 1947, Kuiper remained unable to determine particle sizes within the atmosphere of Venus or to apply the detector to other outstanding problems in solar system astronomy. He desired new, more expensive auxiliary instruments precisely at a time when the NSF seemed more interested in building general telescopes for the discipline. Like Whipple, Kuiper sought secure patronage to maintain his competitiveness within the field, and worried about the instability of federal backers.[56]

Throughout the early 1950s, his most productive period at Yerkes–McDonald, Kuiper had been simply a staff astronomer, first under Struve, then under Strömgren; but in 1957, shortly before the *Sputnik* launch, Kuiper became director of this third largest U.S. observatory. Several years earlier Kuiper and Chandrasekhar had

[54] Kuiper to Gehrels, Aug. 5, 1953, Box 30, and idem to E. C. Abendanon, Feb. 20, 1956, Box 10, both GPK; J. T. Wilson to Kuiper, Dec. 27, 1956, and Kuiper to Allan, June 15, 1956, both Box 16, JTW; see also Kuiper (1959c).

[55] Kuiper to van de Kamp, Sept. 27, 1954, Box 31, GPK.

[56] Quoted from Kuiper to Meinel, Feb. 22, 1956, Box 17, GPK; see also G. P. Kuiper, "[Draft Report on Infrared Stellar Spectrometer]," July 17, 1956, idem to W. Hiltner, Mar. 10, 1956, both Box 11, and idem, "Final Report for NSF Grant G–4140, [Solar System Studies]," June 20, 1961, Box 33, all GPK.

complained to Chicago administrators about Strömgren's management abilities, urging a limited-term reappointment as Yerkes–McDonald director. When Strömgren announced plans to accept an appointment at the Institute for Advanced Studies in Princeton, Chicago president Lawrence Kimpton had sought to appoint Chandrasekhar to the post. After Chandrasekhar had refused, declaring that a theoretician was a poor choice to manage an optical observatory, Kimpton had selected Kuiper, Chandrasekhar's own choice. It is not clear whether Kuiper had initially resisted or sought this appointment. While running Yerkes–McDonald under Struve during 1947–9, Kuiper had complained of feeling like "a Dutch official under German occupation," and developed a painful duodenal ulcer. He realized, however, that returning as director would give him enormous authority to develop solar system astronomy at Chicago, precisely when space research seemed poised to mushroom.[57]

In retrospect, Kuiper's brief second term as head of Yerkes–McDonald seemed a disaster, marked by increased staff squabbles, fights over instrumentation and allocations of resources, disputes over Kuiper's management and hiring decisions, and finally Kuiper's dismissal from the directorship by Kimpton. Some of the blame can be placed squarely on Kuiper's abrupt and autocratic leadership;[58] but between 1958 and 1960 the Yerkes–McDonald observatories were also buffeted by the fundamental changes sweeping American astronomy at large, changes over which Kuiper had little control. He assumed his new responsibilities on September 1, 1957, just thirty-four days before *Sputnik 1*'s launch; he stepped down in January 1960, after NASA had declared reaching the Moon to be a new national priority. During this period, the fragmentation of American astronomy accelerated sharply; this trend was no less apparent in optical observatories like Yerkes–McDonald than in emerging Big Science centers of astronomy such as the Smithsonian Astrophysical Observatory and Harvard.

Kuiper's first challenge on assuming command of Yerkes–McDonald was to decide priorities for instrumentation and fund-raising appeals. Initially he seemed to believe, like Struve, that it was still possible to treat the discipline as an integrated whole and develop research strategies that would appeal to stellar astrophysicists as well as to solar system astronomers. Worried that the rapid growth of world-class telescopes in California and the planned national optical observatory made the McDonald 82-inch less competitive, Kuiper abandoned his 1953 argument that U.S. telescopes were not yet at saturation. Urging Chicago administrators to fund a larger, general-purpose instrument, Kuiper soon warned that "a revolutionary pace is sweeping astronomy, and without constant additions of major and expensive equipment, astronomers get behind so fast and so far as to become rapidly obsolete." Al-

[57] Kuiper to P. Hodges, Oct. 21, 1950, and idem to Oort, Sept. 1, 1950, both Box 18, GPK; and idem to J. Oort, Aug. 28, 1951, Box 47, JHO [quoted]. On the Strömgren–Kuiper transition, see Chandrasekhar and Kuiper to W. Bartky, Feb. 27, 1953, and W. W. Morgan to Bartky, Feb. 10, 1953, both Box 28, UChi; Chandrasekhar to Kuiper, June 11, 1957, Box 18, GPK; Carl E. Sagan OHI (Ronald E. Doel, Aug. 27, 1991, at AIP); and Wali (1990, 69, 201).

[58] See, e.g., Evans and Mulholland (1986).

though aware of the rapid advances in radio astronomy, and the relevance of this technique to solar system studies – he followed with interest the 1955 discovery by Bernard Burke and Kenneth L. Franklin, both of the Carnegie Institution of Washington, that Jupiter emits periodic radio emissions – Kuiper decided against building radio astronomy at Chicago, feeling his institutions could not effectively compete in this field. He retained a faith in the productivity of ground-based optical astronomy and, like many U.S. optical astronomers, worried that federal and military grants were poor bets for institutional development. Advising Kimpton that Yerkes–McDonald researchers had received over $600,000 in federal grants since 1951, but that not a penny was available for permanent upkeep, Kuiper urged his proposed new optical telescope be built with Texas state funds or private money, reflecting the tradition of nonfederal support for American astronomy.[59]

In the aftermath of *Sputnik,* however, Kuiper found it necessary to alter his strategy for running these observatories. A major reason was the flood of funding for solar system research that followed this Soviet achievement, and U.S. efforts to accelerate studies of the Moon and near space as a way to regain lost prestige. Funding for all areas of astronomy rose sharply in the late 1950s: The growth rate for astronomy reached 20 percent by 1959, surpassing that of physics, and NSF funds for astronomy tripled between 1957 and 1960, when it reached $3 million; yet resources targeted for solar system astronomy mushroomed faster still, particularly from organizations not formerly regarded as disciplinary patrons. Almost immediately Kuiper received $61,000 from the Air Force for lunar studies, a tenfold increase over his average annual grants in this field. These patrons requested quick results to meet launch schedules in the emerging space race; this pressure further marked lunar and planetary research as a distinct intellectual area, and heightened the disparity in funding between solar system and other astronomical fields.[60]

By mid-1958 even Kuiper no longer regarded U.S. astronomy as a unified discipline. He now saw his task as observatory director one of balancing the needs of stellar and galactic astronomers – since 1938 the major interest of Yerkes–McDonald staff members – with the needs and demands of solar system researchers. Kuiper took several steps to enhance the competitiveness of his stellar colleagues. Aware that the extremely rapid growth of astronomy was limiting observing time on world-class instruments, including the McDonald 82-inch, Kuiper renewed his campaign to build a new telescope for the benefit of Yerkes–McDonald astronomers. Importantly, Kuiper now abandoned his reservations about approaching federal and mili-

[59] Quoted from Kuiper to Kimpton, Nov. 1, 1959, Box 18, GPK. See also idem to Kimpton, Aug. 26, 1957, and idem to Chandrasekhar, May 1, 1957, both Box 18, GPK; Kuiper grant proposal to Louis Block foundation, Oct. 12, 1957, department file 1957–9, WWM; and Doel (1992, 242–3). On radio astronomy, see Richardson (1957, 89) and Smith and Carr (1964, 44, 55).

[60] Kuiper to Strömgren, Feb. 2, 1954, and idem to C. D. Shane, May 6, 1955, both Box 14, idem to E. C. Abendanon, Feb. 20 and Mar. 12, 1956, both Box 10, and idem to G. Van Doren, Oct. 3, 1957, Box 18, all GPK; idem, "Considerations on a New Photographic Lunar Map," n.d. [ca. 1954–5], Box 11, and idem, grants and proposal folders, Box 15, GPK; see also NSF *Annual Reports* for 1957–60.

tary patrons. He secured a commitment from the Air Force Cambridge Research Center to build a new 60-inch telescope in Chile, primarily for stellar and galactic astronomy (although Kuiper hoped to reserve light-of-the-moon periods for lunar and solar system research). Ultimately this project became the nucleus of the Cerro Tololo Observatory, one of the largest U.S. observatories in the Southern Hemisphere. Kuiper also served as the Yerkes–McDonald representative to the planning group of the national optical observatory, whose construction was beginning atop Kitt Peak in south-central Arizona.[61]

Nevertheless, it is clear that Kuiper began devoting the bulk of his efforts to enhancing solar system astronomy at Yerkes–McDonald, developing new projects that profoundly influenced the character of these facilities. In 1959, Kuiper installed an Air Force satellite-tracking camera at Yerkes. Abandoning his reservations about nonoptical instrumentation, Kuiper also secured NSF funds to build a 28-ft infrared–microwave telescope at McDonald to study the molecular structure of cool stars as well as the thermal emissions of planetary bodies. Moreover, he brought interdisciplinary teams to Yerkes–McDonald, precipitating a significant shift in the institutional and professional character of these institutions. In 1958 he tried to recruit an Air Force unit specialized in celestial mechanics to Yerkes; he also accelerated work on his lunar atlas, a project he had long pursued as an individual effort, by bringing to Yerkes two British researchers experienced in cartography and lunar mapping, A. G. W. (Di) Arthur and Ewen Whitaker. To separate these endeavors from other Yerkes research, and to facilitate interdisciplinary work, Kuiper requested NSF funds to establish an "Institute of Planetary Studies" modeled after the former Kaiser Wilhelm institutes in Germany; in his application Kuiper decried the instability of Air Force grants. Though this plan was rejected, Chicago administrators sanctioned Kuiper's request to grant joint majors in astronomy and geology, a significant departure for both disciplines. These efforts demanded considerable energy. While resisting the temptation to administer full-time, maintaining his lunar research with remarkable determination, Kuiper had little time left for other activities. He missed many meetings of the national observatory planning committee, perhaps resentful that this instrument would largely serve stellar and galactic researchers. Kuiper also raised more funds for his research than did his staff for theirs. In 1959 he secured $55,000 for solar system studies from the NSF, more than all his colleagues in stellar and galactic studies raised for their projects ($46,000).[62]

[61] Kuiper to M. Greenberg, Sept. 16, 1958, idem to M. Alperin, Oct. 4, 1958, idem to C. D. Shane, July 7, 1958, all Box 28, and idem to W. H. van den Bos, Apr. 28, 1960, Box 29, all GPK; see also Edmondson (in press) (I thank the author for advance copies of selected excerpts) and Kuiper (1959a,b, 1717).

[62] Kuiper, "Proposal for a 'Center' or 'Institute' of Planetary and Lunar Studies," July 1958, Box 13, GPK; idem to Jan Oort, July 1, 1958, idem to R. W. Harrison, Apr. 17, 1960, and idem to M. Alperin, Oct. 4, 1958, all Box 18, GPK; "University of Texas, 1960," folder, WWM; "NSF Astronomy Report 1959," Box 32, NSF; also F. K. Edmondson, Jr., private communication, Apr. 18, 1991. Kuiper made no mention of Struve's attempt to create an institute for theoretical astronomy at Yerkes during World War II; noted in Struve to R. M. Hutchins, Apr. 17, 1941, Box 27, Folder 8, UChi.

Kuiper's management of Yerkes–McDonald drew increasing fire from staff members by mid-1959. Several factors contributed to these tensions; among the most important, particularly for isolated astronomical observatories, was Kuiper's style of management. Staff members complained of Kuiper's gruff and autocratic demeanor, his unwillingness to share information on negotiations about the Chilean telescope, and his reluctance to discuss plans to create a joint department of astronomy with the expanding University of Texas (owner of the McDonald Observatory), which would increase resources and researchers at these facilities. Another critical factor, however, was Kuiper's evident partiality to solar system research, and its rapid growth. While Kuiper continued to promote the Chilean observatory, it is difficult to avoid the conclusion that, as lunar and planetary studies became national priorities in the late 1950s, Kuiper devoted ever greater attention to these subjects. Air Force pressure to complete the lunar mapping project prompted Kuiper to assign additional telescope time to Arthur and Whitaker, angering staff researchers. Also in 1959, Kuiper arranged a study of Soviet astronomy for the Central Intelligence Agency, focusing his own contribution on lunar and planetary work. Although he undertook this review for a variety of reasons, including his conviction that understanding Soviet achievements and failures would aid the competitiveness of Yerkes–McDonald, he also desired to position himself as an expert on Soviet planetary science.[63]

Another problem was more fundamental, and unrelated to Kuiper's limitations as a manager of science: By the late 1950s, the core fields of astrophysics and planetary science had little common intersection. This was a natural evolutionary trend within the physical sciences, but one that had enormous implications for solar system research at U.S. observatories. Compared to Russell, Shapley, and Struve, all of whom had regarded the solar system as an integral component of the discipline, the generation that entered American departments of astronomy in the 1950s shared more specialized training and interests. Many junior astronomers hired at Yerkes after 1956, including the Berkeley-trained David N. Limber and the Columbia-educated Kevin H. Prendergast, focused on galactic phenomena; similarly, the husband–wife team of Geoffrey and Margaret Burbidge, hired in 1957, had backgrounds in high-energy physics. They shared few professional interests with Arthur and Whitaker, cartographers without Ph.D.'s. Applied celestial mechanics was equally remote to these scientists, who saw lunar and planetary research aspects of the IGY and of geophysics.[64]

Tensions between these generations of astrophysicists were illuminated by Kuiper's deteriorating relationship with Geoffrey Burbidge, a University of London–

[63] G. Burbidge, [untitled memorandum on loss of confidence in Kuiper], n.d. [ca. Oct. 1959], WWM; and Whitaker (1985, 14–16). On Kuiper's contacts with the CIA, see Doel (1992).

[64] Kuiper to W. G. Whaley, July 29, 1959, Box 18, GPK; and Evans and Mulholland (1986, 133). Kuiper also assigned Arthur and Whitaker to coveted office space at Yerkes, heightening professional resentments; see Whitaker (1985, 18).

trained physicist born in 1919 who replaced the departing Strömgren in 1957. Burbidge had worked on galactic structure since the mid-1940s; he used the 82-inch McDonald reflector to study the rotation curves of peculiar multiple galaxies and was sufficiently regarded by U.S. astronomers to share the 1959 Warner Prize for outstanding junior astronomer. Although Burbidge found congenial colleagues at Chicago, he became increasingly irritated by what he termed Kuiper's "private" projects, including the planned infrared–microwave telescope, for him peripheral to astrophysics. Convinced (accurately) that Kuiper intended to hire researchers sympathetic to solar system astronomy for planned posts at Texas, Burbidge complained directly to Chicago administrators that Yerkes–McDonald astronomers felt "more and more hedged by a vast and vague organization dedicated to the study of the solar system." While Kuiper's proposed institute for planetary studies was "the direction in which he would like the Department of Astronomy to go," Burbidge declared, this was not the wish of the department.[65]

Burbidge's statement underscored what many astronomers came to recognize by 1960: American astronomy was splintering into planetary and stellar branches, divided by unequal patronage, distinct professional mandates, strains over instrumental access, and decreased perception of common intellectual interests. The tensions that played out at Chicago were present at other centers of U.S. astronomy where solar system research had grown vigorous, including UCLA, where Samuel Herrick relocated his research program in applied celestial mechanics from its Department of Astronomy to its College of Engineering in 1961; but in few other places did these frictions reach the intensity they did at Chicago. Kuiper's response to staff discontent fueled animosities. Still smarting from his fallout with Urey, enraged by Burbidge's intolerant dismissal of his research, and peeved that staff astronomers doubted the quick completion of his Chilean telescope project, Kuiper lashed out at his critics in livid letters that worried Chicago administrators.[66] Yet other factors also played a role. Kimpton grew worried about the Chilean project after senior American astronomers, including Shane, argued that Kuiper had seriously underestimated the instrument's cost, and State Department officials reminded the university that the United States wanted warm relations with the democratic Chilean government to support its hemispheric aims. After Kuiper ignored a final warning from Kimpton to put his house in order, Kimpton fired Kuiper as director of Yerkes–McDonald on January 4, 1960. Kimpton urged his appointed successor, staff astronomer

[65] Quoted from Burbidge, [untitled memorandum on loss of confidence in Kuiper], n.d. [ca. Oct. 1959], WWM; W. Bidelman to G. Clemence, Aug. 22, 1959, copy in Box 28, GPK; "Exhibit D: Statement of Kuiper concerning meeting in Chicago of Nov. 9, 1959," and Kuiper to W. Zachariasen, Oct. 31, 1959, both WWM.

[66] Joseph N. Tatarewicz, "Social and Institutional Context of Celestial Mechanics in the Space Age: Style in Celestial Mechanics, Herget versus Herrick," First Biannual History of Astronomy Workshop, Notre Dame, Ind., June 27, 1993; Kuiper to R. W. Harrison, Oct. 18, 1959, Box 18, GPK; F. W. Zachariasen to Kuiper, Oct. 23, 1959, YO; Kuiper to Zachariasen, Dec. 6, 1959, Box 18, GPK; see also Whitaker (1985, 14–16).

William W. Morgan, to carry out an orderly transition with "no further embarrassment of Kuiper." Kimpton's order maintained Kuiper's status as a senior Yerkes astronomer. Indeed, Kimpton and Morgan both expected Kuiper to remain at Chicago after this administrative reshuffling, hardly an uncommon event in academic departments.[67]

Given the depth of animosity between Kuiper and other Yerkes staff, however, this was not a realistic option. In late January, Kuiper announced his plans to transfer his research associates, graduate students, and associated staff – ten people in all – to the University of Arizona. The importance of this move was what it told about the political economy of American science in the post-*Sputnik* era, for Kuiper's new post was not only at a then little-known state institution, but was split between its Department of Astronomy (including the Steward Observatory) and the university's Laboratory of Atmospheric Physics. Although Kuiper later declared that he had long known that "planetary astronomy must in time separate off from stellar astronomy and combine with geophysics," this was at best an exaggeration; through the mid-1950s, as we have seen, Kuiper had resisted mixing geophysics with astronomy. What had changed was the rise of solar system astronomy as a high national priority, the concomitant flood of funding for large-scale planetary research, the emergence of distinct patrons for astronomy's component fields, and the willingness of many universities to accommodate new grant-funded interdisciplinary programs of tremendous scale and apparent permanence. Arizona administrators, led by Richard Harvill, president since 1951, grasped that federal patronage for space exploration and the nearby Kitt Peak National Observatory were a potential means to strengthen universities in the American Southwest, where traditional departments were less firmly entrenched. Harvill's enthusiasm for academic institutes was shared by Richard A. Kassander, Jr., director of the Laboratory of Atmospheric Physics (which, following Kuiper's suggestion, was renamed "Lunar and Planetary Laboratory"). "In the relatively near future," Kassander enthused to Kuiper, even "more interdisciplinary marriages could take place." After Kuiper's arrival in September 1960, the Lunar and Planetary Laboratory became a leading institutional player in solar system research in America;[68] yet the growth of Big Science research programs at Arizona – including the building of a 61-inch telescope solely for solar system studies – marked a changed world for optical astronomers like Kuiper.

Planetary research programs, welcome to Harvill and Kassander for their promise to aid once peripheral state universities, were simultaneously a nemesis to depart-

[67] Kuiper to Harrison, Apr. 17, 1960, Box 18, GPK; G. Mulders to A. Waterman, June 3, 1960, Box 47, NSF; Morgan, "Meeting with Chancellor Kimpton," Jan. 4, 1960, and idem, "Subjects Raised by Dean Whaley," Jan. 6, 1960, both WWM. A chronology in Evans and Mulholland (1986, App. B) contains errors; see "Expanded Version of Statement Made at Faculty Meeting of August 17, [1959]," and Kuiper to W. G. Whaley, Aug. 9, 1960, both Box 18, GPK, and Kuiper (1960).

[68] Quoted from Kuiper to Kimpton, Jan. 25, 1960, Box 18, and Kassander to Kuiper, Jan. 23, 1960, Box 12, both GPK. On university science in the American southwest, see Webb (1983; 1993); on the LPL, see Tatarewicz (1990a) and Whitaker (1985).

mental leaders at these institutions, who saw their autonomy and budgets threatened. The main opponent to Kuiper's appointment at Arizona was Edwin Carpenter, director of its Steward Observatory since 1938. A noted stellar and galactic astronomer, Carpenter had overcome indifferent administrators and severe financial shortfalls to build competitive research programs during the 1930s and 1940s. The planned national optical observatory heartened Carpenter, who arranged to transfer the observatory's 36-inch reflector, hemmed by campus playing-field lights, to Kitt Peak; but word of Kuiper's appointment dismayed him. Carpenter was not antipathetic to solar system research: In the 1930s he had backed Leuschner's asteroid program at Berkeley, and had later encouraged staff member Edwin P. Martz to study the atmosphere of Mars. Carpenter's dismay had to do with the strength of Kuiper's programs, and the prospect that stellar and galactic astrophysics – *his* fields, and the discipline's core studies – would be sidetracked. He responded by portraying Kuiper as outside the discipline. Kuiper's interests "are so far afield from the conventional astronomy in which we are engaged, and indeed are so peripheral with respect to the basic needs of graduate students," Carpenter asserted, "that I have some doubt of his conformation to a new graduate program."[69]

Despite Kuiper's success in creating the Lunar and Planetary Laboratory (LPL) – more apparent in retrospect than to contemporaries – the transformation of American astronomy in the late 1950s was no less traumatic for Kuiper than it was for Whipple and other researchers active in solar system astronomy. Kassander's "interdisciplinary marriage" allowed Kuiper to join a laboratory of atmospheric physics, but it also meant for Kuiper increased separation from the astronomical community where he had worked for nearly three decades. The growing anger, loss, and isolation that Kuiper experienced were reflected in his public writings about solar system astronomy after 1956, as when he declared that since the late nineteenth century the "phenomenal growth of astrophysics, and the exciting explorations of the galaxy and the observable universe, led almost to an abandonment of planetary studies." Like Whipple, Kuiper understood the value of stressing limited resources to new patrons for his field; but it would be unfair to conclude that his claim of "abandonment" was simply a cynical means to aid fund-raising. Seeking funds from new and unfamiliar patrons was necessary precisely because the space race had raised competitiveness in his field to unprecedented new heights, and because the once unified discipline of astronomy (where Kuiper had counted on local and peer-reviewed funds to support his research) no longer existed. The pain of abandonment was real: It occurred, however, only after the mid-1950s, not in the distant past where field members tried to place it. Complaining in 1960 that "major U.S. observatories have been designed and built for stellar work and are used only incidentally and for a small

[69] Quoted from Carpenter to Urey, Jan. 25, 1960, Box 51, HCU. See also George E. Webb, "The Early Years of Astronomy at the University of Arizona," 1986 draft MS (I thank Webb for an advance copy); Leuschner to Whipple, Nov. 11, 1933, FLW; E. P. Martz to Kuiper, Nov. 11, 1953, Box 30, and Kassander to Kuiper, June 4, 1960, Box 12, both GPK; and Whitaker (1985, 19).

fraction of the time for planetary and lunar studies," Kuiper was accurately describing a *new* situation in which competition for telescope time had grown fierce.[70]

Although no evidence suggests that Kuiper deliberately sought to mislead patrons – by the early 1960s the need for major new telescopes to support NASA's planetary programs was beyond dispute – it is also clear that Kuiper painted a very different picture of solar system astronomy in America when funding was not at issue. In 1960, Kuiper was asked to review draft recommendations that Harvard astronomers Hynek and de Vaucouleurs had prepared for NASA's Jet Propulsion Laboratory. Their report, which asserted that twentieth-century astronomers had abandoned solar system research, suggested implementing various ground- and balloon-based observing programs. In a forceful response, which he marked "confidential," Kuiper advised NASA officials that it was "quite wrong to state that professionals as a class have neglected the planets in favor of the stars. While it is true that even more might have been accomplished, this is even more strongly the case for stellar problems." He followed this with a second letter, also confidential. "No good projects on planetary astronomy have lacked basic support" since the end of World War II, he declared. "What has been lacking is a scale of operation that matches *present* needs." Concerned with demonstrating the extent of American expertise in this field, Kuiper stressed the post-1945 vitality of solar system astronomy he increasingly denied in public.[71]

The need to respond to science planners at remote NASA facilities underscored how different solar system astronomy in America had become from what it had been in the 1920s, or even in the early 1950s. Whipple and Menzel had come to favor Big Science approaches to study the solar system, but they failed to retain autonomy over these programs after NASA was established. Optical astronomers like Kuiper, schooled in small-science approaches and wary of the enormity of the IGY, found the changes imposed first by NSF managers and then by the cold-war space race an even greater dislocation. Leaders of solar system astronomy in the 1950s, particularly Kuiper, did become active participants in the NASA era.[72] Yet for Kuiper – and likewise Whipple and others whose careers and research styles were thrown into upheaval in the 1950s – the emerging era of spacecraft exploration was a time of intense transformation and turmoil, heady excitement tinged by lost stability, diminished

[70] Quoted from Kuiper (1956b, 89) and from Kuiper memo, "Need for a Ground-based Lunar and Planetary Observatory," dated June 1960, copy in Box 61, HCU; this 1956 assessment was repeated in Kuiper and Middlehurst (1961, v). Kuiper was not alone in making such claims: In 1959 the former Lowell astronomer James Edson, then a researcher at the White Sands Proving Ground, declared that all contemporary knowledge about the planets had been created by several dozen men "working to a great extent with equipment created out of their private resources." Edson's allusion to the origins of Lowell Observatory was misleading – many Lowell planetary projects after 1945, including the planetary atmospheres project, had been financed by government grants – but Edson was certainly correct that existing patronage systems would not support post-*Sputnik* demands for planetary research; see Edson, "Needs of Planetary Astronomy," n.d. [1959], LVB, and Doel (1990c, 246–53).

[71] Quoted from Kuiper to A. R. Hibbs, Nov. 21, 1960, and idem, "Review of JPL Technical Memo 33-37; March 1, 1961," n.d. [ca. 1961; emphasis in the original], both Box 18, GPK.

[72] See Koppes (1982) and Tatarewicz (1990a, 54–6).

authority over research programs, and above all loss of security within a unitary American astronomical community, retold in tales of disciplinary "abandonment."[73] The permanent institutions that solar system researchers embraced after the late 1950s were not simply enlarged versions of what had existed before, but new entities that now served to define and protect still fragile research communities as the space program reordered the political economy of the physical sciences. They replaced the transient institutions that had sheltered interdisciplinary research in American astronomy before the start of its Big Science era.

[73] Recent examples include Cruikshank (1991) and Young (1989); other instances are noted in Tatarewicz (1990a, xi, 1–25).

Conclusion

In November 1973, one month before his 68th birthday (and just seven weeks before his sudden death in Mexico City), Gerard Kuiper pressed for further expansion of his Lunar and Planetary Laboratory. Writing University of Arizona officials, Kuiper urged that a new spectroscopist and infrared-instrument builder be hired to allow the LPL to compete for access to NASA's new 36-inch Airborne Infrared Observatory and the planned Space Shuttle.[1] His request, modest by 1970s standards, nevertheless revealed how far solar system astronomy in America had grown from its small-science roots. Just thirteen years had passed since Kuiper had established the LPL, emblematic of the emerging age of large-scale planetary science; yet by 1973 the Apollo lunar program was already history. *Mariner 9* was orbiting Mars for a second full year, and *Pioneer 10* was approaching a historic first encounter with Jupiter. Cold-war rivalries and massive outlays of federal funds for space research had built an enormous infrastructure for studying the planets, including the Jet Propulsion Laboratory and Kuiper's LPL.[2] Because the vitality of scientific research is often judged as synonymous with its scale, later generations of planetary scientists came to accept that their field's birth coincided with that of NASA.

Solar system research, however, was already an important element in the landscape of American science well before the 1957 launch of *Sputnik*. Far from a neglected topic, solar system astronomy was a characteristic example of a small-scale, interdisciplinary field, not unlike research programs in physical chemistry and oceanography that similarly blossomed during the first half of the twentieth century.[3] In this book, I have argued that work in planetary astronomy, meteorology, geochemistry, geology, and geophysics formed an interwoven tapestry. None of these distinct intellectual strands can be viewed in isolation from the others. Researchers within these disciplines engaged with one another through complex social networks, including transient institutions; their claims and arguments about fundamental planetary questions were influenced in response to results achieved outside their professional communities. I have also argued that important social, instrumental, profes-

[1] H. P. Larson and G. P. Kuiper to Dean H. K. Bleibtreu, Nov. 3, 1973, Box 1, GPK. The 36-inch instrument was subsequently renamed the Kuiper Airborne Observatory.

[2] Koppes (1982), Smith (1992, 202–8), and Tatarewicz (1990a).

[3] Burstyn (1980) and Servos (1983).

sional and disciplinary transformations affected these fields – first when foundation support for science was eclipsed by federal and military patronage after World War II, and second, when major new federal science programs altered the landscape of American science in the mid-1950s. The rapid rise of NSF patronage for astronomy, the great flood of funds for the IGY, and the sudden political demands for space exploration following *Sputnik*'s launch altered relationships among these disciplines. These changes profoundly recast the boundaries of the physical sciences in America and fragmented the once unified discipline of astronomy. They simultaneously made solar system astronomy professionally, institutionally, and intellectually distinct, and tied its fortunes to the politics of the U.S.–Soviet space race. The transformation of solar system astronomy from an informal interdisciplinary community into an exemplar of Big Science research provides an illuminating example of how cold-war funding practices influenced the character and conduct of science in the United States after 1945, making it more responsive to state interests and more dependent on federal patronage.[4]

One question that this book has posed can be put succinctly: How did changing styles of scientific patronage affect *interdisciplinary* research in America during the early and mid twentieth century?

Solar system research in America was aided by the broad intellectual horizons of U.S. astronomy and the financial and institutional stability of the discipline. One of the most important characteristics of American astronomy was its ecumenical embrace of distinct research programs well into the 1950s. Mathematical astronomers maintained their place within American centers of astronomy – and their focus on planetary bodies – despite the great advance of astrophysics. Astrophysicists who studied meteor origins and planetary atmospheres (initially to understand the interstellar medium and the general abundances of the elements) also saw themselves contributing to disciplinary programs, not as the progenitors of distinct research communities. The inclusiveness of American astronomy, to be sure, was not simply a matter of professional courtesy; rather, it was mandated by the discipline's place in the ecology of academic science institutions. U.S. astronomy departments were small, requiring instructors to be well versed in many fields. Although niches for specialized research did emerge at Harvard, Princeton, Mount Wilson, Lowell, and other major institutions of U.S. astronomy, solar system investigations prospered alongside sustained programs in other fields. Moreover, solar system research also flourished at smaller centers like Cincinnati and Iowa, where astronomers could not easily take part in core disciplinary programs involving stellar or nebular research.

Decentralized academic departments, a hallmark of American higher education by 1900, also stimulated interdisciplinary contacts; but even more important was the faith of leading American scientists, shared by leading patrons of science, that the fertile, untouched plains between cultivated disciplines were the best places to

[4] Dennis (1994), Forman (1992), Galison (1987), Lowen (in press), McDougall (1985), Mukerji (1990), and Smith (1989).

achieve new breakthroughs. By the 1920s "cooperative" institutions dotted the landscape of U.S. science, including Hale's Mount Wilson Observatory and the National Research Council, whose influential committees addressed interdisciplinary problems like the age of Earth. The value that U.S. scientists placed on cooperative approaches as a strategy for doing science cannot be underestimated. In declaring that "[t]he big opportunities are obviously not in pure astronomy or physics or chemistry or geology, but in mixtures," Shapley voiced a conviction shared by many colleagues, shaped by the political economy of American science, that disciplinary boundaries were porous enough to yield profitable new lines of research.[5]

Foundation funding for U.S. science sustained this faith. Leaders of the Rockefeller and Carnegie foundations poured resources particularly into institutions that fostered "cooperative" research among disciplines, aware or their great leverage in stimulating research outside core disciplinary programs. The vast scale of foundation resources made them a particularly influential carrot for academic scientists dreaming of broad interdisciplinary efforts. Shapley's grand ambition to link atomic physics, astronomy, physical chemistry, and mathematical geology within an institute of cosmogony, requiring him to circumvent Harvard's departmental structure and support systems, had little chance to succeed without philanthropic backing, as Shapley himself understood. In this sense, the influence of the Rockefeller and Carnegie foundations extended well beyond the projects they funded. It is impossible to understand the growth of interdisciplinary science programs in the United States during the early twentieth century without considering the flexibility and creativity of philanthropic support for research and transient institutions.[6]

As A. Hunter Dupree has stressed, World War II marked a "Great Instauration" for U.S. science, a watershed where federal and military agencies became the chief patrons of scientific work. The war also exposed scientists to large-scale organizational work and the promise of greatly expanded funding for studies of potential military application.[7] A second point that Dupree makes is often overlooked, however: This transformation did not affect all disciplines equally. While physicists and geophysicists forged close alliances with military agencies during World War II, geologists, long accustomed to fulfilling national service roles through the dominant U.S. Geological Survey, experienced far fewer changes. Wary of federal domination of research, and comfortable with prewar styles of patronage, most astronomers also found little reason to press for abundant federal assistance. Not all boats rose evenly on the flood tide of postwar federal funding; rather, World War II accentuated differences among scientific disciplines. Conflicts over appropriate *scales* of research and support networks after 1945 influenced the development of U.S. science no less than the enormous *growth* of federal funds for scientific work.

[5] Quoted from Shapley to H. Benioff, Feb. 9, 1923, Box 2, DHCOs; see Ben-David (1971), Doel (in press-a), and Kohler (1990).

[6] Kevles (1977, 185–99), Kohler (1991), and Servos (1983).

[7] Dupree (1972) and Leslie (1993).

Government and military funding nevertheless had considerable influence over U.S. astronomy after 1945. Small-scale studies of solar system phenomena, including Kuiper's asteroid programs, were funded by the Office of Naval Research and later by the NSF, which replaced private foundations as the discipline's general patron. Other government agencies, including Navy and Air Force centers, became the principal sponsors for large interdisciplinary programs, such as Whipple's meteor–upper-atmosphere project at Harvard and the Project on Planetary Atmospheres at Lowell (once initial Weather Bureau support petered out). Military patrons also became chief sponsors of transient institutions in this field; with the important exceptions of the Rancho Sante Fe conference of 1950 and the atomic abundances conference in Williams Bay in 1952, many were sustained by military contracts. These funds were critical for American astronomy, helping local institutions like Yerkes–McDonald and Lowell survive the lean years after 1945; but they also made astronomy – which Struve had proudly advertised at the close of World War II as "absolutely useless in any project contemplated to make this nation stronger in a military sense" – more responsive to military goals and needs.[8] This reorientation was not painless. Whipple and Menzel, for example, succeeded in building large, new interdisciplinary projects at Harvard, but at a cost of institutional instability, fragmented research programs, and faculty discord. To the question, Did philanthropic versus federal funding make a difference?, one must point to the greater instability and dislocations that accompanied military contracts, which astronomers came to regard with increasing anxiety and regrets. However, a line cannot be drawn between military and civilian patrons in terms of productivity or disruptive tendencies. ONR funds sustained many areas of U.S. science after 1945 as effectively as many foundation programs had prior to World War II, whereas NSF policies, intentionally or otherwise, helped to splinter the American astronomical community.

Solar system astronomy remained an integral field of American astronomy through the mid-1950s; yet within a few years, this unified discipline began to fragment. Several factors caused this transformation, including the determination of NSF officials to build national optical and radio observatories, sharpening the identity of distinct specialties, and the rapid growth of U.S. astronomy, which increased competition for time on existing world-class optical instruments. The most significant factor, however, was the extremely rapid growth of geophysics, astronomy's close disciplinary neighbor, particularly through the IGY. Clashes arose between proponents of astro-geophysical research, supported by the newly created NASA and other federal-military agencies as part of the burgeoning space program, and traditional stellar astrophysicists, who sought to maintain telescope time for their programs. Transient institutions were no longer sufficient to shelter interdisciplinary projects; to handle them, entrepreneurs created new institutions, such as the Lunar and Planetary Laboratory, or recast older institutions like the Smithsonian Astro-

[8] Quoted from O. Struve to R. M. Hutchins, Aug. 17, 1945, Box 2, UChi.

physical Observatory. This new, Big Science era would later seem a golden age of planetary exploration, but it was nevertheless a difficult time of transition for Kuiper and Whipple, used to different operating scales and greater autonomy over their work. Loss of a unified astronomical discipline was especially painful for them, and both men claimed that American astronomy had abandoned their field. Although they erred in ascribing abandonment to a distant past rather than to the troubled present, their accounts accurately captured the pain of dislocation as disciplinary boundaries and patronage relationships were rapidly redrawn in the late 1950s. The underdog status that this field popularly acquired was a response to, and revealing of, a changed political economy of science. Kuiper's and Whipple's rhetoric defended not science as a cultural activity – a chief worry of U.S. scientists before 1940 – but rather the claims of one scientific *field* against competitors in all other disciplines for federal patronage.[9]

Conflicts in American astronomy during the 1950s indeed may have been unavoidable: Scale is critical in scientific work, and splintering often occurs when research communities grow larger than several hundred people.[10] However, the tumultuous growth of the American space program, driven by cold-war antagonisms, nonetheless illuminated an ominous and often overlooked aspect of federal funding for science that Bowen, Struve, and Kuiper had sought to avoid in U.S. astronomy. Already by 1955, warily watching the rapid growth of Harvard astronomy, many traditional optical astronomers felt ambivalent about yielding disciplinary autonomy to federal patrons. There can be no doubt that much new information was acquired in all fields of astronomy after 1960, and the era of spacecraft exploration had revolutionary impact on planetary science; but given the boom-and-bust cycles that came to affect many domains of U.S. research, including planetary science, their arguments that rapid growth in astronomy would be bought at the price of greater vulnerability to political and economic movements were in retrospect valid. As federal budget cuts began to trim jobs for planetary scientists in the 1970s and 1980s, leading researchers worried that this field was now part of what Chandra Mukerji has aptly termed the government's post–World War II "elite reserve labor force."[11]

Two further points are worth stressing. It is impossible to understand how interdisciplinary bridges in solar system astronomy were built – or left to crumble – without taking into account the role of personality. This was particularly evident in the cases of Urey and Kuiper, who, like most researchers in this field prior to 1960, tried to sustain interdisciplinary work outside formal institutional structures. The particular chemistry of their interactions – Kuiper's sloppiness with citations,

[9] Aronson (1986), Graham, Lepenies, and Weingart (1983), and Morgan (1990); on rhetorical strategies, see Friedman (1989) and Moyer (1992).

[10] Such arguments were not limited to astronomy; see Weart (1992b); on planetary science after 1960, see especially Burrows (1990) and Koppes (1982).

[11] Mukerji (1990, 21).

Urey's fastidiousness about them – caused them to abandon attempts to understand the intellectual and professional standards of each other's communities, and each initially retreated to the less contentious cores of their own disciplines. The intensity of their conflict should not, however, obscure the significant fruits of their initial cooperation. In his extensive, recent study of Antoine-Laurent Lavoisier, the premier chemist of the eighteenth century, Frederic Lawrence Holmes makes the important point that scientific creativity is often the result of collaborative work, not the isolated activities of individual scientists. Both Urey and Kuiper produced their most creative research on planetary evolution during the early, exuberant stage of their relation. Even the incandescence of this controversy, and the real damage to astro-chemical research that it caused, did not deter other astronomers such as Whipple from continuing cooperative work with Urey. Fragile, wild flowers compared to core disciplinary programs, interdisciplinary interactions nevertheless promised benefits sufficient to encourage scientists to risk involvement.[12]

Second, it is also important to observe that many prominent researchers in this emerging interdisciplinary field were simultaneously central figures *in their disciplines.* Russell, Shapley, and Leuschner were influential leaders of the prewar generation of American astronomers, just as Kuiper and Whipple were of theirs. Urey was no less senior in geochemistry, and Harrison Brown, who helped extend the geochemical revolution, had solid standing in physical chemistry. What makes this significant is that leaders of interdisciplinary work are often thought to be social outcasts on the peripheries of their disciplines; in the case of molecular biology in Great Britain during the 1950s, as Pnina Abir-Am has persuasively argued, this was indeed the case. The development of solar system astronomy thus provides an important reminder that interdisciplinary work is not always an adaptive response to difficulties in maintaining core research programs, and that such investigators often have access to significant institutional and professional resources in their *home* communities.[13]

Another significant question that this book has posed is this: What did U.S. researchers learn about the solar system before the dawn of the space age in 1957?

Earlier claims to the contrary, American astronomers clearly retained a fascination with the nature and origin of the solar system. There can be no doubt that reflecting telescopes and their associated spectrographs – the standard tools of astronomy through the first half of the twentieth century – favored studies of stellar and nebular phenomena: Stars, far more abundant than planets, yielded more data than the colder bodies of the solar system. Nevertheless, American astronomers through the 1950s made unprecedented efforts to understand the composition and structure of bodies within the solar system. What is impressive and remarkable is the extent to which U.S. astronomers fashioned links to more distant disciplines. Russell, Kuiper, Whipple, Baldwin, and others united with geochemists to address the problem of

[12] Holmes (1987, 486–91); see also Burchfield (1975, 275).
[13] Abir-Am (1987).

atomic abundances, with physicists and geochemists to address the age of Earth, with geologists to address crater structures, and with nuclear chemists and geophysicists to study the interior structure of planets. At no earlier time were these interdisciplinary connections so finely interwoven or sustained.

Several of their contributions had lasting significance. Russell, Wildt, Adel, Adams, and Menzel, using traditional astrophysical techniques, detected the methane–ammonia components of Jupiter's atmosphere and the oxygen component of Venus's, difficult measurements for their time. Whipple and Wylie were the first to produce definitive evidence that meteors are solar, not interstellar. In 1950 Whipple produced his "dirty snowball" model of comet structure; meanwhile Kuiper, enlarging on his discovery of Titan's atmosphere, produced a new nebular cosmogony that incorporated Oort's theory of long-period comets, supporting the Oort cloud concept and his own proposed comet reservoir, later called the Kuiper belt. Most of these achievements – all major developments in mid-twentieth-century astronomy – involved small members of the solar system. In part the fecundity of this work can be explained by the relative ease of studying these bodies compared to the planets, making such research competitive at many observatories; but credit must also go to Leuschner's active research school at Berkeley, which trained a generation of individuals, including Whipple, to apply physico-mathematical techniques to these problems.

Astronomers also shaped debates over the structure of Earth and other planets, a critical research program in the earth sciences. Planetary diameter measurements influenced theories of Earth's interior during in the early 1940s, and helped convince American geophysicists in the mid- and late 1940s to reject the Kuhn–Rittmann and Ramsey hypotheses, leading alternatives to iron-core models. Frequent contact between astronomers and earth scientists over such issues as ocean formation and the molten Earth also left its mark. Rubey accepted astronomical constraints involving the early Sun's temperature in fashioning his theory of seawater formation, and Kuiper, at least for a time, accepted in his cosmogony constraints imposed by geologists and geochemists concerning Earth's early temperature. Sustained contact between these communities – reinforced by such transient institutions as the Rancho Sante Fe conference – convinced researchers in each that they could not easily dismiss or ignore evidence from the other. Although these research programs were promoted by no single research center, a community – an "invisible college," to use Derek de Solla Price's still relevant term – took shape by the early 1950s.[14] Developments in any of these disciplines cannot be examined in isolation.

Perhaps the most influential contribution from solar system astronomy during the first half of the twentieth century was support for the idea that large bodies had repeatedly struck Earth and the Moon. Baldwin's 1949 argument that massive bombardment had caused most lunar craters and seas – as well as his prediction that numerous impact craters would be discovered on Earth – was accepted by many Amer-

[14] Quoted from Price (1963, 85); see also Crane (1972).

ican astronomers, ending (for this community at least) the centuries-old debate over the origin of lunar craters. Familiarity with atomic bomb results convinced them to support this theory ahead of their European and Soviet colleagues, and the U.S. lunar program after 1958 reflected this orientation. More important, growing support for this theory caused American geologists to reexamine the principle of uniformitarianism, and to debate the role that catastrophic processes had played in planetary evolution. The better-known controversy over continental drift in the 1960s has overshadowed how misgivings over catastrophism caused geologists to reexamine the conceptual roots of their discipline.[15] In this debate Daly and his former Harvard students, including Albritton, played an important but little-recognized role.

Until comprehensive studies of astronomical research in France, Germany, Britain, and the Soviet Union are completed, international comparisons of solar system research in the first half of the twentieth century will be difficult to make. Still, astronomers in other countries clearly entered this field as well. Richard Jarrell has revealed that close to a fifth of the research output of Canadian astronomers was devoted to meteor astronomy, and Loren Graham has emphasized the importance that Soviet astronomers placed on cosmogony. British, Dutch, German, and French researchers also made substantial contributions to planetary, comet, and meteor studies during this time.[16] Nonetheless, published figures from the International Astronomical Union indicate that, through the first half of the twentieth century, U.S. astronomers were particularly well represented in IAU solar system commissions, and in many instances led them (see Appendix D). What does this signify? Undoubtedly the world-class status of U.S. optical telescopes, the damage to Soviet astronomy from the Great Purge and World War II, and the slow postwar recovery of continental European science played a role. Yet U.S. astronomers dominated the most interdisciplinary commissions, including meteoritics, and led four of eight commissions in 1955. This suggests that socio-institutional factors, such as the departmental structure at American universities and historically fluid interactions among U.S. disciplines, were also responsible. Since the creation of sustained interdisciplinary programs is one of the landmark contributions of twentieth-century science, sustained inquiry into this issue is merited.

Solar system astronomy in America before the 1950s was not a branch of geophysics, for it matured as a specialty within an all-embracing astronomical discipline; yet this field was no less part of the process by which the "planetary sciences," as they became known, were created. As environmental issues have loomed in importance, the emergence of the planetary sciences, emphasizing integration of existing scientific disciplines, has come to be recognized as one of the most important scientific developments of the century. Certain problems pursued by researchers in solar system astronomy, including climate change and how massive impacts influenced planetary

[15] This issue awaits critical examination; a helpful introduction remains Albritton (1963).

[16] Graham (1993, 220–4) and Jarrell (1988).

evolution, are now considered aspects of the environmental sciences as well.[17] Solar system astronomers were innovators at the interstices of disciplines, forging links among specialties to address broad planetary questions. If their history has been neglected, it is in part because historians have focused on other fields where military influence over U.S. science after World War II is more evident; but it is also because most histories of twentieth-century science in America are disciplinary histories, wherein such activities appear like blurs in the corners of photographs.[18] Center and periphery are matters of perspective, however, not absolutes; what seems peripheral in one moment can appear far less so in the next.

[17] Bowler (1992, 426), Raup (1991), and Weart (1992a).

[18] Leslie (1993); see especially Forman (1987). Disciplinary approaches to the history of science are aptly defended in Servos (1990, xiii–xvi); on the need for further emphasis on interdisciplinary fields, see Bowler (1992, xvi, 1–31) and Schweber (1993).

Appendixes

Solar System Astronomy as a Community, 1920–1960

In this study, I have argued that solar system astronomy was an active, interdisciplinary component of American astronomy through the first half of the twentieth century, and became a distinct research community only in the late 1950s. Moreover, I have claimed that its emergence as a specialty had to do with important changes that affected the discipline of U.S. astronomy, particularly shifts in patronage and the rise of vigorous research communities along its border with geophysics. Here I substantiate these claims with numerical evidence, placing solar system astronomy between 1920 and 1960 in a more quantitative framework. These measurements also offer new perspectives on the development of American astronomy during the early and mid twentieth century.

Scientific fields evolve into distinct specialties by developing research schools, recruiting new workers, publishing their results, creating new curricula and dedicated institutions, and securing patronage. This study has focused on new research schools within the community of American astronomy, transient and permanent, that stimulated new work in one such field. Awards and significant appointments received by researchers in particular fields indicate in turn their relative standing among colleagues in the discipline, both within national contexts and among the international community at large. Measurements of these factors illuminate the growth and evolution of research communities over time.

Astronomy was the smallest physical science discipline in the United States through the 1950s, and the number of individuals who participated in solar system astronomy was, of course, smaller still. Certain standard statistical yardsticks for measuring the growth of scientific fields, including cocitation indices, are hence unreliable when the population sample is limited to several dozen individuals.[1] In

[1] Although bibliometric data for twentieth-century astronomical research can be gleaned from the annual *Astronomischer Jahresbericht,* its subject categories do not always correlate closely with what contemporary researchers then regarded as the actual intellectual confines of their fields. This makes attempts to use such data to investigate the growth of particular fields problematic; see, e.g., Gieryn (1979, 42), and Gieryn, private communication, Jan. 31, 1991.

addition, the number of awards for which American astronomers competed prior to 1960 was small, and awards were made for overall contributions rather than for work in specific fields.[2] Here I have relied on more prosaic but informative measures. Data are presented on American astronomers who contributed to this field, including their geographical distribution, their publications, their funding, and their participation in solar system commissions of the International Astronomical Union (IAU).

APPENDIX A: AMERICAN CONTRIBUTORS TO SOLAR SYSTEM ASTRONOMY

In undifferentiated disciplines – that is, where scientists share common training, instruments, and resources – it is common for scientists to move in and out of what later become distinct research specialties. American astronomy between 1920 and 1960 was such a discipline, and solar system studies were occasionally made by astronomers better known for their contributions to other fields. Table A.1 shows the distribution of researchers in solar system astronomy, including individuals who also worked on stellar and galactic problems, at American facilities between 1920 and 1960. These researchers also trained graduate students; in Table A.2, I list individuals who took their Ph.D.s on topics involving solar system astronomy at American universities. Table A.3 reports officials of the American Astronomical Society (AAS) who made significant contributions to solar system astronomy.

Such compilations cannot avoid a certain imprecision. To avoid including astronomers who published infrequently on solar system research, I have cited only individuals whose work was familiar to active members of the field. Because World War II was a watershed for many American observatories and university departments, Table A.1 divides the four decades from 1920 to 1960 into prewar and wartime/postwar epochs. Small-capital names signify researchers who devoted at least 50 percent of their aggregate publications to solar system astronomy, including celestial mechanics, during this period. Researchers whose names are bracketed occupied these posts for less than five years (or, for the period 1940–60, began their appointments after 1955). Graduate students are not listed in Table A.1.

[2] Several U.S. astronomers received Bruce Medals from the Astronomical Society of the Pacific for their contributions to solar system astronomy (including V. M. Slipher for planetary spectroscopy in 1935 and A. O. Leuschner for celestial mechanics in 1936). Bruce Medalists, however, were selected for lifetime service to astronomy, making it difficult to analyze disciplinary trends during times of rapid professional flux; see Einarsson (1935, 5) and Tenn (1986, 103–5).

Table A.1. *Participants in solar system astronomy at American astronomical centers, 1920–1960*

Institution	Investigators	Research interests
I. 1920–40		
Berkeley	A. O. LEUSCHNER, H. Thiele	asteroids, celestial mechanics
Chicago	F. R. Moulton	celestial mechanics
Columbia	W. J. Eckert	celestial mechanics
Harvard	H. Shapley, F. L. WHIPPLE, [F. G. WATSON, JR.], [E. ÖPIK], W. J. Fisher	meteors
Iowa	C. C. WYLIE	meteor velocities, meteor craters
Kansas	W. D. ALTER	celestial mechanics, cosmogony
Lick Obs.	W. W. Campbell, R. Trumpler	planetary atmospheres, planetary diameters
Lowell Obs.	C. W. TOMBAUGH, C. O. LAMPLAND, V. M. & E. C. SLIPHER	planetary atmospheres, planetary surfaces, photographic surveys
Mt. Wilson Obs.	W. S. Adams, W. E. Pettit, T. S. Dunham, Jr., F. E. Ross, C. E. St. John, F. E. Wright	planetary atmospheres, lunar surface
Ohio Wesleyan	N. T. BOBROVNIKOFF	comets
Pennsylvania	C. P. OLIVIER[a]	meteors
Princeton	H. N. Russell	planetary atmospheres, cosmogony
Toronto	P. M. MILLMAN	meteors, meteor spectra
U.S. Naval Obs.	C. B. WATTS, G. M. CLEMENCE, J. C. Hammond	celestial mechanics
Yale	D. BROUWER, E. W. BROWN	celestial mechanics
II. 1940–60		
Berkeley	L. E. CUNNINGHAM	celestial mechanics, comets
Chicago (Yerkes–McDonald)	G. P. KUIPER, D. E. Harris, [G. Herzberg], G. van Biesbroeck	planetary atmospheres, asteroids, satellites, Moon, cosmogony
Cincinnati	P. HERGET, E. RABE, P. MUSEN	asteroids, celestial mechanics
Dominion Obs.	C. S. Beals, P. M. MILLMAN	meteor craters, meteors
Griffith Obs. & Planetarium	W. D. Alter	lunar surface
Harvard/Smithsonian Astrophysical Obs.[b]	F. L. WHIPPLE, D. H. Menzel, [R. E. MCCROSKY], [L. Jacchia], R. N. Thomas, [G. de Vaucouleurs], [G. HAWKINS], A. F. Cook	meteors, comets, planetary atmospheres
Indiana	[staff, graduate students]	asteroid reconnaissance
Iowa	C. C. WYLIE	meteors
Lick Obs.	H. M. Jeffers	comets
UCLA	S. HERRICK, F. LEONARD	celestial mechanics, meteors, meteorites
Lowell Obs.	E. C. SLIPHER, H. M. Johnson, H. Giclas, C. O. Lampland	planetary atmospheres, planetary photometry, observations of Mars
Mt. Wilson & Palomar Obs.	E. Pettit, S. B. Nicholson, R. W. Richardson	planetary atmospheres
Northwestern U.	[R. B. Baldwin]	Moon, meteor craters
Ohio State/ Perkins Obs.	N. T. BOBROVNIKOFF	comets
U.S. Naval Obs.	G. CLEMENCE, C. B. WATTS	celestial mechanics
Yale U. Obs.	D. BROUWER, R. Wildt, Harlin J. Smith	celestial mechanics, asteroids, planetary physics

[a]University of Virginia, 1914–28.
[b]The Smithsonian Astrophysical Observatory was moved from Washington, D.C., to the Harvard campus in 1955.
Sources: Biographies, obituaries, observatory reports, curriculum vitae, publications in *Astronomischer Jahresbericht*.

Table A.2. *Known dissertations in solar system astronomy, 1920–1960*

Year	Name	Subject	Institution (advisor if known)
1921	Jeffers, Hamilton M.	comets	Berkeley
1927	Bobrovnikoff, Nicholas T.	comets	Chicago (Frost)
1931	Eckert, Wallace J.	celestial mechanics	Yale (E. W. Brown)
1932	Millman, Peter M.	meteor spectroscopy	Harvard (Shapley)
1932	Bennett, Arthur L.	lunar photometry	Princeton (Russell)
1935	Herget, Paul	celestial mechanics	Cincinnati [trained under Leuschner at Berkeley]
1936	Herrick, Samuel	celestial mechanics	Berkeley (Leuschner)
1938	Watson, Fletcher G., Jr.	meteors, asteroids, cosmogony	Harvard (Shapley)
1941	Hertz, Hans G.	celestial mechanics	Yale (Brouwer)
1942	Grosch, Herbert R. J.	celestial mechanics	Michigan
1946	Cunningham, Leland E.	celestial mechanics	Harvard (Whipple)
1949	Harris, Daniel, III	photometry, celestial mechanics	Chicago (Kuiper)
1949	Bauer, Carl	meteorites	Harvard (Whipple)
1950	Hamid, S. El-D.	meteors	Harvard (Whipple)
1950	Davis, Morris S.	celestial mechanics	Yale (Brouwer)
1951	de Marcus, Wendell C.	planetary physics	Yale (Wildt)
1953	Sinton, William	planetary spectroscopy	Johns Hopkins (Strong)
1956	McCrosky, Richard E.	meteors	Harvard (Whipple)
1956	Gehrels, Tom	asteroids	Chicago (Kuiper)
1956	Duncombe, Raymor	celestial mechanics	Yale (Brouwer)
1958	Wright, Frances W.	meteors	Harvard (Whipple)
1959	Gill, Jocelyn R.	celestial mechanics	Yale (Brouwer)
1960	Sagan, Carl E.	Moon, Venus greenhouse	Chicago (Kuiper)

Note: Several American researchers who made substantial contributions to solar system astronomy in the mid twentieth century, including Fred L. Whipple (Berkeley, 1931), Robert S. Richardson (Berkeley, 1931), Ralph B. Baldwin (Michigan, 1937), Richard N. Thomas (Harvard, 1948), and Elizabeth Roemer (Berkeley, 1955), wrote dissertations outside this field, and are therefore omitted from this list. About 370 dissertations in astronomy were written at U.S. universities between 1920 and 1960; it is important to note that not all dissertation titles for this period could be reviewed, particularly at smaller institutions, and hence graduate work in solar system astronomy is almost certainly underrepresented here.
Sources: Compilations from U.S. observatories and departments of astronomy; biographical directories; dissertation indexes, oral history interviews, and archival materials; Berendzen and Moslen (1972) and Knapp and Goodrich (1952).

Table A.3. *Leadership roles in the AAS among contributors to solar system astronomy, 1920–1*

President					
Wallace W. Campbell	(1922–5)	Henry Norris Russell	(1934–7)	Donald H. Menzel	(1954
Ernest W. Brown	(1928–31)	Harlow Shapley	(1943–6)	Gerald M. Clemence	(1958
Walter S. Adams	(1931–4)				
Vice-President					
Donald H. Menzel	(1946–8)	Dirk Brouwer	(1949–51)	Gerald M. Clemence	(1952
Fred L. Whipple	(1948–50)	Carlyle S. Beals	(1950–1)	Seth B. Nicholson	(1953
Councilor					
Nicholas T. Bobrovnikoff	(1945–8)	Gerard P. Kuiper	(1949–52)	Gerhard Herzberg	(1958
Peter M. Millman	(1947–50)	Paul Herget	(1952–5)		

Source: American Astronomical Society.

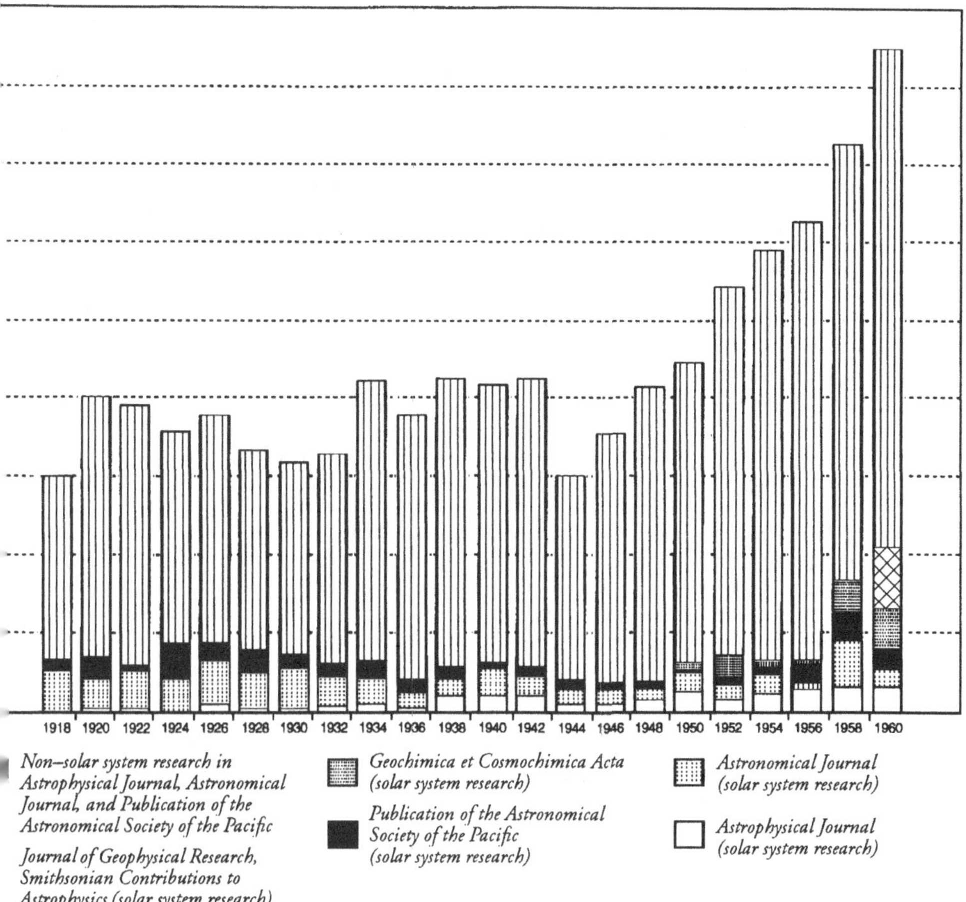

Graph B.1. Publications in U.S. astronomy by research fields, 1918–60.

APPENDIX B: PUBLICATIONS IN SOLAR SYSTEM ASTRONOMY, 1920–1960

Graph B.1 tabulates articles on solar system research published in leading astronomical journals between 1920 and 1960, together with the aggregate totals of publications in other specialties. Rather than examine all astronomical publications, I have focused on the three most prestigious journals for American astronomers in this time: the *Astrophysical Journal*, the *Astronomical Journal*, and the *Publications of the Astronomical Society of the Pacific*. I have also reviewed several publications in related fields that began or expanded in the 1950s and were utilized by astronomers writing on solar system topics – particularly *Geochimica et Cosmochimica Acta, Smithsonian Contributions to Astrophysics,* and the *Journal of Geophysical Research*. Research publications alone are cited; book reviews, descriptions of instrumentation, and general news items are omitted. Sampling was done biannually.

Several important trends are indicated. The overall curve of publications produced by American astronomers in all fields (including solar system astronomy) shows the negative influence of World War II. American astronomers did not regain their prewar publication output until the late 1940s, when federal and military patrons such as the Office of Naval Research and the National Science Foundation made significant investments in the discipline. Equally remarkable, the effect of the Great Depression is hard to discern.[3] This suggests that the stable patronage provided to Mount Wilson by the CIW, new grant-funded projects like the Harvard Arizona Meteor Expedition, and the endowments of major private universities such as Princeton and Harvard made up for the drop in output from smaller facilities.

Solar system astronomy generally kept pace with the expansion of its parent discipline, attesting to its integration within the larger community. Until about 1941, this field contributed between 10 and 15 percent of the total publications of American astronomers. A modest decline appears in the late 1940s – an irony, given the high optimism that Kuiper and Whipple then voiced about the field's growth. This dip may be more apparent than real, for it apparently derives from an increased tendency by mathematical astronomers to publish results in mathematical journals rather than in the *Astronomical Journal.*[4] Field contributions to the *Astrophysical Journal* remained unchanged. In any case, this figure never fell below 10 percent of the discipline's output as a whole.

After the mid-1950s, publications on solar system topics soared, paralleling the curve for the discipline on the whole. (A comparative low point occurred in 1956, perhaps reflecting the Kuiper–Urey controversy.) An important change can be seen nevertheless: New research was now more likely to appear in publications at the periphery of American astronomy, such as the *Journal of Geophysical Research,* rather than in those that carried results from its core research programs, such as *Astrophysical Journal* and the *Astronomical Journal.* These figures offer vivid testimony of the splintering of American astronomy into increasingly distinct specialties in the mid and late 1950s, caused by the rapid growth of IGY-sponsored research and the NSF's rise as astronomy's chief patron (see Chapter 6). Solar system astronomy emerged as an institutionally and professionally distinct specialty during this turbulent period.

APPENDIX C: PATRONAGE FOR SOLAR SYSTEM ASTRONOMY, 1947–1960

The distribution of funds within scientific disciplines offers an important perspective on the relative standing of component fields and research specialties. Graph C.1 shows disciplinary resources provided by its leading external patrons, indicating that portion devoted to solar system astronomy.

It is difficult to begin such a graph prior to 1947. Before World War II, funding for American astronomy was informal and largely decentralized. Excluding large philanthropic grants to build major new observatories, American astronomers secured research support largely through their home universities or through small grants-in-aid programs. Comparing the distribution of research funds among astronomical fields for the period before 1940 is thus difficult, and must await further historical studies.[5]

[3] This is also the case in American physics; see Weart (1979).

[4] Peter Kammeyer, "Celestial Mechanics in 1920," First Biannual History of Astronomy Workshop, Notre Dame, Indiana, June 27, 1993.

[5] See Lankford (1987) and Lankford with Slavings (in press).

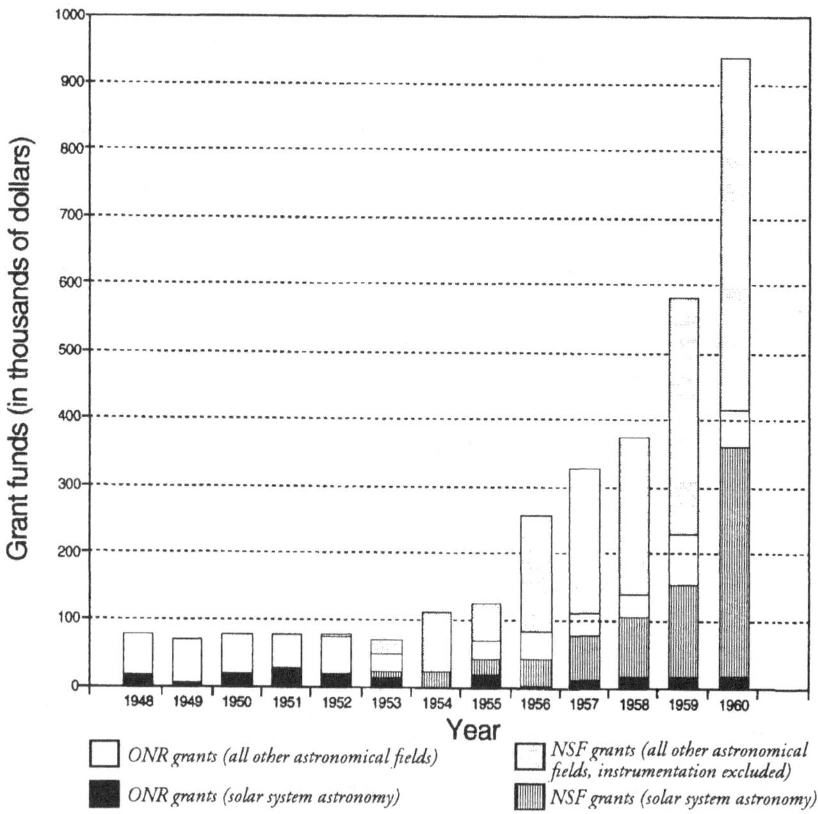

Graph C.1. Funding for U.S. astronomy, 1948–60. *Sources*: ONR figures are drawn from the minutes of meetings of the Committee on Astronomy, Advisory to ONR, National Academy of Sciences. NSF figures were taken from the *Annual Reports* of the National Science Foundation.

The rise of new federal patrons after World War II eases this task for the historian, although local institutional funds (for which no central accounting exists) still played a major role. By the mid-1950s, the two most significant patrons of astronomy in the United States were the Office of Naval Research (ONR) and the National Science Foundation (NSF). Proposals to both agencies were evaluated by panels composed of leading astronomers, including Greenstein, Bowen, and Shane. Decisions about allocating resources thus reflected the views of the discipline's elite.

The ONR and NSF, it is important to note, were not the only patrons for American astronomers involved in solar system research after World War II – nor the most generous. Grants from other military agencies to astronomers often exceeded manyfold the award figures for these steady, discipline-wide patrons; yet too little is known about the distribution, size, and effect of these awards to permit meaningful comparisons, and thus they are not included. The ONR and NSF in any case remained the most significant general patrons for American astronomers through 1960.

This graph confirms archival materials that suggest that solar system astronomy fared well as a field, receiving funds out of proportion to its size. It accounted for a large proportion of funds expended by the ONR, particularly for 1950–3. It also benefited from the explosive growth of federal patronage for astronomy that began in the mid-1950s and accelerated sharply after the *Sputnik* crisis of 1957. By 1960, as new federal agencies – particularly NASA – funded research in lunar and planetary science, the NSF became only one among several important patrons for this field.[6]

Two explanatory notes are in order. First, government fiscal years and academic calendars rarely match with calendar years. I have, however, listed grants as if fiscal and calendar years were aligned, a one-year resolution being sufficient for this purpose. Multiple-year grants are reported evenly over the term of the grant, rather than as a lump sum in the first year of operation, providing a more accurate portrait of funds available in particular years. Second, I have excluded from the NSF totals grants awarded for the construction of major new research instruments, which were not made by the ONR. This eases comparison of NSF with ONR totals; more important, it recognizes that new instruments often address research problems in a variety of fields.

APPENDIX D: AMERICAN PARTICIPATION IN SOLAR SYSTEM COMMISSIONS OF THE IAU, 1922–1958

A final measure of the relative strength of solar system astronomy in the United States can be found in the records of the International Astronomical Union. Membership in IAU commissions dedicated to subfields of solar system astronomy is reproduced in Table D.1. I define solar system here (as throughout this study) as American astronomers themselves did in the 1950s, including planets and satellites, celestial mechanics, comets, meteors, meteorites, and the zodiacal light.[7]

Established in 1919, the IAU quickly became the leading international organization of astronomy. IAU commissions coordinate research programs within particular astronomical fields; commission presidents are elected during the general assemblies of the IAU. Meetings have been held on average every three years, except during World War II and the early years of the cold war.

American astronomers were always present on these commissions between 1922 and 1960, and occasionally dominated them. Eight American researchers were elected commission presidents during this period, many serving successive terms (see Table D.2). Since commission presidents were leaders of these subfields, they often influenced research agendas in these areas. Leuschner used his authority as President of Commission 20 (asteroids) to demand more precise analytical methods for computing asteroid orbits (see Chapter 1), whereas Kuiper brought Urey into Commission 16 (planets), thus increasing contacts between astronomy and planetary geochemistry.

Table D.1, nevertheless, must be used with caution. Commission mandates and activity levels varied over time. Optical astronomy predominated during this period; the IAU

[6] Struve (1960) and Tatarewicz (1990a, 112–14).

[7] G. P. Kuiper, "Solar System Studies," draft proposal, Nov. 12, 1956, Box 33, GPK; Whipple (1956c, iii–ix).

Table D.1. *American participation in the IAU solar system commissions, 1922–1958*

No.	Commission name	1922	1925	1928	1932	1935
7.	Celestial Mechanics	11{2}*	16{3}	17{3}[a]		
15.	Physical Study of Comets[b]					9{2}
16.	Physical Study of Planets[c]	15{3}	21{3}	23{5}	26{8}	24{7}
17.	Lunar[c]					13{5}*
20.	Asteroids	18{3}*	23{4}*	22{3}*	23{4}*	25{5}*
20a.	Subcom. on periodic comets					
21.	Comets	9{2}*[f]				
22.	Meteors	7{1}	12{3}	14{5}*	13{4}*	24{5}
22a.	Subcomm. on meteorites					
IAU membership						
	United States	54	58	75	83	88
	Total	207	248	285	406	495

No.	Commission name	1938	1948	1952	1955	1958
7.	Celestial Mechanics		15{4}*	32{6}*	32{6}*	27{6}*
15.	Physical Study of Comets[b]	12{2}	13{2}	27{4}	31{3}	28{3}
16.	Physical Study of Planets[c]	27{7}	26{6}	37{13}*	33{9}*	21{6}
17.	Lunar[c]	10{1}	13{2}[d]	14{3}*	16{3}*	18{3}
20.	Asteroids	43{6}	52{10}*	67{14}*	70{13}	39{9}
20a.	Subcom. on periodic comets		9{2}[e]	7{2}	8{2}	8{2}
21.	Comets					
22.	Meteors	36{6}	48{11}*	69{15}	74{15}	29{4}
22a.	Subcomm. on meteorites				9{4}*[g]	9{4}*
IAU membership						
	United States	101	170	202	231	281
	Total	554	612	811	887	1,129

Note: Unbracketed numbers indicate total commission membership; bracketed numbers are U.S. astronomers. An asterisk (*) denotes that an astronomer at a U.S. facility served as commission or subcommission president (see Table D.2). In instances where commission members temporarily worked overseas, "home" institutions were tabulated.

[a]Commission 7 was dissolved in 1928 by then-president Willem de Sitter, who found no problems in dynamical astronomy "whose solutions call for combined action and international organization rather than private efforts of mathematicians"; see de Sitter (1933).

[b]Founded in 1935.

[c]Membership in nomenclature committees is not included in this count.

[d]After 1948 this commission also took responsibility for the movement and figure of the Moon.

[e]Founded in 1948.

[f]In 1925 comet orbits were reassigned to Commission 20; physical observations of comets were reassigned to Commission 16.

[g]Founded in 1955.

Source: Transactions of the International Astronomical Union.

Table D.2. *U.S. presidents of IAU commissions for solar system astronomy*

Year	Commission	President
1922	7. Celestial mechanics	A. O. Leuschner
	20. Asteroids	E. W. Brown
	21. Comet orbits	A. O. Leuschner
1925	20. Asteroids	A. O. Leuschner
1928	20. Asteroids	A. O. Leuschner
	22. Meteors	C. P. Olivier
1932	20. Asteroids	A. O. Leuschner
	22. Meteors	C. P. Olivier
1935	17. Lunar motion	E. W. Brown
	20. Asteroids	A. O. Leuschner
1948	7. Celestial mechanics	G. M. Clemence
	20. Asteroids	D. Brouwer
	22. Meteors	F. L. Whipple
1952	7. Celestial mechanics	G. M. Clemence
	16. Planetary	G. P. Kuiper
	17. Lunar motion	C. B. Watts
	20. Asteroids	D. Brouwer
1955	7. Celestial mechanics	D. Brouwer
	16. Planetary	G. P. Kuiper
	17. Lunar motion	C. B. Watts
	22a. Meteorites	F. L. Whipple
1958	7. Celestial mechanics	D. Brouwer
	22a. Meteorites	F. L. Whipple

made little effort to incorporate the rapidly expanding community of radio astronomers within its domain.[8] Moreover, commission memberships were sometimes pared arbitrarily by commission presidents.[9] This table is most accurate as an indicator of relative strength over time.

[8] Edge and Mulkay (1976).
[9] See R. Wildt to B. Lyot, Sept. 29, 1948, Box 1, Folder 5, RW.

Abbreviations

ACRONYMS AND INITIALISMS

AAAS	American Association for the Advancement of Science
AAS	American Astronomical Society
AFCRC	Air Force Cambridge Research Center
AGU	American Geophysical Union
APL	Applied Physics Laboratory, Johns Hopkins
Caltech	California Institute of Technology
CIW	Carnegie Institution of Washington
HCO	Harvard College Observatory
IAU	International Astronomical Union
IBM	International Business Machines
ICBM	intercontinental ballistic missile
IGY	International Geophysical Year
JPL	Jet Propulsion Laboratory
LPL	Lunar and Planetary Laboratory
MICE	Megaton Ice-Contained Project
MIT	Massachusetts Institute of Technology
NAS	National Academy of Sciences
NASA	National Aeronautics and Space Administration
NRC	National Research Council
NSF	National Science Foundation
NTS	Nevada Test Site
ONR	Office of Naval Research
OSRD	Office of Scientific Research and Development
RCAF	Royal Canadian Air Force
RRL	Radio Research Laboratory
SAO	Smithsonian Astrophysical Observatory
UCLA	University of California at Los Angeles
USGS	United States Geological Survey
WPA	Works Progress Administration

ARCHIVAL SOURCES

Note: Box numbers reflect arrangements in place at the time these collections were consulted. Where possible, as with the reprocessed Urey papers, I have updated box citations to match current arrangements.

Abbreviations

AOL	Armin O. Leuschner papers, Bancroft Library, University of California, Berkeley, Calif.
ATW	Alan T. Waterman papers, Library of Congress, Washington, D.C.
BG	Beno Gutenberg papers, California Institute of Technology, Pasadena, Calif.
BL	Bernard Lyot papers, Observatoire de Paris, Paris.
CITA	Department of Astronomy records, California Institute of Technology archives, Pasadena, Calif.
CITG	Division of Geological and Planetary Sciences records, California Institute of Technology archives, Pasadena, Calif.
CSB	Carlyle S. Beals correspondence, Dominion Observatory records, RG 48, National Archives of Canada, Ottawa, Ont., Canada
DB	Dirk Brouwer papers, Yale University Department of Astronomy records, Series V, Yale University archives, New Haven, Conn.
DHCOm	Director's papers, Harvard College Observatory (Donald H. Menzel), Harvard University archives, Cambridge, Mass.
DHCOs	Director's papers, Harvard College Observatory (Harlow Shapley), Harvard University Archives, Cambridge, Mass.
DHM	Donald H. Menzel papers, Harvard University archives, Cambridge, Mass.
ECP	Edward C. Pickering papers, Harvard University Archives, Cambridge, Mass.
EMS	Eugene M. Shoemaker papers, National Archives, Southwest Branch, Laguna Niguel, Calif.
FEW	Frederick E. Wright/Carnegie Moon Committee papers, Huntington Library, San Marino, Calif.
FGW	Fletcher G. Watson, Jr., papers, Harvard University, Cambridge, Mass.
FLW–HU	Fred L. Whipple papers, Harvard University archives, Cambridge, Mass.
FLW–SI	Fred L. Whipple papers, Smithsonian Institution archives, Washington, D.C.
GEB	General Education Board records, Rockefeller Archives Center, North Tarrytown, N.Y.
GPK	Gerard P. Kuiper papers, University of Arizona library archives, Tucson, Ariz.
HAS	H. Alexander Smith papers, Mudd Library, Princeton University, Princeton, N.J.
HCU	Harold C. Urey papers, University of California at San Diego, Mandeville Special Collections, Central University Library, San Diego, Calif.
HHH	Harry H. Hess papers, Firestone Library, Princeton University, N.J.
HNR	Henry Norris Russell papers, Firestone Library, Princeton University, Princeton, N.J.
HW	Harry Wexler papers, Library of Congress, Washington, D.C.
IEB	International Education Board records, Rockefeller Archives Center, North Tarrytown, N.Y.
IGY	International Geophysical Year, U.S. National Committee files, National Academy of Sciences archives, Washington, D.C.
ISB	Ira S. Bowen papers, Huntington Library, San Marino, Calif.
JCM	John C. Merriam papers, Library of Congress, Washington, D.C.
JHO	Jan H. Oort papers, Leiden University archives, Leiden, The Netherlands.
JLG	Jesse L. Greenstein papers, California Institute of Technology archives, Pasadena, Calif.
JTW	J. Tuzo Wilson papers, University of Toronto Library, Toronto, Ont., Canada.
JVV	John Van Vleck papers, American Institute of Physics, College Park, Md.
JWC	Joseph W. Chamberlain papers, Rice University, Houston, Tex.
LAD	Lee A. DuBridge papers, California Institute of Technology archives, Pasadena, Calif.

LC[si]	Leonard Carmichael papers, Smithsonian Institution archives, Washington, D.C.
LOW	Lowell Observatory archives, Flagstaff, Ariz.
LP	Linus Pauling papers, Oregon State University, Corvallis, Ore.
LVB	Lloyd V. Berkner papers, Library of Congress, Washington, D.C.
MT	Merle Tuve papers, Library of Congress, Washington, D.C.
NAS	National Academy of Sciences archives, Washington, D.C.
NASM	National Air and Space Museum historical files, Washington, D.C.
OS	Otto Struve papers, Bancroft Library, University of California, Berkeley, Calif.
OS[mic]	Otto Struve papers, AIP microfilm edition, copy at National Air and Space Museum, Smithsonian Institution, Washington, D.C.
RD	Reginald Daly papers, Harvard University Archives, Cambridge, Mass.
RF	Rockefeller Foundation archives, Rockefeller Archives Center, North Tarrytown, N.Y.
RW	Rupert Wildt papers, Yale University archives, New Haven, Conn.
SAO	Smithsonian Astrophysical Observatory, Assistant Director (Science), 1961–73 Smithsonian Institution archives, Washington, D.C.
SIO	Scripps Institution of Oceanography subject files, 81-16/AC6, Scripps Institution of Oceanography, La Jolla, Calif.
UChi	Office of the President, University of Chicago archives, Chicago, Ill.
USGS	U.S. Geological Survey general files, National Archives, Washington, D.C.
UVa	Office of the President, University of Virginia, Charlottesville, Va.
WSA	Walter Sydney Adams papers, Huntington Library, San Marino, Calif.
WWM	William W. Morgan papers, Yerkes Observatory archives, Williams Bay, Wisc.
YO	Director's Office, Yerkes Observatory, Williams Bay, Wisc.

ORAL HISTORY INTERVIEW COLLECTIONS

OHI	Oral History Interview. The name of the individual interviewed is followed by the interviewer's name, date of interview, and institute where the interview transcript is on deposit. Example: John Smith OHI (Frank Jones, Jan. 1, 1990, SHMA).
SAOHP	Space Astronomy Oral History Project, National Air and Space Museum, Washington, D.C.
SHMA	Sources for the History of Modern Astrophysics oral history interview collection, Center for History of Physics, American Institute of Physics, College Park, Md.
SPHG	Survey Project in the History of Geophysics, Center for History of Physics, American Institute of Physics, College Park, Md.

PUBLICATION TITLES

AJ	*Astronomical Journal*
AJS	*American Journal of Science*
AN	*Astronomical Newsletter*
ApJ	*Astrophysical Journal*
ARHCO	*Annual Report [of the Harvard College Observatory]*
AS	*Annals of Science*

Abbreviations

BAAPG	*Bulletin of the American Association of Petroleum Geologists*
BAIN	*Bulletin of the Astronomical Institutes of the Netherlands*
BMNAS	*Biographical Memoirs of the National Academy of Sciences*
DSB	*Dictionary of Scientific Biography*
GCA	*Geochimica et Cosmochimica Acta*
GSAB	*Geological Society of America Bulletin*
HCO	*Harvard College Observatory [Bulletin, Circular]*
HSPS	*Historical Studies in the Physical [and Biological] Sciences*
JG	*Journal of Geology*
JGR	*Journal of Geophysical Research*
JHA	*Journal for the History of Astronomy*
JRASC	*Journal of the Royal Astronomical Society of Canada*
MNRAS	*Monthly Notices of the Royal Astronomical Society*
PA	*Popular Astronomy*
PAAS	*Publications of the American Astronomical Society*
PASP	*Publications of the Astronomical Society of the Pacific*
PNAS	*Proceedings of the National Academy of Sciences*
PR	*Physical Review*
SA	*Scientific American*
S&T	*Sky and Telescope*
Trans. AGU	*Transactions of the American Geophysical Union*
Trans. IAU	*Transactions of the International Astronomical Union*
VA	*Vistas in Astronomy*

References

A BRIEF NOTE ON SOURCES

The history of solar system research in the twentieth century encompasses studies in physics, astronomy, geology, geochemistry and geophysics, and no single source offers an adequate starting point for relevant scientific literature. For astrophysics and astronomy, the most helpful guide to publications is the *Astronomischer Jahresbericht,* an annual compilation with items catalogued under numerous subject headings. An introduction to geological literature is the *Bibliography of North American Geology,* published as issues of the *Bulletin of the U.S. Geological Survey;* more extensive in scope but regrettably lacking systematic rigor in coverage is William A. S. Sarjeant, *Geologists and the History of Geology: An International Bibliography from the Origins to 1978* (Malabar, Fla.: Krieger, 1980) and his *Supplement 1979–1984 and Additions* (1987). Meteorological literature is covered in *Meteorological Abstracts and Bibliography* (after 1959 *Meteorological & Geoastrophysical Abstracts and Bibliography*). The *Dictionary of Scientific Biography*, ed. Charles C. Gillispie (New York: Scribner's, 16 vols., 1970–80), with subsequent volumes ed. Frederic L. Holmes), is an indispensable source for biographical material. The annual "Observatory Reports" published in *Popular Astronomy* before World War II, and in the *Astronomical Journal* thereafter, are a rich source of information for departments of astronomy in the United States and Canada.

Published items and dissertations cited in the chapter notes are listed below. No attempt is made to distinguish primary and secondary sources, although the distinction should be clear in all cases.

See the List of Abbreviations regarding journal titles and manuscript sources used in this study.

Aaserud, Finn. (1990). *Redirecting Science: Niels Bohr, Philanthropy, and the Rise of Nuclear Physics.* New York: Cambridge University Press.

Abetti, Georgio. (1975). "Schiaparelli, Giovanni Virginio." In Charles C. Gillispie, ed., *DSB* 12: 160–1. New York: Scribner's.

Abetti, G., W. E. Bernheimer, K. Graff, A. Kopff, and S. A. Mitchell. (1929). *Handbuch der Astrophysik. Band IV: Das Sonnensystem.* Berlin: Springer.

Abir-Am, Pnina G. (1987)."The Biotheoretical Gathering, Trans-disciplinary Authority, and the Incipient Legitimation of Molecular Biology in the 1930s: New Perspective on the Historical Sociology of Science." *History of Science* 25: 1–70.

(1988). "The Assessment of Interdisciplinary Research in the 1930s: The Rockefeller Foundation and Physico-chemical Morphology." *Minerva* 26: 153–76.

References

Adams, W. S., and T. Dunham, Jr. (1932). "Absorption Bands in the Infrared Spectrum of Venus." *PASP* 44: 243–7.

Adel, Arthur, and V. M. Slipher. (1934a). "On the Identification of the Methane Bands in the Solar Spectra of the Major Planets." *PR* 46: 240–1.

(1934b). "The Constitution of the Atmospheres of the Giant Planets." *PR* 46: 902–6.

Ahrens, L. H. (1956). "Radioactive Methods for Determining Geologic Ages." In L. H. Ahrens, K. Rankama, and S. K. Runcorn, eds., *Physics and Chemistry of the Earth*, vol. 1, pp. 44–67. London: Pergamon.

Albritton, Claude C. (1989). *Catastrophic Episodes in Earth History.* New York: Chapman & Hill.

ed. (1963). *The Fabric of Geology.* Stanford, Calif.: Freeman, Cooper.

Aldrich, L. B. (1937). "Measurement of the Solar Constant." In W. E. Forsythe, ed., *Measurement of Radiant Energy*, pp. 423–6. New York: McGraw–Hill.

Alter, W. Dinsmore. (1948). "Evolution of the Moon." *Griffith Observer* 12: 114.

(1955). "Nature of the Lunar Rays." *PASP* 67: 257.

Ambartsumian, V. A. (1948). "The Evolution of Stars and Astrophysics" (abst.). *AN* 37: 3.

Angel, Juvenal. (1958). *Careers in Geology.* New York: World Trade Academy Press.

Aronson, Naomi. (1986). "Resistance to Discovery: Vitamins, History, and Careers." *Isis* 77: 630–46.

Aspray, William. (1990). *John von Neumann and the Origins of Modern Computing.* Cambridge, Mass.: MIT Press.

Baker, James G. (1945). "Review [of the telescope designs of Bernhardt Schmidt]." *AN* 33: 3.

Baldwin, Ralph B. (1942). "The Meteoritic Origin of Lunar Craters." *PA* 50, 7: 365–9.

(1949). *The Face of the Moon*. Chicago: University of Chicago Press.

(1980). *The Deadly Fuze: Secret Weapon of World War II.* San Rafael: Presidio Press.

Baldwin, Ralph B., and Don E. Wilhelms. (1992). "Historical Review of a Long-overlooked Paper by R. A. Daly Concerning the Origin and Early History of the Moon." *JGR* 97: 3837–43.

Bashe, C. J., L. R. Johnson, J. H. Palmer, and E. W. Pugh. (1986). *IBM's Early Computers.* Cambridge, Mass.: MIT Press.

Bates, Charles. (1989). "The Formative Rossby–Reichelderfer Period in American Meteorology, 1926–1940," *Weather and Forecasting* 4: 593–603.

Bates, Charles, and John F. Fuller. (1986). *America's Weather Warriors, 1814–1985.* College Station, Tex.: A&M University Press.

Bauer, Carl August. (1948). "On the Origin of Meteorites."*AJ* 53(1171): 196.

(1954). "The Origin and Age of Meteorites." *S&T* 27: 179.

Baum, Werner A. (1985). "Seymour Lester Hess." In Garry E. Hunt, ed., *Recent Advances in Planetary Meteorology*, pp. ix–xi. New York: Cambridge University Press.

Bauschinger, J. (1897). "Vorwort." *Veröffentlichungen des Königlichen Astronomischen Rechen-Instituts zu Berlin* 4: 3–4.

Baxter, James P. (1946). *Scientists against Time.* Boston: Little, Brown.

Beals, Carlyle S. (1957). "A Probable Meteorite Crater of Great Age." *S&T* 16: 526–8.

(1958a). "A Survey of Terrestrial Craters." *Nature* 181: 559.

(1958b). "Fossil Meteorite Craters." *SA* 199(1): 32–9.

Beals, Carlyle S., G. M. Ferguson, and A. Landau. (1956). "A Search for Analogies Between Lunar and Terrestrial Topography on Photographs of the Canadian Shield." *JRASC* 50(5,6): 203–11, 250–61.

Beals, Carlyle S., M. J. S. Innes, and J. A. Rottenberg. (1963). "Fossil Meteorite Craters." In B. M. Middlehurst and G. P. Kuiper, eds., *The Moon, Meteorites and Comets*, pp. 235–84. Chicago: University of Chicago Press.

References

Ben-David, Joseph. (1971). *The Scientist's Role in Society*. Chicago: University of Chicago Press.

Bennett, Arthur L. (1931). "A Photovisual Investigation of the Brightness of 59 Areas on the Surface of the Moon." Ph.D. diss., Dept. of Astronomy, Princeton University, Princeton, N.J.

Berendzen, Richard (1974). "Newton, Hubert Anson." In Charles C. Gillispie, ed., *DSB* 10: 41. New York: Scribner's.

Berendzen, Richard, and Mary Treinen Moslen. (1972). "Manpower and Employment in American Astronomy." In Richard Berendzen, ed., *International Conference on Education in and History of Modern Astronomy. Annals of the New York Academy of Sciences* 198:46–55.

Bernal, J. D. (1949). "The Goldschmidt Memorial Lecture." *Journal of the Chemical Society:* 133: 2108–14.

Beyer, M. (1956). "On the Present Situation in Cometary Research." *VA* 2: 949–57.

Birch, Francis. (1951). "Remarks on the Structure of the Mantle and Its Bearing upon the Possibility of Convection Currents." *Trans. AGU* 32, 533–4.

(1954). "The Earth's Mantle: Elasticity and Constitution." *Trans. AGU* 35, 79–85, 97–8.

Bobrovnikoff, Nicholas T. (1927). "On the Spectra of Comets." *ApJ* 66: 429.

(1951). "Comets." In J. A. Hynek, ed., *Astrophysics: A Topical Symposium*, pp. 302–56. New York: McGraw–Hill.

(1956). "Systematic Variations in the Diameter of Comets." *AJ* 61: 171–2.

Bok, Bart. (1946). "Reports on the Progress of Astronomy: The Time-Scale of the Universe." *MNRAS* 106, 1: 61–75.

(1978). "Harlow Shapley, Nov. 2, 1885–Oct. 20, 1972." *BMNAS* 49: 241–91.

Bondi, Hermann. (1952). *Cosmology*. Cambridge: Cambridge University Press (2d ed., 1961).

Boon, John D., and Claude C. Albritton. (1937). "Meteorite Scars in Ancient Rocks." *Field and Laboratory* 5 (1): 1–9.

Bowler, Peter J. (1992). *The Norton History of the Environmental Sciences*. New York: W. W. Norton.

Bradley, W. H. (1963). "Geologic Laws." In Claude C. Albritton, ed., *The Fabric of Geology*, pp. 12–33. Stanford, Calif.: Freeman, Cooper.

Brickwedde, Ferdinand G. (1982). "Harold Urey and the Discovery of Deuterium." *Physics Today* 35: 34–9.

Brouwer, Dirk. (1935a). "The Minor Planets." *The Telescope* 2: 12–26.

(1935b). "On the Determination of Systematic Corrections to Star Positions from Observations of Minor Planets." *AJ* 44(1022): 57–63.

(1950). "Families of Minor Planets and Related Distributional Problems." *AJ* 55(1186): 162–3.

(1956). "Current Trends in Minor Planet Research." *VA* 2: 943–8.

Brown, Ernest. (1923). "The General Orbits of the Asteroids of the Trojan Group." *AJ* 35: 69–80.

Brown, Harrison. (1949). "A Table of Relative Abundances of Nuclear Species." *Reviews of Modern Physics* 4: 625–34.

(1950a). "The Composition of Our Universe." *Physics Today* 3 (4): 6–13.

(1950b). "On the Compositions and Structures of the Planets." *ApJ* 111: 641–53.

Brown, Harrison, and Edward Goldberg. (1949). "The Neutron Pile as a Tool in Quantitative Analysis: The Gallium and Palladium Content of Iron Meteorites." *Science* 109: 347–53.

Brown, Harrison, Gunnar Kullerud, and Walter Nichoporuk. (1952). *Bibliography on Meteorites*. Chicago: University of Chicago Press.

References

Brown, Harrison, and Claire Patterson. (1947a). "The Composition of Meteoritic Matter: I. The Composition of the Silicate Phase of Stony Meteorites." *JG* 55: 405–11.

(1947b). "The Composition of Meteoritic Matter: II: The Composition of Iron Meteorites and of the Metal Phase of Stony Meteorites." *JG* 55: 508–10.

(1947c). "The Relative Abundance of Some Light Nuclear Species as Determined from the Composition of Stony Meteorites." *PR* 72(6): 456–7.

(1948). "The Composition of Meteoritic Matter: III. Phase Equilibria, Genetic Relationships and Planet Structure." *JG* 56: 85–111.

Brush, Stephen G. (1978a). "A Geologist Among Astronomers: The Rise and Fall of the Chamberlin–Moulton Cosmogony, Parts 1 and 2." *JHA* 9: 1–41; 77–104.

(1978b). "Planetary Science: From Underground to Underdog." *Scientia* 113: 771–87.

(1981). "From Bump to Clump: Theories of the Origin of the Solar System 1900–1960." In Paul Hanle and Von Del Chamberlain, eds., *Space Science Comes of Age: Perspectives in the History of the Space Sciences,* pp. 78–96. Washington, D.C.: Smithsonian Institution.

(1982a). "Chemical History of the Earth's Core." *Eos* 63: 1185–8.

(1982b). "Nickel for Your Thoughts: Urey and the Origin of the Moon." *Science* 217: 891–8.

(1989). "The Age of the Earth in the Twentieth Century." *Earth Sciences History* 8(2): 170–82.

(1990). "Theories of the Origin of the Solar System, 1956–1985." *Reviews of Modern Physics* 62(1): 43–112.

Bucher, Walter H. (1963). "Cryptoexplosion Structures Caused from Without or Within in the Earth? ('Astroblemes' or 'Geoblemes?')." *AJS* 261: 597–649.

Bugos, Glenn E. (1989). "Managing Cooperative Research and Borderland Science in the National Research Council, 1922–1942." *HSPS* 20(1): 1–32.

Bulkeley, Rip. (1991). *The Sputniks Crisis and Early United States Space Policy.* Bloomington: Indiana University Press.

Burchfield, Joe D. (1975). *Lord Kelvin and the Age of the Earth.* New York: Science History Publications.

Burke, John G. (1986). *Cosmic Debris: Meteorites in History.* Berkeley: University of California Press.

Burrows, William E. (1990). *Exploring Space: Voyages in the Solar System and Beyond.* New York: Random House.

Burstyn, Harold L. (1980). "Reviving American Oceanography: Frank Lillie, Wickliffe Rose, and the Founding of the Woods Hole Oceanographic Institution." In Mary Sears and Daniel Merriman, eds., *Oceanography: The Past,* pp. 57–66. New York: Springer.

Cain, Joseph Allen. (1993). "Common Problems and Cooperatives Solutions: Organizational Activity in Evolutionary Studies, 1936–1947." *Isis* 84(1): 1–25.

Campbell, Wallace W. (1920). "Notes on the Problem of the Origin of the Lunar Craters." *PASP* 32: 126–38.

Cassidy, David C. (1992). *Uncertainty: The Life and Science of Werner Heisenberg.* New York: W. H. Freeman.

Chamberlain, Joseph, and Gerard P. Kuiper. (1956). "The Rotational Temperature and Phase Variations of Carbon Dioxide Bands of Venus." *ApJ* 124: 399–407.

Chamberlin, Rollin R. (1949). "Geological Evidence on the Evolution of the Earth's Atmosphere." In Gerard P. Kuiper, ed., *The Atmospheres of the Earth and Planets*, pp. 248–57. Chicago: University of Chicago Press.

Chandrasekhar, S. (1946). "On a New Theory of Weiszäcker on the Origin of the Solar System." *Review of Modern Physics* 18: 94–102.

References

Chant, C. A. (1949). "American Astronomical Society: Eighty-first Meeting." *JRASC* 43: 138–41.

Chao, E. C. T., E. M. Shoemaker, and B. M. Madsen. (1960). "First Natural Occurrence of Coesite." *Science* 132: 220–2.

Charney, Jule. (1959). "On the General Circulation of the Atmosphere." In B. Bolin, ed., *The Atmosphere and the Sea in Motion,* pp. 178–93. New York: Rockefeller Institute Press.

Charters, Alex C. (1960). "High-Speed Impact." *SA* 203(4): 128–40.

Clemence, Gerard M. (1951). "Reports on the Progress of Astronomy: Celestial Mechanics." *MNRAS 111*: 219–31.

(1970). "Dirk Brouwer, 1902–1966." *BMNAS* 41: 69–87.

Clerke, Agnes. (1896). *A Popular History of Astronomy during the Nineteenth Century*. New York: Macmillan.

Coben, Stanley. (1971). "The Scientific Establishment and the Transmission of Quantum Mechanics to the United States, 1919–1932." *American Historical Review* 76: 442–67.

Coblentz, W. W. (1925). "Radiometric Determination of the Temperature of Mars in 1924." *Nature* 116: 472.

Coblentz, W. W., and C. O. Lampland. (1924a). "New Measurements of Planetary Radiation and Planetary Temperatures." *PNAS* 11: 34–6.

(1924b). "Some Measurements of the Spectral Components of Planetary Radiation." *PAAS* 32: 546–7.

Craig, Harmon. (1953). "The Geochemistry of the Stable Carbon Isotopes." *GCA 3*: 36–82.

Craig, Harmon, S. L. Miller, and G. J. Wasserburg, eds. (1964). *Isotope and Cosmic Chemistry: Dedicated to Harold C. Urey on His 70th Birthday, April 29, 1963*. Amsterdam: North-Holland.

Crane, Diana. (1972). *Invisible Colleges: Diffusion of Knowledge in Scientific Communities.* Chicago: University of Chicago Press.

Cressman, Earle R., and Martin C. Noger. (1981). "Geologic Mapping of Kentucky: A History and Evaluation of the Kentucky Geological Survey–U.S. Geological Survey Mapping Program, 1960–1978." *USGS Circular* 801: 1–22.

Crowe, Michael. (1986). *The Extraterrestrial Life Debate: The Idea of a Plurality of Worlds from Kant to Lowell.* New York: Cambridge University Press.

Cruikshank, Dale P. (1991). "A Bright Future for Planetary Astronomy?" *S&T* 82: 4.

(1993). "Gerard Peter Kuiper, 1905–1973." *BMNAS* 62: 259–95.

Cuffey, James. (1956). "Reports of Observatories: Goethe Link Observatory, Indiana University." *AJ* 61: 321–2.

Currie, K. L. (1965). "Analogues of Lunar Craters on the Canadian Shield." In Harold E. Whipple, ed., *Geological Problems in Lunar Research. Annals of the New York Academy of Sciences* 123(2): 915–40.

Dalrymple, G. Brent. (1991). *The Age of the Earth.* Stanford, Calif.: Stanford University Press.

Daly, Reginald A. (1942). *The Floor of the Ocean: New Light on Old Mysteries.* Chapel Hill: University of North Carolina Press.

(1943). "Meteorites and an Earth-Model." *GSAB* 54: 401–56.

(1946). "Origin of the Moon and Its Topography." *Proceedings of the American Philosophical Society* 90(2): 104–19.

(1947). "Vredefort Ring-Structure of South Africa." *JG* 55, 125–45.

Davis, Gordon L. (1950). "Radium Content of Ultramafic Igneous Rocks: III. Meteorites." *AJS* 248: 107–11.

Davis, Gordon L., and Harry H. Hess. (1949). "Radium Content of Ultramafic Igneous Rocks: II. Geological and Chemical Implications." *AJS* 247: 856–82.

Delaporte, E. (1950). "Report [Commission 20]," In J. H. Oort, ed., *Trans. IAU* 7: 215–39.
Dennis, Michael Aaron. (1994). "'Our First Line of Defense': Two University Laboratories in the Postwar American State." *Isis* 85: 427–55.
de Sitter, Willem. (1933). "Report of Commission 7 (Commission de l'Astronomie dynamique et des tables astronomiques)." In F. J. M. Stratton, ed., *Trans. IAU* 47: 27. Cambridge: Cambridge University Press.
de Vaucouleurs, Gérard. (1955). "Mars." In *The New Astronomy*, pp. 187–98. New York: Simon & Schuster.
(1960). "A Survey of Physical Problems of the Nearer Planets and a Review of Observational Techniques Applicable to Balloon-Borne Telescope Systems." In John D. Strong, Gérard de Vaucouleurs, and Fritz Zwicky, *Planetary Astronomy from Satellite-Substitute Vehicles*, pp. 25–148. Research Report 60-6, Holloman AFB, New Mexico, Air Force Missile Development Center, Air Research and Development Command, U.S. Air Force.
de Vaucouleurs, Gérard, and Menzel, Donald H. (1960). "Results of the Occultation of Regulus by Venus, July 7, 1959." *Nature* 188: 28–33.
DeVorkin, David H. (1977). "W. W. Campbell's Spectroscopic Study of the Martian Atmosphere." *Quarterly Journal of the Royal Astronomical Society* 18: 37–53.
(1982). "The Maintenance of a Scientific Institution: Otto Struve, The Yerkes Observatory, and Its Optical Bureau During the Second World War." *Minerva* 18(4): 595–623.
(1984). "The Harvard Summer School in Astronomy." *Physics Today* 37: 48–55.
(1987). "Organizing for Space Research: The V-2 Rocket Panel." *HSPS* 18(1): 1–24.
(1989a). "Along for the Ride: The Response of American Astronomers to the Possibility of Space Research, 1945–1950)." In M. DeMaria and M. Grilli, eds, *The Restructuring of the Physical Sciences in Europe and the USA 1945–1960*, pp. 55–74. Singapore: World Publishing.
(1989b). *Race to the Stratosphere: Manned Scientific Ballooning in America*. New York: Springer.
(1990). "Defending a Dream: Charles Greeley Abbot's Years at the Smithsonian." *JHA* 21: 121–36.
(1992). *Science with a Vengeance: Military Origins of Space Science*. New York: Springer.
(1993). "How Specialties Survive." *Minerva* 31(2) (Summer): 246–52.
(1994). "A Fox Raiding the Hedgehogs: Henry Norris Russell at Mount Wilson." *History of Geophysics* 5: 103–11.
(in press). "Back to the Future: The Response of Astronomers to the Prospect of Government Funding for Research in the Decade following World War II." In David van Keuran and Nathan Reingold, eds., *Science and the Federal Patron.*
DeVorkin, David H., and Ralph Kenat. (1983a). "Quantum Physics and the Stars (I): The Establishment of a Stellar Temperature Scale." *JHA* 14: 102–32.
(1983b). "Quantum Physics and the Stars (II): Henry Norris Russell and the Abundances of the Elements in the Atmospheres of the Sun and Stars." *JHA* 14: 180–222.
(in press). "Consensus Building: Stellar Evolution to 1955." In Owen Gingerich, ed., *The General History of Astronomy,* vol. 4B. New York: Cambridge University Press.
Diamond, Sigmund. (1992). *Compromised Campus: The Collaboration of Universities with the Intelligence Community, 1945–1955.* New York: Oxford University Press.
Dick, Steven J. (1990). "Pulkovo Observatory and the National Observatory Movement: An Historical Overview." In J. H. Lieske and V. K. Abalakin, eds., *Inertial Coordinate Systems on the Sky*, pp. 29–38. Dordrecht: IAU.
(1996). *Biological Universe: The Twentieth Century Extraterrestrial Life Debate and the Limits of Science.* New York: Cambridge University Press.

Dietz, Robert S. (1946a). "The Meteoritic Impact Origin of the Moon's Surface Features." *JG* 54: 359–75.
(1946b). "Geological Structures Possibly Related to Lunar Craters." *PA* 54: 465–7.
(1959). "Shatter Cones in Cryptoexplosion Structures (Meteorite Impact)?" *JG* 67: 496–505.
Doel, Ronald E. (1990a). "Competition and Myth: Solar System Astronomy on the Eve of Space Exploration, 1952–1959" (abst.). *Bulletin of the American Astronomical Society* 22(3): 1038.
(1990b). "Redefining a Mission: The Smithsonian Astrophysical Observatory on the Move." *JHA* 21: 137–53.
(1990c). "Unpacking a Myth: Interdisciplinary Research and Solar System Astronomy, 1920–1958." Ph.D. diss., Dept. of History, Princeton University, Princeton, N.J.
(1992). "Evaluating Soviet Lunar Science in Cold War America." *Osiris* (2d ser.) 7: 238–64.
(1994). "Expeditions at the CIW: Comments and Contentions." *History of Geophysics* 5: 79–87.
(in press-a). "From 'National University' to Local Center of Science: Princeton University, 1918–1932." *Minerva*.
(in press-b). "Geophysics in Universities." In Gregory A. Good, ed., *Garland Encyclopedia for the History of Earth Sciences.* New York: Garland Press.
(in press-c). "Documents and Reports: G. P. Kuiper's Report on Soviet Astronomy to the CIA." *Istoriko-astronomicheskie issledovanniia.*
Doggett, LeRoy E. (in press). "History of Celestial Mechanics." In John Lankford, ed., *Garland History of Astronomy.* New York: Garland.
Dunbar, Carl O. (1960). *Historical Geology*, 2d ed. New York: Wiley.
Dunham, Theodore, Jr. (1938). "Knowledge of the Planets in 1938." In *Cooperation in Research*, pp. 115–34. Washington, D.C.: Carnegie Institution of Washington.
(1956). "Methods in Stellar Spectroscopy." *VA* 2: 1223–83.
Dupree, A. Hunter. (1972). "The Great Instauration of 1940: The Organization of Scientific Research for War." In Gerald Holton, ed., *The Twentieth-Century Sciences: Studies in the Biography of Ideas,* pp. 443–67. New York: Norton.
(1990). "Smithsonian Astrophysical Observatory: From Washington to Cambridge." *JHA* 21: 107–10.
Eckel, Edwin B., ed. (1968). *Nevada Test Site.* Geological Society of America Memoir 110. Boulder, Colo.: Geological Society of America.
Edge, David O., and Michael Mulkay. (1976). *Astronomy Transformed: The Emergence of Radio Astronomy in Great Britain.* New York: Wiley.
Edmondson, Frank K. (1991). "AURA and KPNO: The Evolution of an Idea, 1952–1958." *JHA* 22(1): 68–86.
(in press). *AURA, Kitt Peak, and Cerro Tololo: The Early Years.* New York: Cambridge University Press.
Einarsson, S. (1935). "The Award of the Bruce Gold Medal to Dr. Vesto Melvin Slipher." *PASP* 47: 5–10.
Elsasser, Walter M. (1950). "The Earth's Interior and Geomagnetism." *Reviews of Modern Physics* 22: 1–35.
(1951). "Quantum-Theoretical Densities of Solids at Extreme Compression." *Science* 112: 105–7.
(1963). "The Early History of the Earth." In J. Geiss and E. D. Goldberg, eds., *Earth Science and Meteoritics: Dedicated to F. G. Houtermans on His Sixtieth Birthday,* pp. 1–30. Amsterdam: North-Holland.

Elston, Wolfgang E. (1990). "How Did Impact Processes on Earth and Moon Become Respectable in Geological Thought?" *Earth Sciences History* 9 (1): 82–7.

Engelhardt, Wolf Von. (1982). "Hypotheses on the Origin of the Ries Basin, Germany, from 1792 to 1960." *Geologische Rundschau* 71: 475–85.

England, J. Merton. (1982). *A Patron for Pure Science.* Washington, D.C.: National Science Foundation.

Eucken, Arnold. (1944). "Über den Zustand des Erdinnern." *Die Naturwissenschaften* 32: 112–21.

Evans, David S., and Derral Mulholland. (1986). *Big and Bright: A History of the McDonald Observatory.* Austin: University of Texas Press.

Fisher, D. Jerome. (1963). *The Seventy Years of the Department of Geology at the University of Chicago, 1892–1961.* Chicago: University of Chicago Press.

Fisher, Willard J. (1927). "Projects for Meteor Photography." *PA* 35: 511–513.

Fleming, James Rogers. (1990). *Meteorology in America, 1800–1870.* Baltimore: Johns Hopkins University Press.

Fleming, John A., and Charles S. Piggot. (1956). "Frederick Eugene Wright, 1877–1953." *BMNAS* 29: 317–59.

Forman, Paul. (1987). "Behind Quantum Electronics: National Security as Basis for Physical Research in the United States, 1940–1960." *HSPS* 18 (1): 149–229.

(1992). "Inventing the Maser in Postwar America," *Osiris* (2d ser.) 7: 105–34.

Fowler, William A. (1956). "The Origin of the Elements." *SA* 195 (3): 82–91.

Friedman, Robert Marc. (1989). *Appropriating the Weather: Vilhelm Bjerknes and the Construction of a Modern Meteorology.* Ithaca, N.Y.: Cornell University Press.

Galison, Peter L. (1985). "Bubble Chambers and the Experimental Workplace." In Peter Achinstein and Owen Hannaway, eds., *Observation, Experiment, and Hypothesis in Modern Physical Science,* pp. 309–73. Cambridge, Mass.: MIT Press.

(1987). *How Experiments End.* Chicago: University of Chicago Press.

Gamow, George, and J. Allen Hynek. (1945). "A New Theory by C. F. von Weiszäcker of the Origin of the Planetary System." *ApJ* 101: 249–54.

Gehrels, Tom. (1988). *On the Glassy Sea: An Astronomer's Journey.* New York: American Institute of Physics.

Genuth, Joel. (1987). "Groping Towards Science Policy in the United States in the 1930s." *Minerva* 25(3): 238–68.

Gieryn, Thomas F. (1979). "Patterns in the Selection of Problems for Scientific Research." Ph.D. diss., Dept. of Sociology, Columbia University, New York.

Gilbert, Grove Karl. (1893). "The Moon's Face: A Study of the Origin of Its Features." *Philosophical Society of Washington Bulletin* 12: 241–92.

Gilluly, James W. (1959). *Principles of Geology,* 2d ed. San Francisco: Freeman.

Gilvarry, John J., and Jerald E. Hill. (1956). "The Impact of Large Meteorites." *ApJ* 124: 610–22.

Gold, Thomas. (1955). "The Lunar Surface." *MNRAS* 115: 585–604.

Goldberg, Edward, Aiji Uchiyama, and Harrison Brown. (1951). "The Distribution of Nickel, Cobalt, Gallium, Palladium and Gold in Iron Meteorites." *GCA* 2: 1–25.

Goldberg, Leo, and Donald H. Menzel. (1956). "Solar Physics." In Fred L. Whipple, ed., *Smithsonian Contributions to Astrophysics* 1(1): 103–12. Washington, D.C.: Smithsonian Institution.

Goodman, Clark, and Robley D. Evans. (1941). "Age Measurements by Radioactivity." *GSAB* 52: 491–544.

Goodstein, Judith R. (1991). *Millikan's School: A History of the California Institute of Technology.* New York: Norton.

References

Graham, Loren R. (1987). *Science, Philosophy, and Human Behavior in the Soviet Union.* New York: Columbia University Press.

(1993). *Science in Russia and the Soviet Union: A Short History.* New York: Cambridge University Press.

Graham, Loren R., Wolf Lepenies, and Peter Weingart, eds. (1983). *Functions and Uses of Disciplinary Histories.* Dordecht: Kluwer Academic Publishers.

Grant, Robert. (1852). *History of Physical Astronomy from the Earliest Ages to the Middle of the Nineteenth Century.* London: Robert Baldwin.

Greene, Mott. (1982). *Geology in the Nineteenth Century: Changing Views of a Changing World.* Ithaca: Cornell University Press.

Greenstein, Jesse. (1951). "Interstellar Matter." In J. A. Hynek, ed., *Astrophysics: A Topical Symposium*, pp. 526–97. New York: McGraw–Hill.

(1986). "Harrison Brown and the Astronomers." In Kirk R. Smith, Fereidun Fesharaki, and John P. Holdren, eds., *Earth and the Human Future: Essays in Honor of Harrison Brown,* pp. 18–27. Boulder, Colo.: Westview.

Groeneveld, Ingrid, and Gerard P. Kuiper. (1954). "Photometric Studies of the Asteroids. I." *ApJ* 120: 200–20.

Hager, Dorsey. (1953). "Crater Mound (Meteor Crater), Arizona, a Geological Feature." *BAAPG* 37: 821–57.

Hale, George E. (1920). "Lunar Photography with the Hooker Telescope." *PASP* 32: 112–15.

(1928). "The Possibility of Large Telescopes." *Harper's Monthly Magazine* 156: 639–40.

Halliday, Ian. (1991). "Peter Mackenzie Millman, 1906–1990." *JRASC* 85(2): 67–78.

Hardy, Clyde T. (1954). "Major Craters Attributed to Meteoritic Impact." *BAAPG* 38(5): 917–22.

Harrison, James M. (1954). "Ungava (Chub) Crater and Glaciation." *JRASC* 48: 16–20.

Haurwitz, Bernhard. (1946). "Relations Between Solar Activity and the Lower Atmosphere." *Trans. AGU* 27(11): 161–3.

Hawley, Ellis W. (1992). *The Great War and the Search for Modern Order: A History of the American People and their Institutions,* 2d ed. (New York: St. Martins Press).

Healy, Paul W., Lincoln LaPaz, and Frederick C. Leonard. (1953). "On the Identification of Terrestrial Meteorite Craters." *JRASC* 47: 160–1.

Helmreich, Jonathan E. (1985). *Gathering Rare Ores: The Diplomacy of Uranium Acquisition, 1943–1954.* Princeton: Princeton University Press.

Herget, Paul. (1950). "Minor Planets [Report on the Progress of Astronomy]." *MNRAS* 110(2): 167–9.

(1980). "Armin Otto Leuschner, January 16, 1868 – April 22, 1953." *BMNAS* 49: 129–48.

Herrick, Samuel. (1955). "Obituary Notice: Armin Otto Leuschner." *MNRAS* 114: 295–8.

Herrmann, Dieter B. (1984). *The History of Astronomy from Herschel to Hertzsprung.* Rev. and trans. Kevin Krisciunas. New York: Cambridge University Press.

Hershberg, James G. (1993). *James B. Conant: Harvard to Hiroshima and the Making of the Nuclear Age*. New York: Knopf.

Herzberg, Gerhard. (1938). "On the Possibility of Detecting Molecular Hydrogen and Nitrogen in Planetary and Stellar Atmospheres by their Rotation–Vibration Spectra." *AJ* 37: 428–37.

(1952). "Laboratory Absorption Spectra Obtained with Long Paths." In G. P. Kuiper, ed., *The Atmospheres of the Earth and Planets,* 2d ed., pp. 406–16. Chicago: University of Chicago Press.

Hess, Seymour L. (1948). "A Meteorological Approach to the Question of Water Vapor on Mars and the Mass of the Martian Atmosphere." *PASP* 60: 289–302.

(1950). "Some Aspects of the Meteorology of Mars." *Journal of Meteorology* 7: 1–13.
(1958). "Atmospheres of Other Planets." *Science* 128: 809–14.
(1959). *Introduction to Theoretical Meteorology*. New York: Holt.
Hess, Seymour L., and Hans Panofsky. (1951). "The Atmospheres of the Other Planets." In Thomas F. Malone, ed., *Compendium of Meteorology*, pp. 391–8. Boston: American Meteorological Society.
Hewlett, Richard G., and Oscar E. Anderson, Jr. (1962). *The New World, 1939–1946*. University Park: Pennsylvania State University Press.
Hewlett, Richard G., and Francis Duncan. (1969). *Atomic Shield, 1947/1952. Volume II: A History of the United States Atomic Energy Commission*. University Park: Pennsylvania State University Press.
Hirayama, Kiyotsugu. (1923). "Families of Asteroids." *Japanese Journal of Astronomy and Geophysics* 1 (3): 56–93.
Hoffleit, Dorrit. (1988). "Yale Contributions to Meteoric Astronomy." *VA* 32: 117–43.
Hoffmeister, Cuno. (1922). "Untersuchungen zur Astronomische Theorie der Sternschuppen." *Astronomische Abhandlungen* 4: E1–E33.
(1929). "On the Heliocentric Velocity of Meteors." *ApJ* 69: 157–67.
(1931). "Die tägliche Variation der Sternschnuppenhäufigkeit in der Tropenzone." *Astronomische Nachrichten* 243: 213–26.
(1937a). *Die Meteore*. Leipzig: Akademische Verlagsgesellschaft.
(1937b). "New Cosmic Relationships of Meteors." *PA* 45: 207–9.
Holland, Heinrich D. (1974). "Memorial of Harry Hammond Hess," *American Mineralogist* 59: 415–17.
Holmes, Frederic L. (1987). *Lavoisier and the Chemistry of Life: An Exploration of Scientific Creativity*. Madison: University of Wisconsin Press.
Hooykaas, R. (1963). *The Principle of Uniformity in Geology, Biology, and Theology*. Leiden: Brill.
Howell, Benjamin F., Jr. (1990). *An Introduction to Seismological Research: History and Development*. New York: Cambridge University Press.
Hoyle, Fred. (1949). *Some Recent Researches in Solar Physics*. Cambridge: Cambridge University Press.
(1951). *The Nature of the Universe*. New York: Harper.
Hoyt, William Graves. (1976). *Lowell and Mars*. Tucson: University of Arizona Press.
(1980a). *Planets X and Pluto*. Tucson: University of Arizona Press.
(1980b). "Slipher, Vesto Melvin." *BMNAS* 52: 411–49.
(1987). *Coon Mountain Controversies: Meteor Crater and the Development of Impact Theory*. Tucson: University of Arizona Press.
Hubbert, M. King, Thomas A. Hendricks, and George A. Thiel. (1949). "Report of Committee on Geologic Education of the Geological Society of America." *Geological Society of America Interim Proceedings*, Pt 2: 17–21.
Hufbauer, Karl. (1981). "Astronomers Take up the Stellar-energy Problem, 1917–1920." *HSPS* 11(2): 277–303.
(1991). *Exploring the Sun: Solar Science Since Galileo*. Baltimore: Johns Hopkins University Press.
(1994). "Artificial Eclipses: Bernard Lyot and the Coronagraph." *HSPS* 24(2): 337–94.
Humason, M. L., N. U. Mayall, and A. R. Sandage. (1956). "Redshifts and Magnitudes of Extragalactic Nebulae." *AJ* 61(3): 97–162.
Hynek, J. Allen. (1949). Review of *The Face of the Moon*, by Ralph B. Baldwin. *PA* 57: 257–8.
Index of Selected Publications of the RAND Corporation. (1962). *Vol. 1: 1946–1962*. Santa Monica, Calif.: Rand Corporation.

"International Commission on Meteorites." (1950). *GCA* 1: 71.
Ives, Herbert E. (1919). "Some Large-Scale Experiments Imitating the Craters on the Moon." *ApJ* 50: 245–50.
Jacchia, Luigi G., and Fred L. Whipple. (1956). "The Harvard Photographic Meteor Programme." *VA* 2: 982–94.
Jacobs, John Arthur, R. D. Russell, and J. Tuzo Wilson. (1959). *Physics and Geology*. New York: McGraw–Hill.
Jarrell, Richard A. (1988). *The Cold Light of Dawn: A History of Canadian Astronomy*. Toronto: University of Toronto Press.
Jeans, James H. (1904). *Dynamical Theory of Gases*. Cambridge: Cambridge Universtiy Press.
Jeffreys, Harold. (1923). "The Constitution of the Four Outer Planets." *MNRAS* 83: 350–4.
(1924). *The Earth, Its Origin, History, and Physical Constitution*. Cambridge: Cambridge University Press.
(1935). Review of *Physical and Dynamical Meteorology*, by David Brunt. *Quarterly Journal of the Royal Meteorological Society* 61: 223.
(1952). "The Origin of the Solar System (Bakerian Lecture)." *Proceedings of the Royal Astronomical Society* 214: 281–91.
Jones, H. Spencer. (1953). Review of *The Comets and Their Origin*, by R. A. Lyttleton. *Endeavor* 12: 217.
Jones, Bessie Zaban, and Lyle Gillford Boyd. (1971). *The Harvard College Observatory: The First Four Directorships, 1839–1919*. Cambridge, Mass.: Harvard University Press.
Kargon, Robert H. (1977). "Temple to Science: Cooperative Research and the Birth of the California Institute of Technology." *HSPS* 8: 3–31.
(1977). "Temple to Science: Cooperative Research and the Birth of the CIT." *HSPS* 8: 3–32.
(1982). *The Rise of Robert Millikan*. Ithaca, N.Y.: Cornell University Press.
Keenan, P. C., and W. W. Morgan. (1951). "Classification of Stellar Spectra." In J. A. Hynek, ed., *Astrophysics: A Topical Symposium*, pp. 12–28. New York: McGraw–Hill.
Kellogg, W. W. (1951). "Report of the Standing Committee on Problems of the Upper Atmosphere." *Trans. AGU* 32 (5): 755–9.
Kerr, Frank J., and Fred L. Whipple. (1954). "On the Secular Accelerations of Phobos and Jupiter V." *AJ* 59: 570–3.
Kevles, Daniel J. (1977). *The Physicists: The History of a Scientific Community in Modern America*. New York: Knopf.
Kidwell, Peggy. (1986). "E. C. Pickering, Lydia Hinchman, Harlow Shapley, and the Beginning of Graduate Work at the Harvard College Observatory." *Astronomy Quarterly* 5: 157–72.
(1992). "Harvard Astronomers in World War II." In Clark A. Elliott and Margaret W. Rossiter, eds., *Science at Harvard University: Historical Perspectives*, pp. 275–302. Bethlehem, Pa.: Lehigh University Press.
Kirkwood, Daniel. (1873). *Comets and Meteors: Their Phenomena in All Ages; Their Mutual Relations; and the Theory of Their Origin*. Philadelphia: Lippincott.
(1888). *The Asteroids, or Minor Planets between Mars and Jupiter*. Philadelphia: Lippincott.
Klotz, Irving M. (1949). "On the Calculation of Planetary Temperatures from the Composition of Meteoritic Matter." *Science* 109: 248–51.
Knapp, Robert H., and H. B. Goodrich. (1952). *Origins of American Scientists*. Chicago: University of Chicago Press.
Knopf, Adolph. (1957). "Measuring Geologic Time." *Scientific Monthly* 85 (5): 225–36.

Kohler, Robert E. (1982). *From Medical Chemistry to Biochemistry: The Making of a Biomedical Discipline,* pp. 1–23. New York: Cambridge University Press.
(1990). "The Ph.D. Machine: Building on the Collegiate Base." *Isis* 81: 638–62.
(1991). *Partners in Science: Foundations and Natural Scientists, 1900–1945*. Chicago: University of Chicago Press.
Koppes, Clayton. (1982). *JPL and the American Space Program: A History of the Jet Propulsion Laboratory*. New Haven: Yale University Press.
Krinov, E. M. (1963). "Meteorite Craters on the Earth's Surface." In Barbara M. Middlehurst and Gerard P. Kuiper, eds., *The Moon, Meteorites and Comets*, pp. 183–207. Chicago: University of Chicago Press.
Kuhn, Werner, and Alfred Rittmann. (1941). "Über den Zustand des Erdinnern und seine Entstehung aus einem homogenen Urzustand." *Geologische Rundschau* 32: 215–55.
Kuiper, Gerard P. (1931). "De Planeet Mars, I, II, and III." *Hemel en Dampring* 29: 153–61, 195–208, 221–36.
(1944). "Titan: A Satellite with an Atmosphere." *ApJ* 99(2): 378–83.
(1946). "German Astronomy during the War." *PA* 54: 263–83.
(1948). "Reports of Observatories: Yerkes Observatory, McDonald Observatory." *AJ* 54(1175): 70–6, 225–30.
(1950a). "On the Origin of the Asteroids." *AJ* 55(1186): 164.
(1950b). "Pluto's Diameter." *S&T* 10: 50.
(1951a). "On the Origin of the Solar System." *PNAS* 37(1): 1–14.
(1951b). "On the Origin of the Solar System." In J. A. Hynek, ed., *Astrophysics: A Topical Symposium*, pp. 357–424. New York: McGraw-Hill.
(1951c). "On the Evolution of the Protoplanets." *PNAS* 37(7): 383–93.
(1951d). "On the Origin of the Irregular Satellites." *PNAS* 37(11): 717–21.
(1952a). "Introduction." In G. P. Kuiper, ed., *The Atmospheres of Earth and Planets*, 2d ed., pp. 1–15. Chicago: University of Chicago Press.
(1952b). "Planetary Atmospheres and Their Origin." In G. P. Kuiper, ed., *The Atmospheres of Earth and Planets*, 2d ed., pp. 306–405. Chicago: University of Chicago Press.
(1953a). "Note on the Origin of the Asteroids." *PNAS* 39(12): 1159–61.
(1953b). "Satellites, Comets, and Interplanetary Material." *PNAS* 39(12): 1153–8.
(1953c). *The Solar System. Volume I: The Sun*. Chicago: University of Chicago Press.
(1954). "On the Origin of the Lunar Surface Features." *PNAS* 40: 1096–111.
(1955). "The Lunar Surface: Further Comments." *PNAS* 41: 820–3.
(1956a). "On the Formation of the Planets." *JRASC* 50(2): 57–68; 50(3): 105–21; 50(4): 158–76.
(1956b). "Planets, Satellites, and Comets." *Smithsonian Contributions to Astrophysics* 1(1): 89–93.
(1957). "Report [Commission Pour les Observations Physiques des Planètes et des Satellites]." In P. Th. Oosterhoff, ed., *Trans. IAU* 9: 250–60. Cambridge: Cambridge University Press.
(1959a). "Exploration of the Moon." *Vistas in Aeronautics* 2: 273–313.
(1959b). "The Moon." *JGR* 64(11): 1713–19.
(1959c). "Report of Observatories: Yerkes–McDonald." *AJ* 64: 474.
(1960). "Report of Observatory [Yerkes–McDonald]," *AJ* 65(9): 572.
ed. (1949). *The Atmospheres of Earth and Planets*. Chicago: University of Chicago Press.
Kuiper, Gerard P., Y. Fujita, T. Gehrels, I. Groeneveld, J. Kent, G. van Biesbroeck, and C. J. van Houten. (1958). "Survey of Asteroids." *ApJ Supplement* 3: 289–333.
Kuiper, Gerard P., and Jeannette R. Johnson. (1956). "Dimensions of Contact Surfaces in Close Binaries." *ApJ* 123(1): 90–4.

Kuiper, Gerard P., and Barbara M. Middlehurst, eds. (1961). *Planets and Satellites.* Chicago: University of Chicago Press.

Kuiper, Gerard P., W. Wilson, and R. J. Cashman. (1947). "An Infrared Stellar Spectrometer." *ApJ* 106: 243–50.

Langley, Samuel P. (1888). *The New Astronomy*. Boston: Ticknor.

Lankford, John. (1983). "Photography and the Long-focus Visual Refractor: Three American Case Studies, 1885–1914." *JHA* 14: 77–91.

(1984). "The Impact of Photography on Astronomy." In Owen Gingerich, ed., *The General History of Astronomy,* vol. 4A, pp. 16–39. New York: Cambridge University Press.

(1987). "Private Patronage and the Growth of Knowledge: The J. Lawrence Smith Fund of the National Academy of Sciences, 1884–1940." *Minerva* 25(3): 269–81.

Lankford, John, with the assistance of Rickey Slavings. (in press). *Community, Careers, and Power: American Astronomy 1859–1940 – An Interpretative Essay.* Chicago: University of Chicago Press.

LeGrand, Homer E. (1988). *Drifting Continents and Shifting Theories.* New York: Cambridge University Press.

Leonard, F. C. (1946). "Authenticated Meteorite Craters of the World. A Catalog of Provisional Coordinate Numbers for the Meteorite Falls of the World." *University of New Mexico Publications in Meteorites,* No. 1.

Leslie, Stuart W. (1993). *The Cold War and American Science.* Stanford, Calif.: Stanford University Press.

Leuschner, Armin O. (1935). "Research Surveys of the Orbits and Perturbations of 1091 Minor Planets." *Publications of the Lick Observatory* 19: i–xxi, 1–519.

(1936). "Report [Commission 20]." In F. J. M. Stratton, ed., *Trans. IAU* 5: 136–7. Cambridge: Cambridge University Press.

Leuschner, Armin O., and Holgar Thiele. (1929). "Report on the Progress of the Research Surveys of The Minor Planets." *PASP* 41: 263.

Levin, Alexsey, and Stephen Brush, eds. (1994). *The Origin of the Solar System: Soviet Research, 1925–1991.* College Park, Md.: American Institute of Physics.

Levy, David H. (1991). *Clyde Tombaugh: Discoverer of Planet Pluto.* Tucson: University of Arizona Press.

Lewis, Gilbert N. (1922). "The Chemistry of the Stars and the Evolution of Radioactive Substances." *PASP* 34(202): 309–19.

(1934). "The Genesis of the Elements." *PR* 46: 897–901.

Locke, J. L. (1979). "Carlyle Smith Beals, 1899–1979." *JRASC* 73(6): 325–32.

Lorenz, E. N. (1970). "The Nature of the Global Circulation of the Atmosphere: A Present View." In G. A. Corby, ed., *The Global Circulation of the Atmosphere*, pp. 3–23. London: Royal Meteorological Society.

Lovell, Bernard. (1954). *Meteor Astronomy.* Oxford: Clarendon.

(1990). *Astronomer by Chance.* New York: Basic.

Lowell, Percival. (1908). *Mars as the Abode of Life.* New York: Macmillan Company.

Lowell, Percival. (1911). *Mars and Its Canals.* New York: Macmillan.

Lowell Observatory (1952). "The Study of Planetary Atmospheres, by E. C. Slipher [Project Supervisor] and others." USAF Contract 19(122)-162. Flagstaff, Ariz.: Lowell Observatory. [Available at LOW and U.S. Naval Academy.]

Lowen, Rebecca. (in press). *Creating the Cold War University: Patrons, Scientists and Administrators at Stanford, 1937–1965.* Berkeley: University of California Press.

Luyten, W. J. (1940). Letter to the Editors ["Again the Origin of the Solar System"]. *Observatory* 63: 72–5.

Lyttleton, Raymond A. (1936). "The Origin of the Solar System." *MNRAS* 96: 559–68.

(1948). "On the Origin of Comets." *MNRAS* 108: 465.

(1952). "Note on the Origin of Comets." *ApJ* 115: 333–4.

(1953a). "On the Origin of Comet Tails." *La Physique des Comètes [Communications présentées au Quatrième Colloque international d'astrophysique]*, pp. 351–60. Liège: Institut d'astrophysique.

(1953b). *The Comets and Their Origin*. Cambridge: Cambridge University Press.

(1964). Review of *The Moon, Meteorites, and Comets*, ed. by Barbara M. Middlehurst and Gerard P. Kuiper. *Nature* 202: 526–7.

McBain, Howard Lee. (1931). "Professor Harold C. Urey's Study of Absorption Spectra." *Columbia University Annual Report:* 227–8.

McCutcheon, Robert A. (1991). "The 1936–37 Purge of Soviet Astronomy." *Slavic Review* 50: 100–17.

MacDonald, Thomas L. (1949). Review of *The Face of the Moon*, by R. B. Baldwin. *Journal of the British Astronomical Association* 60: 58–9.

McDougall, Walter. (1985). . . . *The Heavens and the Earth: A Political History of the Space Age*. New York: Basic.

McLaughlin, Dean B. (1955). "Changes on Mars, as Evidence of Wind Deposition and Volcanism." *AJ* 60: 261–70.

McPhee, John. (1974). *The Curve of Binding Energy*. New York: Farrar, Straus & Giroux.

McPherson, J. C. (1984). "Introduction." In W. J. Eckert, *Punched Card Methods in Scientific Computation* (reissue of 1940 ed.), pp. ix–xv. Cambridge, Mass.: MIT Press/ Tomash Publishers.

Manian, Samuel H., Harold C. Urey, and Walker Bleakney. (1934). "An Investigation of the Relative Abundance of the Oxygen Isotopes O^{16}:O^{18} in Stone Meteorites." *Journal of the American Chemical Society* 56: 2601–9.

Mark, Kathleen. (1987). *Meteorite Craters: How Scientists Solved the Riddle of these Mysterious Landforms*. Tucson: University of Arizona Press.

Markov, Alexsander V. (1962). *The Moon: A Russian View*. Chicago: University of Chicago Press.

Marsden, Brian G. (1973). "Lowell, Percival." In Charles C. Gillispie, ed., *DSB* 8: 520–3. New York: Scribner's.

Marvin, Ursula B. (1986). "Meteorites, the Moon, and the History of Geology." *Journal of Geological Education* 34: 140–65.

(1993). "The Meteoritical Society: 1933 to 1993." *Meteoritics* 28: 261–314.

Meadows, A. J. (1972). *Science and Controversy: A Biography of Sir Norman Lockyer*. Cambridge, Mass.: MIT Press.

(1984). "The New Astronomy." In Owen Gingerich, ed., *Astrophysics and Twentieth-century Astronomy to 1950*, pp. 59–72. Cambridge: Cambridge University Press.

"Meeting of the Royal Astronomical Society, Friday, 1948 Dec. 10." (1949). *Observatory* 69): 1–10.

Melosh, H. J. (1989). *Impact Cratering: A Geologic Process*. New York: Oxford University Press.

Menard, Henry W. (1971). *Science: Growth and Change*. Cambridge, Mass.: Harvard University Press.

Menzel, Donald H. (1923). "Water-cell Transmissions and Planetary Temperatures." *ApJ* 58: 65–74.

(1924). "The Atmospheres of the Outer Planets." *PAAS* 32: 225–6.

(1925). "The Atmosphere of Mars." *PAAS* 33: 296.

(1930). "Hydrogen Abundance and the Constitution of the Giant Planets." *PASP* 42: 228–32.

(1939). "The Relationship of Chemistry and Astronomy." *The Telescope* 6(6): 130–2, 142.

Menzel, Donald H., W. W. Coblentz, and C. O. Lampland. (1925). "Planetary Temperatures Derived from Radiation Measurements using Russell's Formula." *PAAS* 33: 297.
Menzel, Donald H., and Fred L. Whipple. (1955). "The Case for H_2O Clouds on Venus." *PASP* 67: 161–8.
Merrill, Paul W. (1950). Review of *The Face of the Moon,* by Ralph B. Baldwin. *PASP* 62: 125–6.
"Meteoritical Activities." (1949). *S&T* 7: 93.
Mikaylov, A. A., M. S. Zverev, P. G. Kulikovskiy, A. G. Masevich, Ye. R. Mustel', V. V. Sobolev, and M. F. Subbotin (1964). *Forty Years of Astronomy in the U.S.S.R., 1917–1957 (Part 1).* Dayton, Ohio: Wright–Patterson AFB Translation Div., Foreign Technical Div.
Miller, Howard S. (1970). *Dollars for Research: Science and Its Patrons in Nineteenth Century America.* Seattle: University of Washington Press.
Miller, Richard L. (1986). *Under the Cloud: The Decades of Nuclear Testing.* New York: Free Press.
Millman, Peter, and Donald W. R. McKinley. (1967). "Stars Fall over Canada." *JRASC* 61: 277–94.
Morgan, Neil. (1990). "The Strategy of Biological Research Programmes: Reassessing the 'Dark Ages' of Biochemistry, 1910–1930." *AS* 47: 139–50.
Moyer, Albert. (1992). *A Scientist's Voice in American Culture: Simon Newcomb and the Rhetoric of Scientific Method.* Berkeley: University of California Press.
Muenger, Elizabeth A. (1985). *Searching the Horizons: A History of Ames Research Center, 1940–1976.* Washington, D.C.: NASA.
Muir, Alex, ed. (1954). *Geochemistry, by V. M. Goldschmidt.* Oxford: Clarendon Press.
Mukerji, Chandra. (1990). *A Fragile Power: Scientists and the State.* Princeton: Princeton University Press.
Nace, R. L. (1958). "Military Use of Geologists in WW II." *GSAB* 69: 615–16.
Nash, Roderick. (1982). *Wilderness and the American Mind.* New Haven: Yale University Press.
National Research Council. (1922). *Celestial Mechanics.* Washington, D.C.: National Research Council of the National Academy of Sciences.
Nebeker, Frederick. (1995). *Calculating the Weather: Meteorology in the 20th Century.* San Diego: Academic Press.
Needell, Allan A. (1987). "Lloyd Berkner, Merle Tuve, and the Federal Role in Radio Astronomy." *Osiris* (2d ser.) 3: 261–88.
(1991). "The Carnegie Institution of Washington and Radio Astronomy: Prelude to an American National Observatory." *JHA* 22(1): 55–67.
(in press-a). *The Horizons of Lloyd V. Berkner: Science and Progress in Cold War America.*
(in press-b). "Rabi, Berkner, and the Rehabilitation of Science in Europe: The Cold War Context of American Support for International Science 1945–1958." In Francis H. Heller and John Gillingham, eds., *The Integration of Europe,* pp. 000–00. New York: St. Martin's Press.
Newcomb, Simon. (1887). *Popular Astronomy.* New York: Harper.
Newell, Homer E. (1980). *Beyond the Atmosphere: Early Years of Space Science.* Washington, D.C.: NASA.
Nölke, Friedrich. (1926). *Entwicklung im Weltall.* Hamburg: H. Grand.
North, John D. (1990). *The Measure of the Universe: A History of Modern Cosmology.* New York: Dover.
"Notes from Observatories [Dominion Observatory, Ottawa]." (1954). *JRASC* 48: 68–9.
Numbers, Ronald L. (1992). *The Creationists.* New York: Knopf.

"Observatory Report [High Altitude Observatory]." (1958). *AJ* 63: 355.

Oftedal, Ivar. (1948). "Memorial to Viktor Moritz Goldschmidt." *Proceedings of the Geological Society of America: 1947 Annual Report*, pp. 149–54.

O'Keefe, John A. (1963). "Two Avenues from Astronomy to Geology." In T. W. Donnelly, ed., *The Earth Sciences: Problems and Progress in Current Research*, pp. 43–58. Chicago: University of Chicago Press.

Olesko, Kathryn M. (1991). *Physics as a Calling: Discipline and Practice in the Köningsberg Seminar for Physics*. Ithaca: Cornell University Press.

Olivier, Charles P. (1925). *Meteors*. Baltimore: Williams & Wilkins.

(1933). "Commission 22 Report." In F. J. M. Stratton, ed., *Trans. IAU* 4: 117. Cambridge: Cambridge University Press.

Oort, Jan H. (1946). "Some Phenomena Connected with Interstellar Matter." *MNRAS* 106: 159–79.

(1950). "The Structure of the Cloud of Comets Surrounding the Solar System, and a Hypothesis Concerning Its Origin." *BAIN* 11(408): 91–110.

Oort, Jan H., and Maarten Schmidt. (1951). "Differences Between Old and New Comets." *BAIN* 11(419): 259–69.

Öpik, Ernst. (1930). "On the Visual and Photographic Study of Meteors." *HCO Bulletin* 879: 5–8.

(1933). "Meteorites and the Age of the Universe." *PA* 41: 71–9.

Oreskes, Naomi. (1990). "American Geological Practice: Participation and Examination." Ph.D. diss., Dept. of Earth Sciences, Stanford University, Stanford, Calif.

Osterbrock, Donald E. (1986). "Nicholas T. Bobrovnikoff and the Scientific Study of Comet Halley 1910." *Mercury* 15(2): 46–50, 63.

(1991). "Ground-based Planetary Science at Lick Observatory 1888–1938." *Bulletin of the American Astronomical Society* 23: 1202.

Osterbrock, Donald E., John R. Gustafson, and W. J. Shiloh Unruh. (1988). *Eye on the Sky: Lick Observatory's First Century*. Berkeley: University of California Press.

Paneth, Friedrich A. (1940). *The Origin of Meteorites*. Oxford: Clarendon Press.

(1946). "Review of Meteoritics." *AN* 36: 9.

(1954). Review of *The Comets and Their Origin,* by R. A. Lyttleton. *GCA* 5: 243.

Pannekoek, Antonie (1961). *A History of Astronomy*. London: G. Allen & Unwin.

Patat, F. (1944). "Arnold Eucken zum 60. Geburtstag." *Die Naturwissenschaften* 32: 101–2.

Patterson, Claire. (1986). "Harrison Brown's Influence on Twentieth-Century Developments in Geochronology." In Kirk R. Smith, Fereidun Fesharaaki, and John P. Holdren, eds., *Earth and the Human Future: Essays in Honor of Harrison Brown*, pp. 10–14. Boulder, Colo.: Westview.

Patterson, Claire, G. Tilton, and M. Inghram. (1953a). "Isotopic Compositions of Quaternary Leads from the Pacific Ocean." *GSAB* 64: 1387–8.

(1953b). "Abundances of Uranium and the Isotopes of Leads in the Earth's Crust." *GSAB* 64: 1461.

(1955). "Age of the Earth." *Science* 121: 69–75.

Pauly, Philip J. (1984). "The Appearance of Academic Biology in Late Nineteenth-Century America." *Journal of the History of Biology* 17(3) (Fall): 369–97.

Pecora, William T. (1960). "Coesite Craters and Space Geology." *GeoTimes* 5(2): 16–20.

Pettijohn, Francis J. (1948) "[Announcement of Award]." *JG* 56(3): 83.

(1984). *Memoirs of an Unrepentant Field Geologist: A Candid Profile of Some Geologists and their Science, 1921–1981*. Chicago: University of Chicago Press.

Pettit, Edson S., and Seth B. Nicholson. (1924a). "Radiation from the Dark Hemisphere of Venus." *PASP* 36: 227–8.

(1924b). "Measurement of the Radiation from Planet Mars." *PASP* 36: 269–272.

Pettit, Edson S., and Robert S. Richardson. (1955). "Observations of Mars Made at Mt. Wilson in 1954." *PASP* 67: 62–73.
Phillips, Norman A. (1956). "The General Circulation of the Atmosphere: A Numerical Experiment." *Quarterly Journal of the Royal Meteorological Society* 82(352): 123–57.
Phillips, Theodore E. R. (1929). "[Report to Commission 16 of the IAU]." In F. J. M. Stratton, ed., *Trans. IAU* 3: 106–10.
Pickering, William H. (1920). "The Origin of the Lunar Formations." *PASP* 32: 116–25.
Plotkin, Howard. (1978). "Edward C. Pickering, the Henry Draper Memorial, and the Beginning of Astrophysics in America." *AS* 35: 365–77.
(1990). "Edward Charles Pickering." *JHA* 21: 47–58.
(1993a). "Harvard College Observatory's Boyden Station in Peru: Origin and Formative Years, 1879-1898." In A. Lafuente, A. Elena, and M. L. Ortega, eds., *Congreso Internacional "Ciencia, Descubrimiento y Mundo Colonial: Mundialización de la Ciencia y Cultura Nacional,"* pp. 689–705. Madrid: Doce Calles.
(1993b). "William H. Pickering in Jamaica: The Founding of Woodlawn and Studies of Mars." *JHA* 24(1–2): 101–22.
Price, Derek de Solla. (1963). *Little Science, Big Science.* New York: Columbia University Press.
Pyne, Stephen J. (1978). "Methodologies for Geology: G. K. Gilbert and T. C. Chamberlain." *Isis* 69, 248: 413–24.
(1980). *Grove Karl Gilbert: A Great Engine of Research*. Austin: University of Texas Press.
(1986). *The Ice: A History of Antarctica.* Iowa City: University of Iowa Press.
Rabbitt, Mary C. (1974). *A Brief History of the U.S.G.S.* Reston, Va.: U.S. Geological Survey.
Rabkin, Yakov M. (1987). "Technological Innovation in Science: The Adoption of Infrared Spectroscopy by Chemists." *Isis* 78(291): 31–54.
Rabe, Eugene. (1950). "Derivation of Fundamental Astronomical Constants from the Observations of Eros during 1926–1945." *AJ* 1184: 112–25.
Ramsey, William H. (1948). "On the Constitution of the Terrestrial Planets." *MNRAS* 108: 406–23.
(1949). "On the Nature of the Earth's Core." *MNRAS* (Geophys. Suppl.) 5: 409–26.
Raup, David M. (1991). *Extinction: Bad Genes or Bad Luck?* New York: W. W. Norton.
(1951). "On the Constitutions of the Major Planets." *MNRAS* 111: 427–45.
Reingold, Nathan. (1972). "American Indifference to Basic Research: A Reappraisal." In George Daniels, ed., *Nineteenth Century American Science: A Reappraisal.* Evanston, Ill.: Northwestern University Press, pp. 38–52 .
(1994). "Science and Government in the United States Since 1945." *History of Science* 32 : 361–86.
Rhodes, Richard. (1986). *The Making of the Atomic Bomb.* New York: Simon & Schuster.
Richardson, Robert S. (1957). "Preliminary Report on Observations of Mars Made at Mount Wilson in 1956." *AJ* 62: 89.
Rosenberg, Charles E. (1979). "Toward an Ecology of Knowledge: On Discipline, Context, and History." In Alexandra Oleson and John Voss, eds., *The Organization of Knowledge in Modern America, 1860–1920,* pp. 440–51. Baltimore: Johns Hopkins University Press.
Rosenberg, Emily S. (1982). *Spreading the American Dream: American Economic and Cultural Expansion, 1890–1945.* New York: Hill & Wang.
Ross, J. E. R. (1957). "Geodetic Surveys in Canada." *JRASC* 51: 341–50.
Rossby, C.-G. (1949). "On the Nature of the General Circulation of the Lower Atmosphere." In Gerard P. Kuiper, ed., *The Atmospheres of the Earth and Planets*, pp. 16–48. Chicago: University of Chicago Press.

(1959). "Current Problems in Meteorology." In Bert Bolin, ed., *The Atmosphere and Sea in Motion*, pp. 9–30. New York: Rockefeller Institute Press.

Rothenberg, Marc. (1985). "History of Astronomy." *Osiris* (2d ser.) 1: 123–9.

Rubey, William W. (1951). "Geologic History of Sea Water: An Attempt to State the Problem." *GSAB* 62: 1111–48.

(1955). "Development of the Hydrosphere and Atmosphere, with Special Reference to Probable Composition of the Early Atmosphere." In A. W. Poldervaart, ed., *Crust of the Earth: A Symposium*, pp. 631–50. New York: Geological Society of America.

Russell, Henry Norris. (1921). "A Superior Limit to the Age of the Earth's Crust." *Proceedings of the Royal Society* 99A: 84–6.

(1929). "On Meteoric Matter Near the Stars." *ApJ* 69: 49–71.

(1935a). "The Atmospheres of the Planets." *Smithsonian Report* 135: 153–68.

(1935b). *The Solar System and Its Origin*. New York: Macmillan.

Russell, Henry Norris, R. S. Dugan, and J. Q. Stewart. (1926–7). *Astronomy: A Revision of Young's Manual of Astronomy* (1st ed., in 2 vols.; 2d ed., 1944). Boston: Ginn.

Russell, Henry Norris, and Donald H. Menzel. (1933). "The Terrestrial Abundance of the Permanent Gases." *PNAS* 19: 997–1001.

Russell, Richard D. (1980). "Isotopes and the Early Evolution of the Earth." In D. W. Strangway, ed., *The Continental Crust and Its Mineral Deposits: The Proceedings of a Symposium Held in Honor of J. Tuzo Wilson*, pp. 49–63. Waterloo, Ont.: Geological Society of Canada.

Sagan, Carl. (1974). "Obituary [of Gerard Peter Kuiper, 1905–1973]." *Icarus* 22: 117–18.

Schlesinger, Frank. (1941). "Reports of Observatories, 1940–41 (Yale University Observatory)." *AJ* 49: 200.

Schlesinger, Frank, and Dirk Brouwer. (1940). "Ernest William Brown, 1866–1938." *BMNAS* 21: 243–73.

Schrödinger, Erwin. (1945). *What Is Life?: The Physical Aspect of the Living Cell.* Cambridge: Cambridge University Press.

Schweber, Silvan S. (1993)."Physics, Community and the Crisis in Physical Theory." *Physics Today* (Nov.): 34–40.

Servos, John. (1983). "To Explore the Borderland: The Foundation of the Geophysical Laboratory of the Carnegie Institution of Washington." *HSPS* 14(1): 147–85.

(1986). "Mathematics and the Physical Sciences in America, 1880–1930." *Isis* 77: 611–29.

(1990). *Physical Chemistry from Ostwald to Pauling: The Making of a Science in America.* Princeton: Princeton University Press.

Shagam, Reginald, R. B. Hargraves, W. J. Morgan, F. B. van Houten, C. A. Burk, H. D. Holland, and L. C. Hollister, eds. (1972). "Preface," in *Studies in Earth and Space Sciences: A Memoir in Honor of Harry Hammond Hess,* pp. xiv–xv. Geological Society of America Memoir 132. Boulder, Colo.: Geological Society of America.

Shapley, Harlow. (1930). *Flights from Chaos: A Survey of Material Systems from Atoms to Galaxies.* New York: Whittlesey.

(1969). *Through Rugged Ways to the Stars*. New York: Scribner's.

ed. (1953). *Climatic Change: Evidence, Causes, and Effects.* Cambridge, Mass.: Harvard University Press.

Shapley, Harlow, and Cecilia H. Payne. (1928). "Spectroscopic Evidence of the Fall of Meteors into Stars." *HCO Circular* 317: 1–11.

Shapley, Harlow, Ernst J. Öpik, and Samuel L. Boothroyd. (1932). "The Arizona Expedition for the Study of Meteors." *PNAS* 18: 16–23.

Shoemaker, Eugene M. (1959). "Structure and Quaternary Stratigraphy of Meteor Crater, Arizona, in the Light of Shock-Wave Mechanics." *GSAB* 70(2): 1748.

(1960a). "Penetration Mechanics of High Velocity Meteorites, Illustrated by Meteor Crater, Arizona." *21st International Geological Conference Report* 18: 418–34.

(1960b). "Brecciation and Mixing of Rock by Strong Shock." Article 192, *U.S. Geological Survey Professional Paper* 400-B: B423–5.

(1961). "Interplanetary Correlation of Geologic Time." *BAAPG* 45, 1: 130.

(1962). "Interpretation of Lunar Craters." In Zdeněk Kopal, ed., *Physics and Astronomy of the Moon*, pp. 282–359. London: Academic Press.

(1963). "Impact Mechanics at Meteor Crater, Arizona." In Barbara M. Middlehurst and Gerard P. Kuiper, eds., *The Moon, Meteorites and Comets*, pp. 301–35. Chicago: University of Chicago Press.

(1981). "Lunar Geology." In Paul A. Hanle and Von Del Chamberlain, eds., *Space Science Comes of Age*, pp. 51–7. Washington, D.C.: Smithsonian Institution.

Shoemaker, Eugene M., Frank M. Byers, and Carl H. Roach. (1958). "Diatremes on the Navajo and Hopi Reservations, Arizona." *U.S. Geological Survey Report* TEI-740: 158–68.

Shoemaker, Eugene M., and E. C. T. Chao. (1961). "New Evidence for the Impact Origin of the Ries Basin, Bavaria, Germany." *JGR* 66, 10: 3371–8.

Shoemaker, Eugene M., and Robert J. Hackman. (1961). "Stratigraphic Basis for a Lunar Time Scale." *GSAB* 71(12): 2112.

Simpson, George Gaylord. (1963). "Historical Laws." In Claude C. Albritton, ed., *The Fabric of Geology*, pp. 24–48. Stanford, Calif.: Freeman, Cooper.

Skinner, Brian J., and Barbara L. Narenda. (1985). "Rummaging through the Attic; or, a Brief History of the Geological Sciences at Yale." *Geological Society of America Centennial Special* 1, pp. 355–76. Boulder, Colo.: Geological Society of America.

Skolovskaya, Z. M. (1976). "Otto Struve." In Charles C. Gillispie, ed., *DSB* 13: 115. New York: Scribner's.

Slichter, Louis B. (1941). "Cooling of the Earth." *GSAB* 52: 561–600.

(1950a). "National Academy of Sciences: The Rancho Santa Fe Conference Concerning the Evolution of the Earth." *PNAS* 36: 511–14.

(1950b). "Rancho Santa Fe Conference Concerning the Evolution of the Earth." *Science* 111: 462.

Smith, Alex G., and T. D. Carr. (1964). *Radio Exploration of the Planetary System*. Princeton: Van Nostrand.

Smith, Kirk R., Fereidun Fesharaki, and John P. Holdren, eds. (1986). *Earth and the Human Future: Essays in Honor of Harrison Brown*. Boulder, Colo.: Westview.

Smith, Robert W. (1982). *The Expanding Universe: Astronomy's 'Great Debate,' 1900–1931*. New York: Cambridge University Press.

(1989). "The Cambridge Network in Action: The Discovery of Neptune." *Isis* 80: 395–422.

(1992). "The Biggest Kind of Big Science: Astronomers and the Space Telescope." In Peter Galison and Bruce Hevly, ed., *Big Science: The Growth of Large-Scale Research*, pp. 184–211. Stanford, Calif.: Stanford University Press.

Smith, Robert W., with contributions by Paul A. Hanle, Robert H. Kargon, and Joseph N. Tatarewicz. (1989). *The Space Telescope: A Study of NASA, Science, Technology, and Politics*. New York: Cambridge University Press.

Spitzer, Lyman. (1939). "The Dissipation of Planetary Filaments." *ApJ* 90: 675–88.

Spurr, Josiah Edward. (1944). *Geology Applied to Selenology*. Lancaster, Pa.: Science Press.

Stock, Chester. (1947). "Memorial to John Campbell Merriam." *Geological Society of America Proceedings* 1947: 183–98.

Strangway, D. W., ed. (1980). *The Continental Crust and Its Mineral Deposits: The Proceed-*

ings of a Symposium Held in Honor of J. Tuzo Wilson. Waterloo, Ont.: Geological Society of Canada.

Strauss, David. (1994). "Percival Lowell, William H. Pickering, and the Founding of the Lowell Observatory." *AS* 51(1): 37–58.

Strong, John D., and William M. Sinton. (1956). "Radiometry of Mars and Venus." *Science* 123: 676.

Struve, Otto. (1942). "Astronomy Faces the War." *PA* 50(9): 465–72.

(1945). Review of *Cosmogony of the Solar System,* by V. G. Fessenkoff. *ApJ* 102(1945): 264–6.

(1947a). "The Story of an Observatory." *PA* 55(5): 227–44, 283–94.

(1947b). "Report of Observatories: Yerkes Observatory, McDonald Observatory." *AJ* 52(1161): 146–51.

(1949). "New Trends in Cosmogony." *S&T* 8: 302–5.

(1950). *Stellar Evolution: An Exploration from the Observatory.* Princeton: Princeton University Press.

(1952). "Comet Theories." *S&T* 11: 269–73.

(1955). "The General Needs of Astronomy." *PASP* 67: 214–23.

(1959). "The Making of the Barringer Meteorite Crater." *S&T* 18: 187–9.

(1960). "Astronomers in Turmoil." *Physics Today* 13(9): 18–23.

Struve, Otto, and Velta Zebergs. (1962). *Astronomy of the Twentieth Century*. New York: Macmillan.

Stuewer, Roger H. (1972). "Gamow, George." In Charles C. Gillispie, ed., *DSB* 5: 271–3. New York: Scribner's.

Suess, Hans, and H. C. Urey. (1956). "Abundance of the Elements." *Reviews of Modern Physics* 28(1): 53–74.

Sullivan, Walter. (1961). *Assault on the Unknown: The International Geophysical Year.* New York: McGraw–Hill.

Sullivan, Woodruff T., III, ed. (1984). *The Early Years of Radio Astronomy: Reflections Fifty Years after Jansky's Discovery.* New York: Cambridge University Press.

Sylves, Richard T. (1987). *The Nuclear Oracles: A Political History of the General Advisory Committee of the Atomic Energy Commission, 1947–77.* Ames: Iowa State University Press.

Tatarewicz, Joseph N. (1990a). *Space Technology and Planetary Astronomy.* Bloomington: Indiana University Press.

(1990b). "Urey, Harold C." In Frederic L. Holmes, ed., *DSB* 18: 943–8. New York: Scribner's.

Tayler, R. J., ed. (1987). *History of the Royal Astronomical Society, Volume 2: 1920-1970.* Oxford: Blackwell Scientific.

Taylor, June H. (1979). *Yellowcake: The International Uranium Cartel.* New York: Pergamon.

Tenn, Joseph S. (1990). "Simon Newcomb: The First Bruce Medalist." *Mercury* 19: 18, 30.

ter Haar, Dirk. (1948). "Recent Theories about the Origin of the Solar System." *Science* 107: 405–11.

(1950). "Further Studies on the Origin of the Solar System." *ApJ* 111: 179–90.

Thomas, Richard N., and Fred L. Whipple. (1951). "Physical Theory of Meteors, II. Astroballistic Heat Transfer." *ApJ* 114: 448–65.

Tilton, G. R., and G. L. David. (1959). "Geochronology." In Philip Abelson, ed., *Researches in Geochemistry*, vol. 1, pp. 190–216. New York: John Wiley.

Todd, David. (1897). *A New Astronomy*. New York: American Book Company.

Tombaugh, Clyde W. (1956). *Interim Report on Search for Small Earth Satellites for the Period 1953–1956*. State College: New Mexico College of Agriculture and Mechanic Arts.

Tombaugh, Clyde W., and Patrick Moore. (1980). *Out of the Darkness: The Planet Pluto.* Harrisburg, Pa.: Stackpole.
Turner, H. H. (1904). "Some Reflections Suggested by the Application of Photography to Astronomical Research." *Smithsonian Institution Annual Report:* 171–84.
Urey, Harold C. (1924). "The Distribution of Electrons in the Various Orbits of the Hydrogen Atom." *AJ*: 1–10.
(1949). "A Hypothesis Regarding the Origin of the Movements of the Earth's Crust." *Science* 110: 445–6.
(1950). "The Structure and Chemical Composition of Mars." *PR* 80: 295.
(1951a). "The Origin and Development of the Earth and Other Terrestrial Planets." *GCA* 1: 209–77.
(1951b). "Condensation Processes and the Origin of the Major and Terrestrial Planets." In A. Farkas and E. P. Wigner, eds., *L. Farkas Memorial*, pp. 3–12. Jerusalem: Research Council of Israel.
(1952a). "Chemical Fractionation in the Meteorites and the Abundance of the Elements." *GCA* 2: 269–82.
(1952b). "The Origin and Development of the Earth and Other Terrestrial Planets: A Correction." *GCA* 2: 263–8.
(1952c). *The Planets.* New Haven: Yale University Press.
(1953). "On the Concentration of Certain Elements at the Earth's Surface." *Proceedings of the Royal Society of London* A219: 281–92.
(1955a). "The Origin of Tektites." *PNAS* 41: 27–31.
(1955b). "Some Criticisms of 'On the Origin of the Lunar Surface Features' by Gerard P. Kuiper." *PNAS* 41: 423–7.
(1956). "Diamonds, Meteorites, and the Origin of the Solar System." *ApJ* 124: 623–36.
(1959). "Primary and Secondary Objects." *JGR* 64(11): 1721–37.
(1960). "Lines of Evidence in Regard to the Composition of the Moon." In Hilde Kallmann Biji, ed., *Space Research: Proceedings of the First International Space Science Symposium*, pp. 1114–21. New York: Interscience.
Urey, Harold C., and Harmon Craig. (1953). "The Composition of the Stone Meteorites and the Origin of the Meteorites." *GCA* 4: 36–82.
Urey, Harold C., Samuel Epstein, Heinz A. Lowenstam, and Charles McKinney. (1950). "Paleotemperatures of the Upper Cretaceous." *Science* 111: 462.
Van Helden, Albert. (1984). "Telescope Building, 1850–1900." In Owen Gingerich, ed., *The General History of Astronomy,* vol. 4A, pp. 40–58. New York: Cambridge University Press.
van Woerkom, A. J. J. (1948). "On the Origin of Comets." *BAIN* 10: 445–72.
von Niessl, Gustav, and Cuno Hoffmeister. (1925). "Katalog der Bestimmungsgrößen für 611 Bahnen großer Meteore." *Denkschrift der Akadamie Wissenschaften Mathematik-Naturwissenschaften*.
Wali, Kameshwar C. (1990). *Chandra: A Biography of S. Chandrasekhar.* Chicago: University of Chicago Press.
Walker, Mark. (1989). *German National Socialism and the Quest for Nuclear Power.* New York: Cambridge University Press.
Watson, Fletcher, Jr. (1935). "Origin of Tektites." *Nature* 136: 105–6.
(1936a). "Distribution of Meteoric Mass in Interstellar Space." *Harvard Annals* 105: 623–32.
(1936b). "Meteor Craters." *PA* 44(1): 2–17.
(1938). "Small Bodies and the Origin of the Solar System." Ph.D. diss., Dept. of Astronomy, Harvard University, Cambridge, Mass.
(1940). "Colors and Magnitudes of Asteroids." *HCO Bulletin* 913: 3–4.

(1941). *Between the Planets*. Philadelphia: Blakiston.
(1953). Review of *Comets and Meteor Streams,* by J. G. Porter. *S&T* 12 (4): 105.
Weart, Spencer R. (1979). "The Physics Business in America, 1919–1940: A Statistical Reconnaissance." In Nathan Reingold, ed., *The Sciences in the American Context: New Perspectives*, pp. 295–358. Washington, D.C.: Smithsonian Institution Press.
(1992a). "From the Nuclear Frying Pan into the Global Fire." *Bulletin of the Atomic Scientists* 48 (5): 18–27.
(1992b). "Solid State as Community." In Lillian Hoddeson, Ernest Braun, Jürgen Teichmann, and Spencer Weart, eds., *Out of the Crystal Maze: Chapters from the History of Solid State Physics,* pp. 617–69. New York: Oxford University Press.
Webb, George E. (1983). *Tree Rings and Telescopes: The Scientific Career of A. E. Douglas.* Tucson: University of Arizona Press.
(1993). "Leading Women Scientists in the American Southwest: A Demographic Portrait, 1900–1950." *New Mexico Historical Review* 68 (Jan.): 41–61.
Weissman, Paul R. (1993). "Comets at the Solar System's Edge." *S&T* 84: 26–9.
Whipple, Fred L. (1938). "Photographic Meteor Studies." *Proceedings of the American Philosophical Society* 79: 499–548.
(1939). "Upper Atmospheric Densities and Temperatures from Meteor Observations." *PA* 47: 419–25.
(1941). *Earth, Moon and Planets.* Philadelphia: Blakiston.
(1943). "Meteors and the Earth's Upper Atmosphere." *Reviews of Modern Physics* 15(4): 246–64.
(1948a). "Kinetics of Cosmic Clouds." *Harvard Observatory Monographs No. 7: Centennial Symposia, December 1946,* pp. 109–42. Cambridge, Mass.: HCO.
(1948b). "The Dust Cloud Hypothesis." *SA* 178: 34–45.
(1949a). "Comets, Meteors and the Interplanetary Complex." *AJ* 54: 179–80.
(1949b). "The Harvard Photographic Meteor Program." *S&T* 8(4): 90–3.
(1949c). Review of *The Face of the Moon,* by Ralph B. Baldwin. *S&T* 8: 258–9.
(1950a). "A Comet Model: I. The Acceleration of Comet Encke." *ApJ* 111 (2): 375–94.
(1950b). "On Tests of the Ice Conglomerate Model for Comets." *AJ* 55: 83.
(1951). "A Comet Model: II. Physical Relations for Comets and Meteors." *ApJ* 113: 464–74.
(1952a). "Results of Rocket and Meteor Research." *Bulletin of the American Meteorological Society* 33(1): 13–25.
(1952b). "On Meteor Masses and Densities." *AJ* 57: 28–9.
(1952c). Review of *The Planets,* by Harold C. Urey. *S&T* 12(3): 279–81.
(1954). "Density, Pressure, and Temperature Data Above 30 Kilometers." In Gerard P. Kuiper, ed., *The Earth as a Planet,* pp. 491–513. Chicago: University of Chicago Press.
(1955a). "Meteors." *PASP* 67: 367–86.
(1955b). "On the Mass–Luminosity Relationship for Meteors." *AJ* 60: 182–3.
(1956a). "The Scientific Value of Artificial Satellites." *Journal of the Franklin Institute* 262: 95–109.
(1956b). "Meteors." *Smithsonian Contributions to Astrophysics* 1: 83–6.
(1959a). "Solid Particles in the Solar System." *JGR* 64: 1653–64.
(1959b). "On the Lunar Dust Layer." *Vistas in Astronautics* 2: 267–72.
ed. (1956c). "New Horizons in Astronomy." *Smithsonian Contributions to Astronomy* 1 (1): iii–x, 1–181.
Whipple, Fred L., Luigi Jacchia and Zdenêk Kopal. (1952). "Variations in the Density of the Upper Atmosphere." In G. P. Kuiper, ed., *The Atmospheres of the Earth and Planets*, 2d ed., pp. 149–58. Chicago: University of Chicago Press.

Whitaker, Ewen A. (1985). *The University of Arizona's Lunar and Planetary Laboratory: Its Founding and Early Years*. Tucson: University of Arizona.

White, Robert M., ed. (1953). "Contributions to the Study of Planetary Atmospheric Circulation." *AFCRC Technical Report 53-25* (Nov.): 3.

Whitley, Richard. (1976). "Umbrella and Polytheistic Disciplines and Their Elites." *Social Studies of Science* 6: 471–97.

Whitnah, Donald R. (1961). *A History of the United States Weather Bureau.* Urbana: University of Illinois Press.

Wildt, Rupert. (1932). "Ammoniakgas in der Atmosphäre des Planeten Jupiter." *Forschungen und Fortschritte* 8: 233.

(1940). "Notes on the Surface Temperature of Venus." *ApJ* 91: 266–8.

(1942). "On the Chemistry of the Atmosphere of Venus." *ApJ* 96(1): 312–14.

(1947). "Reports on the Progress of Astronomy: The Constitution of the Planets." *MNRAS* 107: 84–102.

(1958). "Inside the Planets." *PASP* 70: 237–50.

Wilford, John Noble. (1995). "Scientists Find Source of Comets on Outer Edges of Solar System." *New York Times,* June 15, pp. 1, A26.

Wilhelms, Don E. (1987). "Presentation of the G. K. Gilbert Award to Ralph B. Baldwin." *GSAB* 99: 150–3.

(1993). *To A Rocky Moon: A Geologist's History of Lunar Exploration*. Tucson: University of Arizona Press.

Willett, Hurd C. (1948). "Patterns of World Weather Changes." *Trans. AGU* 29(6): 803, 809.

(1949). "Long-period Fluctuations of the General Circulation of the Atmosphere." *Journal of Meteorology* 6(1): 34–50.

Wilson, Charles W., Jr. (1953). "Wilcox Deposits in Explosion Craters, Stewart Country, Tennessee, and Their Relation to the Origin and Age of Wells Creek Basin Structure." *GSAB* 64: 753–68.

Wilson, J. Tuzo. (1954). "The Development and Structure of the Crust." In G. P. Kuiper, ed., *The Earth as a Planet*, pp. 138–214. Chicago: University of Chicago Press.

Wolf, Max, 1863–1932: Ein Gedenkblatt. (1933). Berlin: de Gruyter.

Wood, Robert Muir. (1985). *The Dark Side of the Earth*. Boston: George Allen & Unwin.

Wright, Frederick E. (1927). "Gravity on the Earth and on the Moon." *Scientific Monthly* 24: 448–62.

(1935). "The Surface Features of the Moon." *Smithsonian Institution Annual Report:* 169–82.

(1938). "The Surface of the Moon." In *Cooperation in Research,* pp. 59–74. Washington, D.C.: Carnegie Institution of Washington.

Wylie, Charles C. (1922). "The Cepheid Variable η Aquilae." *ApJ* 56: 217–31.

Wylie, Charles C. (1933). "Real Paths for Five Meteors." *Contributions of the University of Iowa Observatory* 4: 115.

(1937). "On von Niessl's Velocities for Meteors (Second Paper)." *PA* 45: 209–15.

Yochelson, Ellis L. (1985). "The Role and Development of the Smithsonian Institution in the American Geological Community." *Geological Society of America Centennial Special* 1: 337–53.

Young, Andrew T. (1989). "Planetary Studies and U.S. Space Program." In M. Capaccioli and H. G. Corwin, Jr., eds., *Gérard and Antoinette de Vaucouleurs: A Life for Astronomy*, pp. 31–5. Singapore: World Scientific.

Young, Charles. (1902). *Manual of Astronomy.* Boston: Ginn.

Zaslow, Morris. (1975). *Reading the Rocks: The Story of the Geological Survey of Canada.* Toronto: Macmillan.

Index

Index

Index

Index

Index

Index

Index

Index

Index

Index